The Human Microbiome Handbook

HOW TO ORDER THIS BOOK

BY PHONE: 877-500-4337 or 717-290-1660, 9AM–5PM Eastern Time

BY FAX: 717-509-6100

BY MAIL: Order Department
DEStech Publications, Inc.
439 North Duke Street
Lancaster, PA 17602, U.S.A.

BY CREDIT CARD: American Express, VISA, MasterCard, Discover

BY WWW SITE: http://www.destechpub.com

The Human Microbiome Handbook

Edited by

Jason A. Tetro
Visiting Scientist
Department of Molecular and Cellular Biology
University of Guelph

Emma Allen-Vercoe, Ph.D.
Associate Professor
Department of Molecular and Cellular Biology
University of Guelph

DES*tech* Publications, Inc.

The Human Microbiome Handbook

DEStech Publications, Inc.
439 North Duke Street
Lancaster, Pennsylvania 17602 U.S.A.

Copyright © 2016 by DEStech Publications, Inc.
All rights reserved

No part of this publication may be reproduced, stored in a retrieval system, or transmitted, in any form or by any means, electronic, mechanical, photocopying, recording, or otherwise, without the prior written permission of the publisher.

Printed in the United States of America
10 9 8 7 6 5 4 3 2 1

Main entry under title:
    The Human Microbiome Handbook

A DEStech Publications book
Bibliography: p.
Includes index p. 359

ISBN: 978-1-60595-159-1

Contents

Preface   ix

1. **Some Historical Notes on Bowel Microflora** 1
   SYDNEY M. FINEGOLD

2. **Ecology of the Human Microbiome** 9
   KALUDYNA BOREWICZ and HAUKE SMIDT
   2.1. Overview   9
   2.2. Microbiota of the Gastrointestinal Tract   9
   2.3. Microbial Composition in the GI Tract of
        Healthy Adults   11
   2.4. Microbial Ecosystem Function in the GI Tract of
        Healthy Adults   16
   2.5. Selected Diseases Associated with Dysbiosis of the
        Intestinal Microbiota   21
   2.6. Acknowledgments   28
   2.7. References   28

3. **From Birth to Old Age: Factors that Shape the
   Human Gut Microbiome** 35
   ALEXANDRA NTEMIRI, CATHERINE STANTON,
   R. PAUL ROSS and PAUL W. O'TOOLE
   3.1. Introduction   35
   3.2. Establishment of the Gut Microbiota   38

3.3. Shaping Factors of Gut Microbiota Composition   42
3.4. Functional Gastrointestinal Disorders: Inflammatory Bowel Disease and Irritable Bowel Syndrome   52
3.5. Ageing and Microbiota Alterations   55
3.6. Concluding Remarks   58
3.7. References   60

4. Microbial Biochemical Processes Critical to Human Health 73
VICKY DE PRETER and KRISTIN VERBEKE

4.1. Introduction   73
4.2. Metabolic Function of the Colonic Microbiota   73
4.3. Carbohydrate Metabolism   74
4.4. Physiological Effects of Short-chain Fatty Acids   77
4.5. Short-chain Fatty Acids as Signaling Molecules   79
4.6. The Role of Short-chain Fatty Acids in the Immune Response   81
4.7. Fermentation of Proteins   82
4.8. Microbial Metabolism of Polyphenols   92
4.9. Conclusion   96
4.10. References   97

5. The Gut Microbiome: Pathways to Brain, Stress, and Behavior 109
AADIL BHARWANI, JOHN BIENENSTOCK and PAUL FORSYTHE

5.1. Introduction   109
5.2. Evidence of Microbiota–CNS Interactions   111
5.3. Biological Underpinnings of Neural Communication   118
5.4. Conclusion   124
5.5. Acknowledgements   124
5.6. References   124

6. Effects on Immunity 131
LEANDRO A. LOBO, ROSANA B.R. FERREIRA and L. CAETANO M. ANTUNES

6.1. Introduction   131
6.2. Germ Free Mice   132

    6.3. Galt Formation   134
    6.4. Small Molecules, the Microbiome, and the
         Immune System   145
    6.5. The Impact of Other Microbiomes on Host Immunity   153
    6.6. The Impact of the Microbiota on Immune Disorders   155
    6.7. Conclusion   159
    6.8. References   160

7. **Microbiota-Related Modulation of Metabolic Processes in the Body** 171
    TINGTING JU, JIAYING LI and BENJAMIN P. WILLING
    7.1. Introduction   171
    7.2. Short-chain Fatty Acids   172
    7.3. Microbiota and Bile Acid Metabolism   182
    7.4. Gut Microbiota, TMAO, and Atherosclerosis   189
    7.5. Metabolic Inflammation   194
    7.6. Early Life Microbiome Programming
         Metabolic Outcomes   199
    7.7. Conclusion   200
    7.8. References   200

8. **An Overview of Microbiota-Associated Gastrointestinal Diseases** 213
    CLAUDIA HERRERA, VIRGINIA ROBLES-ALONSO and
    FRANCISCO GUARNER
    8.1. Host-microbes Interactions in the
         Gastrointestinal Tract   213
    8.2. Antibiotics and Risk of Disease   216
    8.3. The Gut Microbiota in Inflammatory
         Bowel Disease   219
    8.4. The Gut Microbiota in Functional
         Bowel Disorders   223
    8.5. The Gut Microbiota in Liver Diseases   225
    8.6. References   229

9. **An Overview of Microbiota-Associated Epigenetic Disorders** 235
    DAWN D. KINGSBURY and HOLLY H. GANZ
    9.1. Introduction   235

9.2. Microbial Influence on Human Health   241
9.3. Conclusions   256
9.4. References   256

10. Fecal Microbiota Transplantation in Gastrointestinal Disease 267
ROWENA ALMEIDA and ELAINE O. PETROF

10.1. Introduction   267
10.2. Basic Concepts in Intestinal Microbiome Function   267
10.3. Introduction to Fecal Microbiota Transplantation   270
10.4. Applications of Fecal Microbiota Transplantation   271
10.5. Summary   298
10.6. References   299

11. Probiotics and the Microbiome 327
GREGOR REID

11.1. A Long History of Time   327
11.2. Rationale for Modulating Health   328
11.3. The Constantly Growing List and How the Strains Function   330
11.4. Probiotics and Microbiome Mysteries   338
11.5. References   339

12. Considering the Microbiome as Part of Future Medicine and Nutrition Strategies 347
EMMA ALLEN-VERCOE

12.1. Introduction   347
12.2. Mining the Human Microbiota for New Drugs   347
12.3. Protecting the Gut Microbiota from Collateral Damage during Antibiotic Exposure   348
12.4. Microbial Ecosystem Therapeutics   350
12.5. Predicting the Influence of Xenobiotics on the Human Microbiota   352
12.6. Leveraging Microbiome Knowledge to Optimize Nutrition Strategies   353
12.7. Summary   354
12.8. References   355

Index   359

Preface

THE term "microbiome" has been in use for over 50 years but only in the last 15 years has it gained popularity in the health community. The word describes the totality of microorganisms in and on a particular environment. In humans, this totality includes the gastrointestinal tract (including the mouth), the skin, the respiratory tract, the genitalia, and even the ocular surface. But while this singular concept has garnered significant attention, our understanding of the scope in terms of public health and medicine continues to be enigmatic.

For over a century we have known microbes play a role in our lives, although for the majority of this span, the focus has been on infection or, ecologically speaking, parasitism. We now know the number of pathogens amounts to only a tiny fraction of the entirety of the microbial species on earth and less than one-tenth of the microbes associated with the human body. The rest have been primarily studied outside of the realm of human health with discoveries limited to journals focusing on microbiology rather than medicine.

Over the last 40 years, we have seen a burgeoning increase in the number of scientific articles examining the interaction of microbes and humans in terms of "commensalism" as well as "mutualism"; ecological terms that now also apply in the field of medicine and public health because of a deeper appreciation of the microbial ecology of the body. We are not solely made up of 37 trillion human cells; we also have microbes totalling up to three times that number. Through observation at the lab bench, in animal models, and clinical trials, we are learning how these two very different organisms—mammal and microbe—interact. More importantly, we have a growing understanding of how this interkingdom interface affects acute as well as chronic health outcomes.

The Human Microbiome Handbook was conceived as an examination of our knowledge about the microbial influence in public health. Though the amount of data continues to increase at a staggering rate, many trends of microbe-human interaction have become solidified. These are duly explored within the pages of this book. The range of topics encompasses many branches of medicine from gastroenterology to metabolism to immunology and mental health. In each chapter, the authors, all of whom are experts in their individual microbiome fields, provide the latest findings and, where applicable, mechanism-based explanations. All told, this compilation will provide any medical or health professional with the necessary knowledge and applicable references to ensure a well-rounded appreciation of the microbiome and its impact on our health.

Many health professionals have only a rudimentary understanding of the microbiome. This book has been designed to ensure all individuals can access the most pertinent information in the field. This has been accomplished by separating the book into three sections, beginning with a general overview of the microbiome and gradually moving to specific mechanisms, including discussions on disease and possible therapeutics. In this way, it is our hope that any reader, regardless of academic background, will be able to gain enough information for use in their future work and practice.

The first section provides an introductory perspective on the microbiome in which a more general observation of the knowledge is provided. Chapter 1, by one of the pioneers of microbiome research, Sydney Finegold, is historical in nature, taking us through his journey in the field over five decades. Chapter 2, by Dutch researchers Kaludyna Borewicz and Hauke Smidt, provides an overview of the microbiome as a part of the human body. This chapter also introduces the concept of ecology in which microbial populations, not solely singular species, are now the focus of research. The final section provides an overview of the concept of our microbiome as more than a static entity. Chapter 3, headed by Paul O'Toole from Ireland, provides a longitudinal examination of the nature of the gut microbiome from birth to death.

The second section of this book examines the trends of microbial influence on our bodily processes. Vicky De Preter and Kristin Verbeke from Belgium examine first the microbial side of the interaction.

Chapter 4 takes a look at the life cycle of bacteria and how certain by-products can act not as waste but as useful stimuli for several associated biological systems. The effect of microbes and mental health is next examined in Chapter 5 by Canadian scientists, Aadil Bharwani, John Bienenstock, and Paul Forsythe. These researchers are forging the path to our understanding of how microbes in the gastrointestinal tract can affect our mental state and influence pathologies such as depression. The key to this may lie in immune system interactions, and Chapter 6 by Leando Lobo, Rosana Ferreira, and Caetano Antunes, from Brazil, explores this concept. Although much has already been learned as a result of traditional, infection-based work, incorporation of the microbiome into this field of study may lead to the development of microbially-mediated immune therapies. Finally in Chapter 7, Tinting Ju, Jiaying Li, and Benjamin Willing, from Canada, provide an examination of how microbes can modulate our metabolism. In the context of human health, microbes have a significant influence and may be the key to several chronic illnesses such as obesity, diabetes, and cardiovascular disease.

The third section deals specifically with disease and therapies. The theme in this section is "balance". As in all ecological environments, equilibrium of species is needed in order to attain harmony, and when this balance is disrupted, problems may ensue. We now understand the same applies to the human body and several diseases once thought to be mysteries have been elucidated on the basis of this lack of ecological balance. In addition, when the ecology is restored, balance can be re-established and health can be returned.

In Chapter 8, Spanish researchers, Claudia Herrera, Virginia Robles-Alonso, and Francisco Guarner examine the effects of microbes on our gastrointestinal health and how a change in ecology may lead to chronic health problems including inflammatory bowel disease, liver diseases, and antibiotic-mediated illnesses. In Chapter 9, Holly Ganz and Dawn Kingsbury, from the United States, explore one of the most hotly debated topics in microbiome research: epigenetics. Though this field is still relatively new, we are beginning to appreciate how microbes are not only influencing our cellular world, but also our genes. This chapter will examine what is already known and as well will explore several hypotheses to explain potential mechanisms behind some of our most problematic diseases.

In contrast to disease, Rowena Almeida and Elaine Petrof, from Canada, provide an in-depth look at one of the most discussed medical procedures today. Known as fecal microbiota transplantation, or FMT, this process of restoring a balanced ecology in the gastrointestinal tract has of late gained significant notoriety. Chapter 10 will unveil the mechanisms, reveal the benefits and drawbacks, and dispel the myths. Apart from FMT, the other major interest for health professionals is the realm of probiotics. Canadian scientist Gregor Reid, in Chapter 11, will provide an examination of the nature of probiotics—what they are, what they are not—and will explore the beneficial properties of these special microbes. He will also provide a critical perspective on questions associated with their use and where gaps in our understanding may be filled.

The end of this book offers a positive outlook for the future. We are still only beginning to understand the scope of microbial influence on our health and illness. As we continue to explore the once-hidden ecology within our bodies, we will unveil even more incredible mechanisms and possibly routes to novel and perhaps even revolutionary therapies. Although we have come far in the short period of time since Lederberg introduced the microbiome terminology to the world, we also know the journey will extend long into the future and change the face of health practice. *The Human Microbiome Handbook* will enable anyone to join the journey, if only as a witness, and to gain awareness and readiness for the marvels that undoubtedly will come. For those in pursuit of medical and health degrees or simply wishing to learn more about the involvement of microbes in their field, understanding the impact of the microbiome now will make for an even richer practice down the road.

We wish you a good read and a very balanced microbiome.

JASON TETRO
EMMA ALLEN-VERCOE

CHAPTER 1

Some Historical Notes on Bowel Microflora

SYDNEY M. FINEGOLD, M.D., MACP, D (ABMM)

SINCE so much of the bowel flora is anaerobic, it makes sense to start with what was known about anaerobic bacteria in the "olden days".

I graduated from UCLA in 1943 as a Bacteriology major. This school is one of good reputation. Still, I learned virtually nothing about anaerobes; just that clostridia were anaerobic bacteria and were responsible for some serious and often fatal infections, or intoxications, such as tetanus, botulism, and gas gangrene. There were laboratory sessions for most of the courses we took as bacteriology majors, but we didn't do anything with any clostridia and did not even see pictures of these organisms or of the serious clinical illnesses related to them in our textbooks. There might well have been concern about handling such bacteria in the laboratory since penicillin was only available for the military in 1943 and was in such short supply that urine was saved from patients receiving it so that penicillin could be recovered from it and used again, but there are many benign anaerobes that could have been used in college courses. (As a Navy Corpsman assigned to the clinical microbiology lab at Long Beach Naval Hospital from 1943 to 1945, I was assigned the task of collecting all urine from patients treated with penicillin.)

In medical school (1945 to 1949), I worked part time in the surgical research laboratory of Dr. Edgar Poth who was well known for his studies on so-called "intestinal sulfonamides", used prophylactically in patients having bowel surgery. These compounds were tested initially in dogs and my job was to obtain fecal samples and study the fecal flora using a protocol that was set up previously. For anaerobic flora, we used Brewer plates (special Petri dishes whose lids came down to a very short distance from the agar surface so that the air space was quite limited) with Brewer thioglycollate agar which supported the growth of many anaerobes. What was not known at that time (and I didn't know until sometime later) was that virtually all clinically significant aerobic and microaerophilic bacteria are facultative and grow well (often better) under anaerobic conditions. We did not know to test all organisms recovered on these Brewer agar plates for the ability to grow under nonanaerobic conditions. In fact, there was no identification of anything growing on those plates; we simply determined the "anaerobic counts" by counting colonies on these plates, not even counting different colony types. They had used these procedures for many years before I was involved.

In my postgraduate work in Minneapolis I worked with Dr. Wesley Spink and Dr. Wendell Hall. There was no specialty of Infectious Diseases yet, but I chose their program because they worked with brucellosis and other bacterial infections and I was still very interested in microbiology. During my clinical training, I had a patient with pleural empyema. I removed purulent pleural fluid by thoracentesis; it was putrid and I was surprised when the laboratory told me they didn't grow any bacteria from it. I looked at the Gram stain with the Chief of the Clinical Microbiology Laboratory and we couldn't decide that there were any bacteria present, just pink-staining pleomorphic "globs". I presented this patient at a conference attended by Faculty and students from several teaching hospitals in the city and no one had any suggestions as to what the cause of this infection was. Finally, one of my colleague Fellows, Gordon Riegel, from the University and VA hospitals in Minneapolis, timidly asked whether this might be an anaerobic empyema. Gordon had trained earlier at Johns Hopkins and remembered one professor talking about anaerobic infections and noting that the discharges were often foul smelling and it was difficult to grow these organisms. No one knew how to respond to the Fellow. I discussed the case further with the head of the Clinical Microbiology Lab and she had no other suggestions.

I had another period in military service from 1951 to 1953. Then I got my first real faculty position 62 years ago as a staff physician at the VA Hospital in Los Angeles and on the faculty of the UCLA School of Medicine in the Department of Medicine and the Department of Microbiology, Immunology, and Molecular Genetics. As luck would have it, we had another case of putrid empyema which did not grow any organisms. I recalled the patient from Minneapolis and I discussed the two cases with Vera Sutter, Ph.D., head of the Clinical Microbiology Lab at the VA hospital where the patient was being treated. We looked at the Gram stains together and found the same questionable pleomorphic bacteria. This time I decided I needed to pursue these anaerobes. Vera said she remembered seeing an anaerobic jar in the basement somewhere and searched until she found it. We again cultured pus from this patient, both on plates in the anaerobic jar and on aerobic plates. We again grew no aerobes but recovered two different gram-negative anaerobic bacilli from the plates incubated in the anaerobic jar. I was very lucky that no bacteria grew from either of the two putrid empyema patients. Anaerobic infections very commonly are mixed with aerobes as well as anaerobes. For that reason, anaerobic infections are often overlooked because the aerobic bacteria grow and the infection is attributed to them. I was also unlucky because if there had been gram-positive anaerobes (cocci, for example) present, I would have seen them on the Gram stain and with negative cultures I would have realized there was some kind of fastidious organism present.

I was finally launched on a many-years-long study of anaerobic bacteria. This was no easy task as it required classification, optimum methods of growing and preserving cultures, unique features of the bacteria, and clinical presentations of anaerobic infections. I was amazed to find anaerobes in so many different settings. Early on I found a small green book by Louis D.S. Smith of Montana on nonspore-forming anaerobic bacteria and their activities. As I got into literature searches, I became aware of centuries-old studies by French and German microbiologists in particular; I was amazed at how much they knew in the 1800s. I published *Anaerobic Bacteria in Human Disease* in 1977 summarizing our studies and those of others. My laboratory, with some outside collaborators, published the *Wadsworth Anaerobic Bacteriology Manual* in 1972, now in its sixth edition and called the *Wadsworth-KTL Anaerobic Bacteriology Manual*.

Early in my academic career, and overlapping my new-found major interest in anaerobic bacteria, I also became interested in bowel flora. Neomycin was a newly introduced antibiotic and it was noted that there was little absorption by the oral route, so the levels achieved in the gut were relatively huge. This led to an interest in using this and similar drugs for preoperative preparation of patients for bowel surgery. With my background from Dr. Poth's laboratory, I was very much interested in studying this compound. I started by determining what the impact of oral neomycin was on the bowel microflora. This was so early in my career that I still was not using the anaerobic jar routinely. I made serial 10-fold dilutions of feces and planted them onto various agar plates that would permit recovery of various known colonic bacteria and also planted them into a set of thioglycollate broths. The appearance of the cultures at 48 hours was really striking. There was no growth on any of the plates incubated aerobically, but the thioglycollate broths were turbid all the way out to 10^{12}/ml! Aerobic subcultures from these broths were sterile, but subcultures incubated in anaerobic jars yielded many anaerobic bacteria of various types.

We subsequently learned about other systems for growing anaerobes, including watch glasses placed on the surface of inoculated plates by Professor Haenel of Potsdam, East Germany. These watch glasses were close to the agar surface and early growth of aerobes soon converted the space to an anaerobic environment. It was tedious working with this setup but Haenel managed to do excellent studies of bowel flora with it. Initially we used line gas (methane) in our anaerobic jars; fortunately, it was not so toxic to anaerobes in Los Angeles and we could grow some of them (but didn't know what we might be missing). Later, commercial kits to provide an anaerobic atmosphere with carbon dioxide in jars became available, as did catalysts to remove traces of oxygen. We ultimately switched to anaerobic chambers when these became available, and to tanks of pure nitrogen, hydrogen, and carbon dioxide gases, individually and in appropriate mixtures. Learning to identify anaerobes, even by the crude techniques available at that time, was a problem. Initially, we called them "gray colonies" (the *Bacteroides fragilis* group, it turned out), "clear colonies" (some of these were *Fusobacterium* we later found out), and brown or black "pigmented" colonies on blood or hemoglobin-containing media.

In comparison to the rapid, wide spectrum of analyses performed on a day-to-day basis, this work may seem minimal. Yet, back then, everything was. Take the mere concept of sharing results and/or communicating with colleagues. Today, the communication possibilities are great and one can phone or e-mail anyone and expect to usually get responses that are very helpful and save much time. At present, one can usually easily arrange to visit other laboratories briefly or even arrange to spend several months or even years studying with someone who has perfected techniques and procedures to deal with problems you have not yet coped with yourself. And textbooks and current literature are presently readily available. One can travel to scientific meetings to listen to and even meet leaders in various fields that may be of interest. When I was starting out, these communication benefits were not so readily available. I did write to and subsequently briefly visited several leaders in the United Kingdom, France, and Germany when I had the opportunity to do so. I was fortunate to meet such notable Professors as Garrod, Beerens, and the grand master of anaerobes, Professor André Prévot, who unfortunately was ill on the day I met him and couldn't meet with me for more than half an hour. But in that short period of time, one could gain a wealth of knowledge and even find a direction for future work. Also, it is so much more personable than any electronic media; you have to exist in order to communicate.

Of course, a half-hour talk does little in the context of the second generation systems currently used to study the microbiome. Using a machine such as the Illumina permits rapid detection and identification of complex microbial floras. These can then be catalogued in databases and analyzed using a number of different software methods. This has indeed helped us to better understand the microbes such as those seen in the human colon. But we were able to do many important studies older methods combined with a DNA sequencer and real-time PCR machine. I will comment on some of these studies in the remainder of Chapter 1.

We studied small bowel fluid from a patient with blind loop syndrome and found six different anaerobes and a total anaerobic count one log higher than the total aerobic count. We developed and evaluated several selective media; we improved gas liquid chromatographic procedures for quantitation of fatty acids and alcohols; and we compared the efficiency of anaerobic jars, the Anoxomat system, and anaerobic chambers. We found that antibiotic susceptibility patterns of various anaerobes were useful as guides to classification and characterization of certain anaerobes and studied these patterns with various anaerobes as a guide to therapy of infections with these organisms. We studied the effect of various antimicrobial drugs on the normal bowel flora of patients. We studied the toxins of *Clostridium difficile* and the epidemiology of disease due to this organism in the hospital setting. We studied an outbreak of enterocolitis in our hospital due to phage type 54 staphylococci resistant to kanamycin, neomycin, paromomycin, and chloramphenicol. We studied the normal flora of ileostomy and transverse colostomy effluents and the flora of the maternal cervix and the newborn's gastric fluid and conjunctivae.

We were the first to isolate *Acidaminococcus fermentans* and *Megasphaera elsdenii* from normal human feces. Our group studied the bacteriology of infections in patients undergoing head and neck cancer surgery that provided guidance for the type of antimicrobial prophylaxis that would be most effective in prophylaxis for such patients. We studied the impact of a partially chemically defined diet on the bowel flora of humans. We also had the opportunity to study stool specimens from two patients presenting with d-lactic acidosis; one patient had previously had most of the small bowel removed because of mesenteric thrombosis and the other patient had previously undergone a jejunoileal bypass. The stool flora of both patients was quite abnormal on admission with predominantly gram-positive anaerobic bacilli, *Eubacterium*, *Lactobacillus*, and *Bifidobacterium*, which produced primarily d-lactic acid. The patients responded well to oral vancomycin therapy.

A very important study that we did in collaboration with Dr. Ernst Drenick, an internist and nutritionist, and Dr. Edward Passaro, Jr., a general surgeon, concerned patients undergoing jejunoileal bypass surgery for obesity. The really unique approach of this study was to obtain specimens from patients in the operating room who did not receive any preoperative antimicrobial prophylaxis. Specimens were obtained during surgery from the proximal jejunum and distal ileum. The plan was to obtain similar specimens from any patients who might require surgery for complications relating to the original surgical procedure. We could also compare the data from patients who had only specimens from after the bypass procedure since they were all processed in the same way.

Among eight patients from whom we had baseline studies, the proximal jejunum was sterile in five. The other three had a predominantly aerobic flora with low counts. Only one patient had anaerobes in the jejunum and counts were low. Ileal contents were sterile in two patients; the other six had variable counts. The ileal contents had higher counts than the jejunal contents; the flora resembled fecal flora qualitatively but with lower counts and a higher ratio of aerobes to anaerobes. Only one of the original patients required repeat surgery; he had a sterile jejunum at the first surgery but at re-operation the functioning small bowel was colonized with fecal-type organisms with a total count of $10^{7.5}$/ml.

Looking at the three patients with no baseline studies, one had a high total bacterial count of $>10^9$/ml., another had *Fusobacterium varium* outnumbering the *B. fragilis* group in both the functioning small bowel and in the blind loop. The third case yielded only *E. coli* from the excluded loop. This latter patient, despite a sparse flora, had severe complications suggesting that perhaps toxin production or metabolic behavior might account for some complications. The various complications that may be seen in these bypass patients include an inflammatory bypass enteritis, pneumatosis cystoides intestinalis, impaired liver function, and even fatal hepatic coma, polyarthritis, skin lesions, eye complications, etc. Metronidazole typically was quite effective therapeutically.

We also did microbiology studies in 10 patients undergoing so-called biliopancreatic bypass (Scopinaro procedure). Collection of specimens, only from the bypassed segment (biliopancreatic bowel segment), was done in the operating room at the start of the procedure and with no antibiotic bowel preparation preoperatively. Counts of organisms recovered were relatively low (10^2 to 10^7/ml. Three subjects developed diarrhea that was moderate to severe which responded promptly to metronidazole given orally.

The final notable study we performed was a comparison of bowel flora in different populations with different incidences of colon cancer—Japanese with their traditional diet, Seventh Day Adventists with variable incidences of meat consumption, people on the standard American or Western diet, and people with colonic polyps. This study went on for years thorough bacteriologic studies on their stools as we could in the 1970s. This important study, however, really should be done again with second generation sequencing techniques.

In the past 15 years, we have been studying the fecal flora of children with regressive autism,of autistic children in comparison with that of normal control children, and with that of their siblings. Our first publication (with Sandler *et al.* 2000) was a small open-label study of oral vancomycin but it was important because of the dramatic improvement in virtually all the 10 children treated. All the subjects relapsed after the short treatment course was stopped, but this study established that the clostridia recovered from their stools played a key role in the disease.

A study published in *Clin. Infect. Dis.* in 2002 showed the importance of clostridia and included small bowel aspirates as well as stool. We documented the presence of clostridia by quantitative culture, real-time PCR, and analysis of 16-23 S space region. Bacteria found that were much more frequently found in autistic children than in the control patients were *Clostridium bolteae*, sp. nov., and perhaps some closely related species. A pyrosequencing study was performed and published in 2010. This study led to recognition of five *Desulfovibrio* spedies as role players in autism, the findng that *Bifidobacterium* counts were low in stools of autistic children as compared to controls. We have confirmed the work of others as to the importance of certain *Sutterella* species in autism and of a protective role for *Akkermansia*, as well as *Bifi-*

dobacterium, but have not published this as yet. We have recently found that an unusual clostridial toxin plays a role in autism.

As to where we stand with the colon and indeed the microbiome, even after all the years of work, I have realized we are only at the beginning. Our laboratory has detected a number of novel taxa and studied, named, and reported them, with various colleagues. Included were: *Bilophila wadsworthia, Sutterella wadsworthensis, Clostridium bolteae, Cetobacterium somerae, Anaerotruncus colihominis, Anaerofustus stercorihominis, Clostridium bartlettii, Porphyromonas uenonis, Bacteroides nordii, Bacteroides salyersae, Fastidiosipila sanguinis, Parabacteroides goldsteinii, Porphyromonas somerae, Alistipes onderdonkii, Alistipes shahii, Peptoniphilus duerdenii, Peptoniphilus koeneneniae, Peptoniphilus gorbachii, Peptoniphilus olsenii, Anaerococcus murdochii, Blautia wexlerae, Porphyromonas bennonis, Murdochiella asaccharolytica, Gemella asaccharolytica*, and *Corynebacterium pyruviproducens*. Along with Paul Lawson and others, we have even recommended reclassification of a few organisms, notably the *Ruminococcus* group. This group is now being regarded as one of the three major enterotypes of the gut microbiome. That means all this work is only one-third of the information we have now. As we continue to learn more with even higher levels of analysis, this fraction may diminish even further. Although this may appear at first to be disheartening after over six decades of work, I am happy. While the microbiome continues to expand in its scope, much of which will be discussed in this book, it all started with a general look at the colon and the belief there was much more to the picture. As we continue to learn, that picture is larger than we might have ever imagined.

CHAPTER 2

Ecology of the Human Microbiome

KALUDYNA BOREWICZ and HAUKE SMIDT

2.1. OVERVIEW

RECENT technological and conceptual developments in culture independent approaches targeting bacterial 16S ribosomal RNA (rRNA) genes have offered a new way of looking at microbial ecosystems. This in turn has contributed to the current expansion in the number of research projects aiming at characterizing microbiota composition and function in health and disease. Healthy human microbiota is composed of many complex and diverse microbial ecosystems, with estimated 10^{14} microbial cells inhabiting the human body (Savage 1977). These microbial ecosystems are also unique between different body sites and between individuals, and this variation in microbial composition can be attributed to many factors including host genetics, environment, diet, and early life microbial exposure (Human Microbiome Project 2012). Despite taxonomic differences in microbial community structure, the core metabolic and functional pathways carried out by these ecosystems seem to be relatively stable, suggesting that the role of microbiota in health and disease may be largely due to disturbances in microbial function, rather than changes in microbiota composition alone (Human Microbiome Project 2012).

2.2. MICROBIOTA OF THE GASTROINTESTINAL TRACT

The human gastrointestinal (GI) tract is by far the most densely colonized and best studied microbial ecosystem found in the human body. It is estimated that 1,000–1,500 species of bacteria can inhabit an average adult GI tract, but this number could be even higher (DiBaise 2008). Each person carries approximately 160 bacterial species and about 10 million microbial genes, which give each individual a unique microbial make up (Li 2014). Host genetics may contribute to these individual variations in microbiota, and it has been shown to be an important factor affecting bacterial community composition and function (Moreno-Indias 2014).

Microbial colonization of the GI tract in healthy humans starts at birth and is influenced mainly by the mode of delivery (vaginal versus Caesarean section) and the method of feeding (breast milk versus formula) during infancy (Moreno-Indias 2014). An adult-like microbiota becomes established with introduction of solid foods and begins to resemble microbiota of adults during the first year of life, after which it remains relatively stable throughout adulthood. Diet, infections, antibiotic use, and other environmental conditions can temporarily disturb the normal gut microbial ecosystem, however, these disturbances tend to be temporary and in most cases, the microbiota is able to recover back to its former state. Microbial composition changes in elderly, as the diversity and stability of gut microbiota decrease with age (Moreno-Indias 2014).

Despite the individual variation in microbial composition, the majority of bacterial species found in the human gut belong to two phyla: *Bacteroidetes* and *Firmicutes* (Mariat *et al.* 2009). Most species in the phylum *Bacteroidetes* belong to the class *Bacteroidetes*, and more specifically to the genera *Bacteroides* and *Prevotella*. Most species in the phylum *Firmicutes* belong to *Clostridium* clusters IV and XIVa, which include genera *Clostridium*, *Eubacterium*, and *Ruminococcus*. Other detected phyla include *Proteobacteria, Actinobacteria, Fusobacteria, Spirochaetes, Verrucomicrobia*, and *Lentisphaerae* (Gerritsen *et al.* 2009). In addition to bacterial groups, *Archaea* (methanogens) and eukaryotic microorganisms (fungi) are also part of healthy human gut microbiota.

Metagenomic sequencing data suggests that even with individual differences in microbiota composition, the metabolic pathways remain stable in the GI tract of healthy subjects (Human Microbiome Project 2012). This collection of microbes forms a dynamic ecosystem which is known to exert important metabolic, physiological, and immunological functions on its host, as well as to provide protection from pathogens through so-called colonization resistance (Wade 2013). The host, on the other hand, offers the microbes a stable environment and nutrients necessary for their survival. The general understanding of the microbial ecosystem function has increased tremendously in the recent years, however, the details are still largely unknown. It is becoming clear that the network of interactions, whether these are positive or negative, is very complex and we are now only at the beginning of understanding the roles of different bacterial groups, and how their functions influence the host.

In order to understand how microbial ecosystems contribute in health and disease, we should first know which microbes comprise the healthy human microbiota. More importantly, we need to ascertain the specific roles they perform and how their presence can impact the host. In the following sections we will first give an overview of the key microbial groups and their functions in different regions of a GI tract of healthy adults. Later, we will discuss how changes in microbiota correlate with selected types of diseases.

2.3. MICROBIAL COMPOSITION IN THE GI TRACT OF HEALTHY ADULTS

The human GI tract can be divided in anatomical regions, each characterized by a different set of physicochemical conditions which create a unique environment for microbial growth. The most important factors influencing intestinal microbiota include pH, redox potential, nutrient content, motility, and presence of host secretions such as digestive enzymes, bile, and mucus. The environment at each anatomical region can be further divided into the luminal content and the mucosal layer. The mucosal layer forms a lining along the GI tract and consists of a single sheet of epithelial cells and an irregular coating of mucus that protects the cells from direct action of host secretions, food, and pathogens found in the lumen. The mucosal layer also provides a site of attachment for commensal microbiota. In the following sections, we will describe microbial ecosystems with respect to different regions of the GI tract.

2.3.1. The Oral Cavity

The oral cavity comprises many different niches that provide unique conditions for microbial growth. Most microbes are associated with the mucosal surfaces on the cheeks or tongue, and hard surfaces of teeth, braces, or dentures, and there is no resident microbiota in the lumen, because the passage time of food in the mouth is very short. The oral microbial ecosystem is very diverse, with about 10^{12} bacterial cells of about 1,000 different species belonging to phyla *Actinobacteria, Bacteroidetes, Firmicutes, Proteobacteria, Spirochaetes, Synergistetes,* and *Tenericutes*, candidate phylum TM7, and the uncultured divisions GN02 and SR1 (Wade 2013; Tlaskalova-Hogenova et al. 2011; Soro et al. 2014; He et al. 2015). The relative distribution of each microbial phylum differs between individuals and between location in the mouth (Zaura et al. 2009). The most predominant genera include *Actinomyces, Streptococcus, Neisseria, Veillonella, Porphyromonas*, and *Selenomonas*. In addition, viruses, protozoa, fungi, and a small number of methanogenic *Archaea* are also members of the normal microbiota. The microbial composition at the species level is highly variable between individuals and can be influenced by factors such as age, diet, oral health, and hygiene (Wade 2013).

2.3.2. The Upper Gastrointestinal Tract

The upper gastrointestinal tract includes the esophagus, stomach, and duodenum. In humans, microbial ecosystem composition and function in the upper GI tract are still largely unknown, due to poor accessibility of these areas and the need for invasive procedures to obtain samples. In the surveys on microbiota of the distal esophagus, members of six phyla, namely *Firmicutes, Bacteroides, Actinobacteria, Proteobacteria, Fusobacteria*, and *TM7*, were found in the mucosal layer, and most common genera included *Streptococcus, Prevotella*, and *Veillonella* (Pei et al. 2004; Fillon et al. 2012). Research shows that the distal esophagus is inhabited by a complex but conserved microbial community, with composition resembling the oral microbiota of the host (Pei et al. 2004). Similar to the oral cavity, food does not stay in the esophagus long enough to allow for establishment of resident microbiota. The stomach is the first part of the GI tract that holds food for longer periods of time. Thus, the microbial distribution in the stomach, and in the descending regions of the GI tract, is spatially specific, with different microbes associated with the gastric content and with the mucosal layer (Wang and Yang 2013). Because of its low pH which can only be tolerated by certain acid-resistant bacteria, the bacterial counts in the stomach content are generally low, with about 10^3–10^4 bacterial cells per mL (Tlaskalova-Hogenova et al. 2011). The microbiota of gastric content can vary depending on diet or influx of bacteria from the mouth, esophagus, and duodenum, however, these factors affect to a lesser degree the mucosa-associated microbiota which is protected in the mucus and much more stable (Wang and Yang 2013). Culture independent studies on stomach microbiota showed that in the mucosal layer *Firmicutes, Proteobacteria, Bacteroidetes*, and *Fusobacteria* were the most abundant phyla, and *Streptococcus, Prevotella, Porphyromonas, Neisseria, Haemophilus*, and *Veillonella* were common genera, but the distribution of taxa at genus level was highly variable between individuals (Stearns *et al.* 2011; Bik *et al.* 2006; Li *et al.* 2009). One of the important, and certainly most well-studied species found in about 50% of the human population is *Helicobacter pylori*, which has been associated with gastric diseases such as gastritis and cancer (Wang and Yang 2013). The duodenum is the last part of the upper GI tract and the first part of the small intestine, and it is discussed in Section 2.3.3.

2.3.3. The Small Intestine

The small intestine is the site where most of the host enzymatic digestion and absorption of nutrients, in particular lipids and simple carbohydrates, takes place. Studies on microbial composition are again very limited with the majority of findings being based on biopsy specimens in association with various GI disorders. The duodenal lumen forms a unique environment characterized by a low pH, fast passage time, and the presence of antimicrobial bile and digestive enzymes, making it an unfavourable place for microbial growth. No culture independent studies up to date focused on resident microbiota in human duodenal content. On the other hand, biopsy samples provided insight in microbiota in the duodenal mucosa. In a recent study using 16S rRNA gene-targeted HITChip analysis of duodenal biopsies from children, 13 phylum-like level bacterial groups were detected, and *Proteobacteria, Bacilli*, and *Bacteroidetes* were the most abundant taxa, with each individual subject showing a different and unique microbial profile (Jing Cheng *et al.* 2013). The predominant genus-like groups included *Sutterella wadsworthensis et rel., Streptococcus mitis et rel., Aquabacterium, Streptococcus intermedius et rel.*, and *Prevotella melaninogenica et rel.* (Jing Cheng *et al.* 2013). In a study using sequencing of 16S rRNA gene clone libraries, the most abundant phyla detected in biopsies from children and adult subjects were *Firmicutes, Proteobacteria, Bacteroidetes*, and also *Actinobacteria, Fusobacteria*, and *Deinococcus-Thermus* (Nistal *et al.* 2012). Most sequences were classified as *Streptococcus* and *Prevotella* spp. in both age groups, and 5% of sequences that were found only in healthy children could not be assigned to any known genus. Bacterial community richness was higher in the adult group as compared to the juvenile group, with members of *Veillonella, Neisseria, Haemophilus, Methylobacterium*, and *Mycobacterium* present in adult mucosa. It is interesting to note that overall duodenal microbiota composition seems to resemble the microbiota found in the oral cavity and esophagus, and less so the microbiota found in the lower GI tract (Wacklin *et al.* 2013). The number of bacterial cells and diversity increase along the intestine, and it is estimated that the jejunum harbors 10^5–10^6 bacteria per mL of content (Tlaskalova-Hogenova *et al.* 2011). An earlier study examining mucosa biopsies of human jejunum showed that *Streptococcus* and *Proteobacteria* were the most abundant taxa and contributed respectively to 68% and 13% of all microbiota detected (Wang *et al.* 2005). A more recent study showed that ileostomy effluent samples can provide a good representation of microbial composition in the human jejunum/proximal-ileum without the need for invasive sampling (Zoetendal 2012). The most predominant (common core) taxa in ileostoma-effluent and in jejunum included *Bacilli* (*Streptococcus* spp.), *Clostridium* cluster IX (*Veillonella* spp.), *Clostridium* cluster XIVa, and *Gammaproteobacteria* (Zoetendal 2012). Similar findings came from an earlier study on ileostoma-effluent where the most abundant species were members of the *Lactobacillales* and *Clostridiales*, mainly *Streptococcus bovis*-related species and the *Veillonella* group, as well as species belonging to *Clostridium* cluster I and *Enterococcus* (Booijink *et al.* 2010). However, the ileum-associated *Bacteroidetes* and *Clostridium* clusters III, IV, and XIVa were reduced in ileostoma-effluent samples. Bacterial numbers increase to about 10^8–10^9 cells per mL of ileal digesta. Biopsies and catheter-collected lumen samples revealed that the bacterial community in the human ileum is dominated by species belonging to *Bacteroidetes* and *Clostridium* clusters IV and XIVa and resembles the microbiota found in the colon (Tlaskalova-Hogenova *et al.* 2011; Wang *et al.* 2005). Similar to the ileostomy-effluent samples, ileum microbiota is also characterized by short and long term fluctuations in microbial profiles within individuals and large interindividual variability between patients (Booijink *et al.* 2010).

2.3.4. The Large Intestine

The large intestine is separated from the small intestine by the ileocecal valve, and it can be divided into the cecum; the ascending, transversing, and descending colon; the rectum, and the anal canal. The cecum is the first region of the large intestine that receives food from the small intestine. It is also connected with the appendix—a small and rudimentary projection, which in humans has no function in food digestion, but it may play an important role as a reservoir of microbiota and in stabilizing and restoring the colon microbial ecosystem, especially after disturbance, for example due to antibiotic use (Laurin *et al.* 2011; Bollinger *et al.* 2007). Unlike the small intestine, microbial composition and function of the human large intestine has been studied to great extent, mostly because of the ease of collecting fecal samples, and because of the high density of microbial cells, estimated to be around 10^{11}–10^{12} per mL (Tlaskalova-Hogenova *et al.* 2011). The most predominant microbial groups found in the human large intestine include *Bacteroides*, members of the various *Clostridium* clusters, *Bifidobacterium, Enterobacteriaceae*, and *Eubacterium*. Even though the large intestine can be divided into five anatomical regions, the microbial composition is very uniform, and fecal material seems to represent well the microbiota in the entire region (Gerritsen *et al.* 2011). However, just like in other parts of the GI tract, in the large intestine there is a large difference between microbial ecosystems found in the lumen and mucosal layer. Fecal samples represent the luminal fraction only, and the mucosal layer is much less explored due to the need for more invasive methods in collecting biopsy samples. Large intestinal microbiota is very diverse, highly unique to each individual, and relatively stable over time (Lahti *et al.* 2014). Factors such as age, disease, or the use of antibiotics may permanently alter the microbial composition (Lahti *et al.* 2014). Recent studies utilizing large cohorts of subjects suggested that the fecal microbiota composition in healthy adults can be categorized into three major enterotypes dominated by different bacterial populations, in particular *Bacteroides, Prevotella*, and *Ruminococcus* (Arumugam *et al.* 2011; Benson *et al.* 2010). These enterotypes are independent of age, ethnicity, gender, and body mass. However, this division is still controversial, and some studies failed to detect presence of enterotypes in both the elderly (Claesson 2012) and in adult research populations (Huse *et al.* 2012).

Another large study suggested an alternative to the enterotype theory (Lahti *et al.* 2014). The authors noted that in fecal samples of Western adults, certain bacterial groups, namely *Dialister* spp., *Bacteroides fragilis, Prevotella melaninogenica, P. oralis*, and two groups of uncultured

Clostridiales cluster I and II, were bimodally distributed in the healthy human population, representing so called "tipping elements" (Lahti *et al.* 2014). These bistable bacterial groups were either very abundant or almost absent, and unstable at their intermediate abundance levels (Lahti *et al.* 2014). In addition, the condition of the bistable groups, especially the *Bacteroides* and *Prevotella*, seemed to correlate with the shifts in other bacteria, and as a result they were believed to be driving the overall composition of the colonic ecosystem towards specific enterotypes (Lahti *et al.* 2014).

2.4. MICROBIAL ECOSYSTEM FUNCTION IN THE GI TRACT OF HEALTHY ADULTS

Metagenomic studies provide insight on the functional potential of microbiota by analyzing microbial genes, collectively known as the microbiome. A recent study reported that each person carries about 10 million bacterial genes in their GI tract, the majority of which are involved in bacterial metabolism (Li *et al.* 2014; Turnbaugh *et al.* 2009). Additional information about microbial activity can be obtained from metatranscriptomics, metabolomics, and metaproteomics analyses. These approaches provide insight about microbial gene regulation and expression, as well as the production of metabolites, proteins, vitamins, and regulatory elements. Similar to compositional diversity, there is a large functional variation in different microbial ecosystems, but the core metabolic and functional pathways carried out by the same types of ecosystems seem to be relatively conserved and stable (Human Microbiome Project 2012). It is also common for the same metabolic functions to be carried out by different bacterial groups, meaning that correlating the compositional and functional changes in the ecosystem maybe less straightforward because changes in the composition and the function of a given microbial ecosystem can be independent from each other (Zoetendal 2008).

2.4.1. The Oral Cavity

The oral cavity is the first point of contact between microbiota, diet, and host. Despite regular influx of food ingested by the host, the majority of nutrients for the oral commensal microbes are derived from glycoproteins present in saliva and gingival crevicular fluid (Homer *et al.* 1999). Complete breakdown of these glycoproteins requires cooperation between different species of bacteria. For example, oral streptococci (e.g., *S. oralis, S. sangiunis*) remove oligosaccharide side chains and break down the protein core by their proteolytic, endopeptidase, and glycosidic activity, while other Gram-negative anaerobes (e.g., *Porphyromonas gingivalis, Prevotella intermedia, Prevotella nigrescens*, and *Peptostreptoccus micros*) further break down proteins into peptides and amino acids (Homer *et al.* 1999; Wickstrom *et al.* 2009; Bao *et al.* 2008). Amino acids can then be fermented to short chain fatty acids (SCFA), including branched chain fatty acids, which are further degraded by other bacteria and by methanogenic Archaea (Wade 2013). Certain food components, such as gluten or nitrate can also be degraded/transformed by microbial enzymes, and the processes and products are crucial for the health and well-being of the host, while breakdown of these functions can be linked with host diseases (Hezel and Weitzberg 2013; Helmerhorst 2010; Zamakhchari 2011). As already mentioned, the mouth is an open environment and commensal bacteria create a barrier against colonisation with transient microbes and any opportunistic pathogens that can enter with food or water. An *in vitro* study on oral microbiota from mice provided a good illustration of how the cooperation of different commensal species can leverage a community response to pathogen invasion. The study proposed that cooperation of three different species of oral streptococci were involved, with *S. saprophyticus* sensing the presence of an invader, and initiating the defence pathway, *S. infantis* acting as a mediator, and *S. sanguinis* producing hydrogen peroxide and acting as a killer (He *et al.* 2014). Besides colonization resistance, oral microbiota plays an important role in maintaining host-microbe homeostasis, by interacting with host mucosal cells and training the host's immune system to recognize and destroy pathogens, while down-regulating the proinflammatory immune response towards the commensal bacteria normally present in the mouth (Srinivasan 2010).

2.4.2. Upper Gastrointestinal Tract

Upper gastrointestinal tract microbiota function is still not well understood, and most studies to date focused on specific pathogens and their role in the aetiology of different diseases and to a lesser extent on the microbial interactions in a healthy ecosystem. Little is known about the ecology of microbiota inhabiting the esophagus and stomach, but its role in colonization resistance and protection from pathogens is likely to be an important one. Normal microbiota generates a microenvironment that can inhibit growth of pathogens by competing for substrates and binding sites, stimulating host immune responses against invaders and production of antimicrobial substances. For example, *in vitro* and *in vivo* studies using animal models showed that stomach colonization with *H. pylori* is inhibited by the normal commensal microbiota and by probiotic strains of *Lactobacillus, Bifidobacterium*, and *Saccharomyces*, suggesting the importance of microbial interaction in pathogen resistance (Wang and Yang 2013). Other studies using human biopsy samples also reported changes in intestinal microbiota associated with gastric cancer, however, the exact function and causality of this association is still being investigated (Tlaskalove-Hogenova 2011). It is likely that microbial metabolites, bacterial lipopolysaccharides (LPS), lipoproteins, lipoteichoeic acids (LTA), flaggellins, and bacterial nucleic acids can interfere with the normal function of gastric mucosa, causing chronic inflammation, changes in mucin production, metaplasia, and eventually can lead to diseases (Tlaskalove-Hogenova 2011; Jing Cheng *et al.* 2013). The functions of the microbiota in the duodenum are still not well understood, but changes in microbial composition between Celiac disease patients and healthy controls suggest that the microbiota plays a role in immune response, inflammation, and maintaining gut homeostasis (Jing Cheng *et al.* 2013; Wacklin *et al.* 2013). The homeostasis of gut epithelia relies to a large extent on adequate activation of toll-like receptors (TLRs), which recognize microbe-associated motifs, regulate the immune response to pathogens, and affect the epithelial barrier by regulating the expression of tight junction proteins, mucin, and antimicrobial peptides by the host's intestinal cells (Jing Cheng *et al.* 2013).

2.4.3. The Small Intestine

The small intestine is the site where most of the host enzymatic digestion and absorption of energy from the diet takes place. Thus, diet is an important factor modulating microbial function, by selecting bacterial groups that are better equipped to break down different dietary substrates (Moreno-Indias 2014). For example, certain *Lactobacillus* spp. found in duodenum and jejunum have been associated with weight gain and leanness, and differ in their metabolic capacities to break down dietary carbohydrates and fats supplied by the host (Moreno-Indias 2014). The transit time in the small intestine is very short, and *Streptococcus* and *Veillonella* spp., which dominate the microbial ecosystem in the jejunum and ileum, are well adapted to quickly metabolize a variety of available carbohydrates, first to lactate (*Streptococcus*) and then to acetate and propionate (*Veillonella*) (Booijink *et al.* 2010). Recent metatranscriptome analysis of the ileostoma effluent confirmed a high abundance of genes involved in the transport and metabolism of diet-derived simple carbohydrates and linked the task mainly to *Streptococcus* groups (Aidy *et al.* 2015). In addition to its function in carbohydrate metabolism, it was concluded that small intestine microbiota could also play a key role in immune system development and homeostasis. For example, the ileum is connected with a large mass of gut associated lymphoid tissue (GALT) and Peyer's patches, and commensal bacteria, such as different strains of streptococci, were shown to induce specific immune responses in the host (Aidy *et al.* 2015). The close contact between the microbiota and the host cells in the small intestine underlines the current hypothesis that microbially derived metabolites or toxins also modulate gene expression via the gut-brain neural circuit and may influence endocrine function (e.g., secretion of glucagon and incretins) and even show an effect on mood or behavior of the host (Moreno-Indias 2014; Aidy *et al.* 2015).

2.4.4. The Large Intestine

Large intestine microbial ecosystem function has been well studied, mainly due to the ease of collecting fecal samples, but also because it has been known for a long time that colonic microbial processes play an important role in human health. The most direct role is in the digestion and metabolism, as the large intestinal microbiota breaks down indigestible food components and provides the host with an otherwise inaccessible source of energy. It also produces SCFA which are the main source of energy for colonocytes (Leser and Molbak 2009). In addition, the colonic microbiota is a main source of vitamins K and B12, it prevents colonization by pathogens, and it plays an important role in regulating the host's immune responses (Moreno-Indias *et al.* 2014; Leser and Molbak 2009). A study on the fecal microbiome of healthy Japanese subjects was among the first to explore microbial ecosystem function in the human colon using culture-independent methods. The study revealed that a high proportion of genes present were related to carbohydrate metabolism and transport. The authors also noted an enrichment of peptidases and enzymes for anaerobic pyruvate metabolism and reduction in genes involved in fatty-acid metabolism. There were also high levels of enzymes involved in energy storage, antimicrobial peptide transport, and multidrug efflux pump peptides (Kurokawa *et al.* 2007). The authors concluded that these enzymes may help certain commensal microbes to compete with each other and thus may be essential for maintenance of ecosystem balance. Enzymes for DNA repair were also enriched. On the other hand, there was a low abundance of genes involved in biosynthesis of flagella and chemotaxis and in oxygen take-up (Kurokawa *et al.* 2007). Interestingly, these patterns in gene distribution were not observed in unweaned infants, suggesting that infant microbiota is less complex and thus microbial ecosystem function is less stable, more dynamic, and highly adaptable. In adult microbiota a higher diversity of bacterial species exists with large interindividual variability in microbial composition, yet there is a shared functional core, which is believed to be stable and much more uniform between individuals (Turnbaugh *et al.* 2009; (Kurokawa *et al.* 2007). Recently, more in depth analyses showed that there could be functional differences correlating with different enterotypes found in the colon (Arumugam *et al.* 2010). For example, the *Bacteroides*-rich type has more bacterial species that are capable of producing vitamins C, B2, B5, and H. This group is dominated by species that utilize carbohydrate fermentation as the main energy source. On the other hand, the *Prevotella* type showed higher numbers of species producing vitamin B1 and folic acid, and included species that use mucin glycoproteins as a source of energy, similarly to the *Ruminococcus* type (Arumugam *et al.* 2010).

One of the important functions of colonic microbiota that received a lot of attention in recent years is the production of SCFA, and in particular butyrate, by bacteria from *Clostridium* clusters IV and XIVa. The main butyrate-producing species are believed to be *Eubacterium rectale* and *Faecalibacterium prausnitzii*, in addition to others in the genera *Coprococcus* and *Roseburia* (Louis and Flint 2009). The process provides a great example of synergic interaction between diet, microbes, and host, and the presence of butyrate producers in the colon has been shown to be negatively correlated with functional dysbiosis, reduction of the risk of infections with opportunistic pathogens, and the decrease in oxidative stress (Moreno-Indias *et al.* 2014). Butyrate producers can respond to different environmental conditions, such as diet or pH, and engage different fermentation pathways in which the final products are lactate, formate, hydrogen, and carbon dioxide. It has been shown that cross-feeding between bifidobacteria and butyrate producers is also possible: bifidobacteria break down polysaccharides and produce lactate and acetate which are further utilized by butyrate-producers to form butyrate (Louis and Flint 2009). Butyrate is known to play an important role in maintaining homeostasis of the intestine. It is the main source of energy for colonocytes and it inhibits expression of proinflammatory cytokines in the mucosal layer of intestine (Segain *et al.* 2000). In addition, butyrate has a positive effect on integrity of the mucosal layer by stimulating the expression of tight junction proteins and by inducing the production of mucin and antimicrobial peptides (Fava and Danese 2011).

2.5. SELECTED DISEASES ASSOCIATED WITH DYSBIOSIS OF THE INTESTINAL MICROBIOTA

2.5.1. Inflammatory Bowel Disease

Inflammatory Bowel Disease (IBD) is a chronic inflammatory disorder affecting the mucosal layer of the intestines. The two main types of IBD are Crohn's disease and ulcerative colitis. Factors such as genetics, diet, gut permeability, stress, and microbiota changes seem to be contributing factors in development of IBD. Both Crohn's disease and ulcerative colitis are linked with a decrease in microbial diversity (Hansen *et al.* 2010), reduced levels of *Firmicutes*, especially *Faecalibacterium prausnitzii* (Benjamin *et al.* 2012; Varela *et al.* 2013), *Ruminococcaceae* and *Roseburia* (Morgan *et al.* 2012), and increased levels of *Enterobacteriaceae* (Hansen *et al.* 2010), *Bacteroides, Prevotella* (Benjamin *et al.* 2012), adherent-invasive *Escherichia coli, Campylobacter concisus*, and enterohepatic *Helicobacter* (Sanchez *et al.* 2013). One of the characteristics of IBD is a decrease in microbial SCFA production. This can be due to reduction in the abundance of the two main butyrate producing bacteria: *Faecalibacterium*—a commensal bacterium with anti-inflammatory properties—and *Roseburia*, which also produces butyrate from *Ruminococcaceae* derived acetate (Morgan *et al.* 2012). The decrease in butyrate-producing bacteria is often accompanied by an increase in sulphate-reducing bacteria, which produce toxic hydrogen sulphide. Hydrogen sulphide blocks butyrate utilization by colonocytes and interferes with innate immune defence (Fava and Danese 2011). In both Crohn's disease and ulcerative colitis, the leaky epithelial barrier and reduced innate immune defence lead to increased translocation of bacteria through the lamina propria and an increase in an inflammatory reaction and the formation of ulcers. Despite growing evidence that

IBD is linked with dysbiosis of intestinal microbiota and with changes in metagenomic pathways, it is still unclear if the changes in microbiota are the cause or the consequence of the intestinal inflammation (Fava and Danese 2011; Morgan *et al.* 2012). Recent studies suggest that mutations in certain host genes coding for receptors involved in bacterial recognition and killing of pathogens, such as TLR4 D299G, TLR1L80P, NOD 2/CARD15, Arg702Trp, Gly908Arg, and Leu1007, might be an important risk factor in aetiology of IBD (Cario 2005; Pierik *et al.* 2006; Ogura *et al.* 2001).

2.5.2. Obesity

Obesity has many risk factors which together lead to perturbations in energy balance and weight gain and may result in development of other metabolic and cardiovascular diseases (Turnbaugh *et al.* 2009). High calorie diet is a main risk factor for obesity, but recent studies show that gut microbiota may also play a role, for example by breaking down indigestible dietary substrates and providing additional energy to the host (Turnbaugh *et al.* 2006). Surplus energy is stored in a form of fat, and excessive fat accumulation leads to weight gain and obesity. For example, it has been shown that germ-free (GF) mice fed diets rich in fat and sugar (the "Western diet") did not develop obesity (Backhed *et al.* 2007). In turn, transferring gut microbiota of obese mice into GF mice led to rapid increase in body fat, despite of restricted calorie intake (Turnbaugh *et al.* 2006). Diet not only provides energy to the host, but it also may affect gut microbiota composition and function, selecting for species that are best adapted to utilize different dietary compounds (Scott *et al.* 2008; Zhang *et al.* 2010). These diet-induced changes in the microbial composition and function have been linked with obesity, but the specific effect on different bacterial groups and the causality are less clear (Scott *et al.* 2008; Zhang *et al.* 2010). Earlier studies on microbiota of obese humans and mice reported an increased ratio of *Firmicutes* to *Bacteroidetes* (Armougom *et al.* 2009), however, recent studies contradict these findings and suggest that obesity might be linked with a decreased ratio of *Firmicutes* to *Bacteroidetes* (Schwiertz *et al.* 2010; Duncan *et al.* 2008), the increase in other bacterial phyla, for example *Actinobacteria* (Turnbaugh *et al.* 2009), or *Proteobacteria* (Zhang *et al.* 2009), or the overall decrease in microbial diversity (Turnbaugh *et al.* 2009). The inconsistencies in these findings might be partially due to differences in the research methods used, but could also relate to the differences in host genotypes which had been shown to influence microbial composition and could predispose certain individuals to developing metabolic conditions (Zoetendal *et al.* 2001). It is also likely that the obesity-associated changes in the gut microbial ecosystem structure and function are more refined than the phylum level. Reports on obesity and microbial changes at a genus or species level are still limited, but a recent study found that obese people had higher fecal levels of *Lactobacillus reuteri, Bifidobacterium animalis*, and *Methanobrevibacter smithii* and lower levels of *B. animalis, L. paracasei*, and *L. plantarum* than the lean controls (Million *et al.* 2012). Studies on microbiota function in obese and lean subjects show that shifts in relative abundance of microbial phyla are accompanied by changes in metabolic pathways involved in carbohydrate metabolism and SCFA production (Turnbaugh *et al.* 2006, 2009; Schwiertz *et al.* 2010). The increase in bacterial enzymes involved in the degradation of indigestible polysaccharides increases the levels of monosaccharides which become readily available to the host. In addition, there is an increase in the production of SCFA which become converted into triglycerides in the liver (Schwiertz *et al.* 2010). SCFA can activate G-protein coupled receptors (GPR41 and GPR43) in the gut and induce the secretion of PYY peptide which decrease intestinal transit, allowing longer time for nutrient uptake (Erejuwa *et al.* 2014). At the current state of research, it is still unclear whether changes in microbiota are a contributing factor causing obesity, or whether microbial dysbiosis is the result of obesity. A recent prospective study suggested that changes in microbiota, such as a decrease in *Staphylococcus aureus* and an increase in *Bifidobacterium* spp., preceded development of obesity in a group of children (Kalliomaki *et al.* 2008). Transplantation studies on GF mice provide another line of evidence that the composition and function of the intestinal microbiota is an important factor in aetiology of obesity.

2.5.3. Diabetes Mellitus Type 1 and Type 2

Diabetes mellitus type 1 and type 2 are metabolic diseases characterized by insufficient production of insulin and insulin resistance, respectively, which lead to high blood glucose levels. Recent study on rats and children with type 1 diabetes (T1D) showed that changes in gut microbiota may play a role in the aetiology of these diseases (Roesch *et al.* 2009; Murri *et al.* 2013). Diabetic children had higher numbers of *Clostridium, Bacteroides*, and *Veillonella* and lower numbers of *Lacto-*

bacillus, Bifidobacterium, C. coccoides-E. rectale group, and *Prevotella*. Type 2 diabetes (T2D) has been associated with changes in gut microbiota composition and function. In T2D, the ratio of *Firmicutes* to *Bacteroidetes* was reduced and the *Bacteroides-Prevotella* to *C. coccoides-E. rectale* group ratio was increased in patients with elevated plasma glucose levels (Larsen *et al.* 2010). In addition, *Clostridia* levels were lower and *Betaproteobacteria* levels were higher in the diabetic group, as compared to nondiabetic controls (Larsen *et al.* 2010). The mechanism by which altered gut microbiota and diabetes interact is complex and is likely to proceed through a cascade of events. Members of the *C. coccoides/E. rectale* group are the main butyrate producers in the human colon, and butyrate is important for maintaining the integrity of the intestinal barrier and protecting the host against invasion by opportunistic pathogens and the transfer of endotoxins into plasma. Thus, the decrease in *C. coccoides/E. rectale* and the corresponding increase in the level of *Bacteroidetes* could be linked with metabolic endotoxemia (Qin *et al.* 2012). *Bacteroidetes* are Gram-negative bacteria containing lipopolysaccharides (LPS) in their outer membrane. LPS are endotoxins, and increased LPS levels in the colon, as well as changes in gut permeability, result in higher LPS levels in blood serum. LPS had been shown to induce production of proinflammatory molecules by macrophages and to disrupt the function of pancreatic β-cells and insulin secretion (Rodes 2013). The same study concluded that specific probiotic strains of *Lactobacillus* and *Bifidobacterium* could decrease LPS levels in the colon and inflammation (Rodes 2013). Similarly, in another study higher levels of *Bifidobacterium* spp. were associated with the increase in the production of YY and glucagon-like peptide (GLP), reduced gut permeability, and decreased LPS in plasma (Cani *et al.* 2009).

2.5.4. Colorectal Cancer

Colorectal cancer (CRC) is one of the most common forms of cancer (Chan and Giovannucci 2010). High-calorie diets rich in animal fat, red and processed meat, or alcohol, and low in dietary fiber, whole grains, and vegetables are considered a main risk factor (Chan and Giovannucci 2010; Huycke and Gaskins 2004). Gut microbiota structure and function is largely dependent on diet, and certain bacterial metabolites are known to be proinflammatory and tumor inducing. Normal colonic microbiota are composed of members of *Clostridium* clusters IV and XIVa, *Lacto-*

bacillales, Bifidobacteriales, and *Actinomycetales* which are believed to have a protective effect in the development of CRC because of their role in the synthesis of butyrate and other SCFA, as well as conjugated linoleic acids (Louis and Flint 2009; Devillard *et al.* 2007). It has been noted that in CRC patients there is a significant decrease in *Firmicutes*, in particular *Roseburia* spp. and *Eubacterium* spp., and as a consequence, a reduction in butyrate production (Wang *et al.* 2012). Butyrate has an important role in reducing inflammation, suppressing precancerous cells, and inducing apoptosis of tumor cells (Wang *et al.* 2012). Changes in microbial composition allow opportunistic pathogens, such as *Enterococcus, Streptococcus*, and *Escherichia/Shigella* to proliferate and cause damage to the gut epithelial cells (Martin *et al.* 2004). There is a wide range of bacterial metabolites which cause damage to the DNA in the host's gut epithelial cells and may lead to chromosomal instability and the development of CRC. For example, fecapentaenes produced by *Bacteroides* spp., heterocyclic amines produced by *Salmonella typhimurium* and *Enterococcus* spp., and hydrogen sulphide produced by sulphate-reducing bacteria such as *Desulfovibrio, Desulfomonas, Desulfotomaculum, Desulfobulbus, Desulfobacter, Desulfococcus, Desulfosarcina*, and *Desulfonema* have all been implicated to have a role in development of CRC (Huycke and Gaskins 2004). Other mechanisms include high superoxide production by *Enterococcus faecalis* (Huycke and Abrams 2002), the metabolism of 7α-dehydroxylating bacteria of the genera *Eubacterium* and *Clostridium* (Wells 2000), and the formation of colonic lesions by *Streptococcus bovis/gallolyticus* (Abdulamir 2011). For each of the associations listed above, the exact mechanism of action in the formation of CRC is still being investigated.

2.5.5. Celiac Disease

Celiac Disease (CD) is an autoimmune, inflammatory disorder of the small intestine triggered by a diet containing gluten proteins found in wheat. It has a genetic component and seems to affect mainly individuals who carry the leukocyte antigen alleles HLA-DQ2 or HLA-DQ8 [83]. Since not all genetically predisposed individuals develop the disease, it has been proposed that changes in gut microbiota may also play a role in aetiology of CD (Jing Cheng *et al.* 2013; Nadal *et al.* 2007). Earlier studies using duodenal biopsy samples reported increased microbial diversity in CD patients, decreased ratio of *Lactobacillus–Bifidobacterium/Bacteroides–E. coli* (Nadal *et al.* 2007), and higher levels of *Bacteroides vulgatus* and *Escherichia coli* (Schippa *et al.* 2005). Similarly, a more recent study found higher diversity in duodenal mucosa and a lower ratio of *Firmicutes/Proteobacteria* in children with active CD, as compared to those with the nonactive disease and healthy controls (Sanchez *et al.* 2013). The same study reported that CD patients had a higher abundance of the families *Enterobacteriaceae* and *Staphylococcaceae*, particularly the species *Klebsiella oxytoca, Staphylococcus epidermidis*, and *Staphylococcus pasteuri*, and controls had more *Streptococcus anginosus* and *Streptococcus mutans* (Sanchez *et al.* 2013). Studies on fecal microbiota in CD-predisposed and healthy infants noted major differences between the two groups, however, the conclusions regarding changes in specific bacterial groups were contradicting, with one study finding very low levels of *Bacteroidetes* in CD infants (Sellitto *et al.* 2012), and the other study reporting a reduction in *Bifidobacterium* spp. and *B. longum*, but an increase in *B. fragilis* group and *Staphylococcus* spp. (De Palma *et al.* 2012). Finally, few studies reported no changes in the microbial composition in relation to the disease status, but noted changes in the TLR signalling pathways, which are involved in inflammatory responses and the expression of tight junction proteins important in maintaining the integrity of the intestinal mucosa (Jing Cheng *et al.* 2013; Kalliomaki *et al.* 2012). The causality of the association is still unknown but just like in other metabolic diseases, decrease in commensal populations and the increase in levels of Gram-negative or pathogenic bacteria could contribute to the pathogenesis of CD by altering intestinal permeability and inducing inflammation (Nadal *et al.* 2007). In addition, recent studies suggest the dysbiosis in oral microbiota could also play a role in the aetiology of this disease (Helmerhorst *et al.* 2010; Zamakhchari *et al.* 2011).

2.5.6. Dental Caries and Periodontal Disease

The main risk factors in the formation of dental caries are frequent sugar intake and low saliva production, both of which promote growth of aciduric and acidogenic strains of lactate producing streptoccoci and lactobacilli. On the other hand, certain strains had been linked with carries-free status. These include *Streptococcus oligofermentas* which inhibits growth of *S. mutans* (Liu *et al.* 2012), and *Porphyromonas catoniae* (Tong 2003). *Porphyromonas gingivalis, Aggregatibacter actinomycetemcomitans, Prevotella intermedia, Tannerella forsythia,*

Parvimonas micra, Fusobacterium nucleatum, and *Treponema denticola* have all been associated with periodontal disease. However, lactic acid-producing bacteria, such as *Streptococcus cristatus, S. salivarius, S. mitis*, and *S. sanguinis*, as well as probiotic *Lactobacillus brevis* and *L. reuteri*, were shown to attenuate inflammatory markers associated with periodontitis, to produce antimicrobial agents, and to reduce inflammation in periodontal cases (Riccia *et al.* 2007; Sliepen *et al.* 2009; Teughels *et al.* 2013).

2.5.7. Other Health Conditions

Other health conditions have been associated with changes in gut microbiota. Diseases such as Irritable Bowel Syndrome (IBS) (Collins 2014; Ohman and Simren 2013), antibiotic-associated diarrhoea (Young and Schmidt 2004), pouchitis (Angriman *et al.* 2014), necrotizing enterocolitis (Stewart *et al.* 2012), gastric ulcers, esophagitis (Jensen *et al.* 2013), Barrett's oesophagus (Wang and Yang 2013) and malnutrition (Tilg and Moschen 2013) are just some examples of diseases linked with changes in gut microbiota composition and function.

However, the impact of gut microbiota on hosts goes beyond its direct effect on the function of the digestive tract. Many other health conditions are now being associated with changes in the structure and function of microbial ecosystems in the gut, but also in other body sites. An interesting example comes from the studies on hypertension and the role of oral microbiota. Recent studies on nitrate supplementation and hypertension suggested that facultative anaerobic oral bacteria (in particular *Streptococcus salivarius, S. mitis, S. bovis, Veionella* spp., *Staphylococcus aureus, S. epidermidis, Nocardia* spp., and *Corynebacterium* spp.) may play an important role in nitrate metabolism, by reducing nitrate to nitrite (Hezel and Weitzberg 2013). Nitrite can be absorbed and converted to nitric oxide, which is essential for maintaining vascular health by reducing hypertension and lowering blood pressure. In a study on healthy subjects, the use of antimicrobial mouth rinse eliminated the beneficial effect of nitrate supplements, suggesting that oral microbiota may contribute to maintaining cardiovascular health (Govoni *et al.* 2008). As discussed earlier, the development of the immune system also seems to largely depend on microbiota. The new "hygiene hypothesis" claims that limiting early-life infection impedes natural immune system development and causes predisposition to allergic disease (Wold 1998), atopic eczema, allergic rhinoconjunctivitis, and asthma (Trompette *et al.* 2014). Reduced microbial diversity during infancy has been associated with an array of allergic diseases later in life (Bisgaard *et al.* 2011). Also, other autoimmune diseases, such as multiple sclerosis, lupus, and rheumatoid arthritis have been shown to be correlated with changes in gut microbiota (Tlaskalova-Hogenova *et al.* 2011; Kamada *et al.* 2013). Finally, recent studies suggest that certain mental conditions, such as depression, anxiety, and autism may all have a microbiota dysbiosis component in their aetiology (Collins 2014; Malkki 2014; Hsiao *et al.* 2013; Foster and McVey Neufeld 2013).

2.6. ACKNOWLEDGMENTS

The authors are grateful for financial support through the TKI Agri&Food program as coordinated by the Carbohydrate Competence Center (CCC3; www.cccresearch.nl).

2.7. REFERENCES

Abdulamir, A.S., R.R. Hafidh, and F. Abu Bakar 2011. The association of *Streptococcus* bovis/gallolyticus with colorectal tumors: The nature and the underlying mechanisms of its etiological role. *J Exp Clin Cancer Res. 30*: p. 11.

Aidy, S.E., B.V.D. Bogert, and M. Kleerebezem. 2015. The small intestine microbiota, nutritional modulation and relevance for health. *Current Opinion in Biotechnology. 32*: p. 14–20.

Angriman, I., M. Scarpa, and I. Castagliuolo. 2014. Relationship between pouch microbiota and pouchitis following restorative proctocolectomy for ulcerative colitis. *World J Gastroenterol. 20*(29): p. 9665–74.

Armougom, F., *et al.* 2009. Monitoring bacterial community of human gut microbiota reveals an increase in *Lactobacillus* in obese patients and Methanogens in anorexic patients. *PLoS One. 4*(9): p. e7125.

Arumugam, M., *et al.* 2011. Enterotypes of the human gut microbiome. *Nature. 473*(7346): p. 174–80.

Backhed, F., *et al.* 2007. Mechanisms underlying the resistance to diet-induced obesity in germ-free mice. *Proc Natl Acad Sci USA. 104*(3): p. 979–84.

Bao, G.J., *et al.* 2008. Proteolytic activities of oral bacteria on ProMMP-9 and the effect of synthetic proteinase inhibitors. *Open Dent J. 2*: p. 96–102.

Benjamin, J.L., *et al.* 2012. Smokers with active Crohn's disease have a clinically relevant dysbiosis of the gastrointestinal microbiota. *Inflamm Bowel Dis. 18*(6): p. 1092–100.

Benson, A.K., *et al.* 2010. Individuality in gut microbiota composition is a complex polygenic trait shaped by multiple environmental and host genetic factors. *Proc Natl Acad Sci US. 107*(44): p. 18933–8.

Bik, E.M., et al. 2006. Molecular analysis of the bacterial microbiota in the human stomach. *Proc Natl Acad Sci USA. 103*(3): p. 732–7.

Bisgaard, H., et al. 2011. Reduced diversity of the intestinal microbiota during infancy is associated with increased risk of allergic disease at school age. *J Allergy Clin Immunol. 128*(3): p. 646–52 e1-5.

Bollinger, R.R., et al. 2007. Biofilms in the large bowel suggest an apparent function of the human vermiform appendix. *J Theor Biol. 249*(4): p. 826–31.

Booijink, C.C., et al. 2010. High temporal and inter-individual variation detected in the human ileal microbiota. *Environ Microbiol. 12*(12): p. 3213–27.

Cani, P.D., et al. 2009. Changes in gut microbiota control inflammation in obese mice through a mechanism involving GLP-2-driven improvement of gut permeability. *Gut. 58*(8): p. 1091–103.

Cario, E. 2005. Bacterial interactions with cells of the intestinal mucosa: Toll-like receptors and NOD2. *Gut. 54*(8): p. 1182–93.

Chan, A.T. and E.L. Giovannucci 2010. Primary prevention of colorectal cancer. *Gastroenterology. 138*(6): p. 2029–2043 e10.

Claesson, M.J., et al. 2012. Gut microbiota composition correlates with diet and health in the elderly. *Nature. 488*(7410): p. 178–84.

Collins, S.M. 2014. A role for the gut microbiota in IBS. *Nat Rev Gastroenterol Hepatol. 11*(8): p. 497–505.

De Palma, G., et al. 2012. Influence of milk-feeding type and genetic risk of developing coeliac disease on intestinal microbiota of infants: The PROFICEL study. *PLoS ONE. 7*(2): p. e30791.

Devillard, E., et al. 2007. Metabolism of linoleic acid by human gut bacteria: Different routes for biosynthesis of conjugated linoleic acid. *J Bacteriol. 189*(6): p. 2566–70.

DiBaise, J.K., et al. 2008. Gut microbiota and its possible relationship with obesity. *Mayo Clin Proc. 83*(4): p. 460–9.

Duncan, S.H., et al. 2008. Human colonic microbiota associated with diet, obesity and weight loss. *Int J Obes (Lond). 32*(11): p. 1720–4.

Erejuwa, O.O., S.A. Sulaiman, and M.S. Ab Wahab. 2014. Modulation of gut microbiota in the management of metabolic disorders: The prospects and challenges. *Int J Mol Sci. 15*(3): p. 4158–88.

Fava, F. and S. Danese. 2011. Intestinal microbiota in inflammatory bowel disease: Friend of foe? *World J Gastroenterol. 17*(5): p. 557–66.

Fillon, S.A., et al. 2012. Novel device to sample the esophageal microbiome--the esophageal string test. *PLoS One. 7*(9): p. e42938.

Foster, J.A. and K.A. McVey Neufeld. 2013. Gut-brain axis: How the microbiome influences anxiety and depression. *Trends Neurosci. 36*(5): p. 305–12.

Gerritsen, J., et al. 2011. Intestinal microbiota in human health and disease: The impact of probiotics. *Genes Nutr. 6*(3): p. 209–40.

Govoni, M., et al. 2008. The increase in plasma nitrite after a dietary nitrate load is markedly attenuated by an antibacterial mouthwash. *Nitric Oxide. 19*(4): p. 333–7.

Hansen, J., A. Gulati, and R.B. Sartor. 2010. The role of mucosal immunity and host genetics in defining intestinal commensal bacteria. *Curr Opin Gastroenterol.* 26(6): p. 564–71.

He, X., *et al.* 2015. Cultivation of a human-associated TM7 phylotype reveals a reduced genome and epibiotic parasitic lifestyle. *Proc Natl Acad Sci USA.* 112: p. 244–249.

He, X., *et al.* 2014. The social structure of microbial community involved in colonization resistance. *ISME.* 8(3): p. 564–74.

Helmerhorst, E.J., *et al.* 2010. Discovery of a novel and rich source of gluten-degrading microbial enzymes in the oral cavity. *PLoS One.* 5(10): p. e13264.

Hezel, M. and E. Weitzberg. 2013. The oral microbiome and nitric oxide homoeostasis. *Oral Dis.*

Homer, K.A., R.A. Whiley, and D. Beighton. 1999. Proteolytic activity of oral streptococci. *FEMS Microbiology Letters.* 67(3): p. 257–260.

Hsiao, E.Y., *et al.* 2013. Microbiota modulate behavioral and physiological abnormalities associated with neurodevelopmental disorders. *Cell.* 155(7): p. 1451–63.

Human Microbiome Project. 2012. C., Structure, function and diversity of the healthy human microbiome. *Nature 486*(7402): p. 207–14.

Huse, S.M., *et al.* 2012. A core human microbiome as viewed through 16S rRNA sequence clusters. *PLoS One.* 7(6): p. e34242.

Huycke, M.M., Abrams V., and M. D.R. 2002. *Enterococcus faecalis* produces extracellular superoxide and hydrogen peroxide that damages colonic epithelial cell DNA. *Carcinogenesis.* 23: p. 529–536.

Huycke. M. and H.R. Gaskins. 2004. Commensal bacteria, redox stress, and colorectal cancer: Mechanisms and models. *Exp Biol Med.* 229: p. 586–597.

Jensen, E.T., *et al.* 2013. Early life exposures as risk factors for pediatric eosinophilic esophagitis. *J Pediatr Gastroenterol Nutr.* 57(1): p. 67–71.

Jing Cheng, *et al.* 2013. Duodenal microbiota composition and mucosal homeostasis in pediatric celiac disease. *BMC Gastroenterolog.* 13: p. 1-13.

Kalliomaki, M., *et al.* 2008. Early differences in fecal microbiota composition in children may predict overweight. *Am J Clin Nutr.* 87: p. 534–538.

Kalliomaki, M., *et al.* 2012. Expression of microbiota, Toll-like receptors, and their regulators in the small intestinal mucosa in celiac disease. *J Pediatr Gastroenterol Nutr.* 54(6): p. 727–32.

Kamada, N., *et al.* 2013. Role of the gut microbiota in immunity and inflammatory disease. *Nat Rev Immunol.* 13(5): p. 321–35.

Kurokawa, K., *et al.* 2007. Comparative metagenomics revealed commonly enriched gene sets in human gut microbiomes. *DNA Res.* 14(4): p. 169–81.

Lahti, L., *et al.* 2014. Tipping elements in the human intestinal ecosystem. *Nat Commun.* 5: p. 4344.

Larsen, N., *et al.* 2010. Gut microbiota in human adults with type 2 diabetes differs from non-diabetic adults. *PLoS ONE.* 5(2): p. e9085.

Laurin, M., M.L. Everett, and W. Parker. 2011. The cecal appendix: One more immune component with a function disturbed by post-industrial culture. *Anat Rec.* 294: p. 567–579.

Leser, T.D. and L. Molbak. 2009. Better living through microbial action: The benefits of the mammalian gastrointestinal microbiota on the host. *Environ Microbiol. 11*(9): p. 2194–206.

Li, J., *et al.* 2014. An integrated catalog of reference genes in the human gut microbiome. *Nat Biotechnol. 32*(8): p. 834–41.

Li, X.X., *et al.* 2009. Bacterial microbiota profiling in gastritis without *Helicobacter pylori* infection or non-steroidal anti-inflammatory drug use. *PLoS One. 4*(11): p. e7985.

Liu, Y., *et al.* 2012. Interspecies competition and inhibition within the oral microbial flora: environmental factors influence the inhibition of *Streptococcus* mutans by *Streptococcus* oligofermentans. *Eur J Oral Sci. 120*(3): p. 179–84.

Louis, P. and H.J. Flint. 2009. Diversity, metabolism and microbial ecology of butyrate-producing bacteria from the human large intestine. *FEMS Microbiol Lett. 294*(1): p. 1–8.

Malkki, H. 2014. Neurodevelopmental disorders: Human gut microbiota alleviate behavioural symptoms in a mouse model of autism spectrum disorder. *10*(2): p. 60–60.

Mariat, D., *et al.* 2009. The Firmicutes/Bacteroidetes ratio of the human microbiota changes with age. *BMC Microbio. 9*: p. 123.

Martin, H.M., *et al.* 2004. Enhanced *Escherichia coli* adherence and invasion in Crohn's disease and colon cancer. *Gastroenterology. 127*(1): p. 80–93.

Million, M., *et al.* 2012. Obesity-associated gut microbiota is enriched in *Lactobacillus reuteri* and depleted in *Bifidobacterium animalis* and *Methanobrevibacter smithii*. *Int J Obes (Lond). 36*(6): p. 817–25.

Moreno-Indias, I., *et al.* 2014. Impact of the gut microbiota on the development of obesity and type 2 diabetes mellitus. *Front Microbiol. 5*: p. 190.

Morgan, X.C., *et al.* 2012. Dysfunction of the intestinal microbiome in inflammatory bowel disease and treatment. *Genome Biol. 13*(9): p. R79.

Murri, M., *et al.* 2013. Gut microbiota in children with type 1 diabetes differs from that in healthy children: A case-control study. *BMC Med. 11*: p. 46.

Nadal, I., *et al.* 2007. Imbalance in the composition of the duodenal microbiota of children with coeliac disease. *J Med Microbiol. 56*(Pt 12): p. 1669–74.

Nistal, E., *et al.* 2012. Differences of small intestinal bacteria populations in adults and children with/without celiac disease: Effect of age, gluten diet, and disease. *Inflamm Bowel Di. 18*(4): p. 649–56.

Ogura, Y., *et al.* 2001. A frameshift mutation inNOD2 associated with susceptibility to Crohn's disease. *Nature. 411*: p. 603–606.

Ohman, L. and M. Simren. 2013. Intestinal microbiota and its role in irritable bowel syndrome (IBS). *Curr Gastroenterol Rep. 15*(5): p. 323.

Pei, Z., *et al.* 2004. Bacterial biota in the human distal esophagus. *Proc Natl Acad Sci USA. 101*(12): p. 4250–5.

Pierik, M., *et al.* 2006. Toll-Like receptor-1, -2, and -6 polymorphisms influence disease extension in inflammatory bowel diseases. *Inflamm Bowel Dis. 12*: p. 1–8.

Qin, J., *et al.* 2012. A metagenome-wide association study of gut microbiota in type 2 diabetes. *Nature. 490*(7418): p. 55–60.

Riccia, D.N., et al. 2007. Anti-inflammatory effects of *Lactobacillus* brevis (CD2) on periodontal disease. *Oral Dis.* 13(4): p. 376–85.

Rodes, L. 2013. Effect of probiotics *Lactobacillus* and Bifidobacterium on gut-derived lipopolysaccharides and inflammatory cytokines: An *in vitro* study using a human colonic microbiota model. *Journal of Microbiology and Biotechnology.* 23(4): p. 518–526.

Roesch, L.F., et al. 2009. Culture-independent identification of gut bacteria correlated with the onset of diabetes in a rat model. *ISME J.* 3(5): p. 536–48.

Sanchez, E., et al. 2013. Duodenal-mucosal bacteria associated with celiac disease in children. *Appl Environ Microbiol.* 79(18): p. 5472–9.

Savage, D.C. 1977. Microbial ecology of gastrointestinal tract. *Annu Rev Microbiol.* 31: p. 107–133.

Schippa, S., et al. 2010. A distinctive 'microbial signature' in celiac pediatric patients. *BMC Microbiol.* 10: p. 175.

Schwiertz, A., et al. 2010. Microbiota and SCFA in lean and overweight healthy subjects. *Obesity (Silver Spring).* 18(1): p. 190–5.

Scott, K.P., S.H. Duncan, and H.J. Flint. 2008. Dietary fibre and the gut microbiota. *Nutrition Bulletin.* 33: p. 201–211.

Segain, J.P., et al. 2000. Butyrate inhibits inflammatory responses through NFκB inhibition: Implications for Crohn's disease. 47: p. 397–403.

Sellitto, M., et al. 2012. Proof of concept of microbiome-metabolome analysis and delayed gluten exposure on celiac disease autoimmunity in genetically at-risk infants. *PLoS One.* 7(3): p. e33387.

Sliepen, I., et al. 2009. Microbial interactions influence inflammatory host cell responses. *J Dent Res.* 88(11): p. 1026–30.

Soro, V., et al. 2014. Axenic culture of a candidate division TM7 bacterium from the human oral cavity and biofilm interactions with other oral bacteria. *Appl Environ Microbiol.* 80(20): p. 6480–9.

Srinivasan, N. 2010. Telling apart friend from foe: Discriminating between commensals and pathogens at mucosal sites. *Innate Immun.* 16(6): p. 391–404.

Stearns, J.C., et al. 2011. Bacterial biogeography of the human digestive tract. *Sci Rep.* 1: p. 170.

Stewart, C.J., et al. 2012. The preterm gut microbiota: Changes associated with necrotizing enterocolitis and infection. *Acta Paediatr.* 101(11): p. 1121–7.

Teughels, W., et al. 2013. Clinical and microbiological effects of *Lactobacillus* reuteri probiotics in the treatment of chronic periodontitis: A randomized placebo-controlled study. *J Clin Periodontol.* 40(11): p. 1025–35.

Tilg, H. and A.R. Moschen. 2013. Gut microbiota: Malnutrition and microbiota—A new relationship? 10(5): p. 261–262.

Tlaskalova-Hogenova, H., et al. 2011. The role of gut microbiota (commensal bacteria) and the mucosal barrier in the pathogenesis of inflammatory and autoimmune diseases and cancer: Contribution of germ-free and gnotobiotic animal models of human diseases. *Cell Mol Immunol.* 8(2): p. 110–20.

Tong, H. 2003. *Streptococcus oligofermentans* sp. nov., a novel oral isolate from caries-free humans. *International Journal of Systematic and Evolutionary Microbiology.* 53(4): p. 1101–1104.

Trompette, A., *et al.* 2014. Gut microbiota metabolism of dietary fiber influences allergic airway disease and hematopoiesis. *Nat Med.* 20(2): p. 159–66.

Trynka, G., C. Wijmenga, and D.A. van Heel. 2010. A genetic perspective on coeliac disease. *Trends Mol Med.* 16(11): p. 537–50.

Turnbaugh, P.J., *et al.* 2009. A core gut microbiome in obese and lean twins. *Nature.* 457(7228): p. 480–4.

Turnbaugh, P.J., *et al.* 2009. The effect of diet on the human gut microbiome: A metagenomic analysis in humanized gnotobiotic mice. *Sci Transl Med.* 1(6): p. 6ra14.

Turnbaugh, P.J., *et al.* 2006. An obesity-associated gut microbiome with increased capacity for energy harvest. *Nature.* 444(7122): p. 1027–31.

Varela, E., *et al.* 2013. Colonisation by Fecalibacterium prausnitzii and maintenance of clinical remission in patients with ulcerative colitis. *Aliment Pharmacol Ther.* 38(2): p. 151–61.

Wacklin P, *et al.* 2013. The duodenal microbiota composition of adult celiac disease patients is associated with the clinical manifestation of the disease. *Inflamm Bowel Dis.* 19(5): p. 934–941.

Wade, W.G. 2013. The oral microbiome in health and disease. *Pharmacol Res.* 69(1): p. 137–43.

Wang, Z.K. and Y.S. Yang. 2013. Upper gastrointestinal microbiota and digestive diseases. *World J Gastroenterol.* 19(10): p. 1541–50.

Wang, M., *et al.* 2005. Comparison of bacterial diversity along the human intestinal tract by direct cloning and sequencing of 16S rRNA genes. *FEMS Microbiol Ecol.* 54(2): p. 219–31.

Wang, T., *et al.* 2012. Structural segregation of gut microbiota between colorectal cancer patients and healthy volunteers. *ISME J.* 6(2): p. 320–9.

Wells, J.E.A.P.B.H. 2000. Identification and characterization of a bile acid 7α-dehydroxylation operon in *Clostridium* sp. strain TO-931, a highly active 7α-dehydroxylating strain isolated from human feces. *Appl Environ Microbiol.* 66(3): p. 1107–1113.

Wickstrom, C., *et al.* 2009. Proteolytic degradation of human salivary MUC5B by dental biofilms. *Microbiology.* 155(Pt 9): p. 2866–72.

Wold, A.E. 1998. The hygiene hypothesis revised: Is the rising frequency of allergy due to changes in the intestinal flora? *Allergy.* 53(46): p. 20–25.

Young, V.B. and T.M. Schmidt. 2004. Antibiotic-associated diarrhea accompanied by large-scale alterations in the composition of the fecal microbiota. *Journal of Clinical Microbiology.* 42(3): p. 1203–1206.

Zamakhchari, M., *et al.* 2011. Identification of Rothia bacteria as gluten-degrading natural colonizers of the upper gastro-intestinal tract. *PLoS One.* 6(9): p. e24455.

Zaura, E., *et al.* 2009. Defining the healthy "core microbiome" of oral microbial communities. *BMC Microbiol.* 9: p. 259.

Zhang, C., *et al.* 2010. Interactions between gut microbiota, host genetics and diet relevant to development of metabolic syndromes in mice. *ISME J.* 4(2): p. 232–41.

Zhang, H., *et al.* 2009. Human gut microbiota in obesity and after gastric bypass. *Proc Natl Acad Sci USA. 106*(7): p. 2365–70.

Zoetendal, E.G., *et al.* 2012. The human small intestinal microbiota is driven by rapid uptake and conversion of simple carbohydrates. *ISME J. 6*(7): p. 1415–26.

Zoetendal, E.G., M. Rajilic-Stojanovic, and W.M. de Vos. 2008. High-throughput diversity and functionality analysis of the gastrointestinal tract microbiota. *Gut. 57*(11): p. 1605–15.

Zoetendal, E.G., *et al.* 2001. The host genotype affects the bacterial community in the human gastrointestinal tract.

CHAPTER 3

From Birth to Old Age: Factors that Shape the Human Gut Microbiome

ALEXANDRA NTEMIRI, CATHERINE STANTON, R. PAUL ROSS
and PAUL W. O'TOOLE

3.1. INTRODUCTION

THE term microbiome generally refers to the microbiota and the collective genetic information those organisms carry. This genetic information supplements the genomic and metabolic potential of the host. A common approach to analyzing the microbiome is to monitor compositional and functional changes in the microbiota and then to seek to correlate those changes with health or disease symptoms. This is usually a prelude to attempting to identify mechanisms and effectors.

Largely based on culture-independent techniques, the core composition of the intestinal microbiota has been identified. The same phyla, taxa, and genera are generally found in the microbiota of the gastrointestinal tract of humans regardless of the condition of the subjects under testing. Both healthy and nonhealthy young or older subjects carry this core, although proportions may vary significantly. Thus selective pressure and coevolution between the host and the intestinal commensals have seemingly established a functionally stable microbiota composition in the gut. Imbalance in the microbial populations in the gut, such as an imbalance in diversity known as dysbiosis, indicate changes in the health status of the host or major environmental perturbations such as diet or antibiotics.

The changes in the microbiota from early colonisation events to those that occur during ageing are summarised here. Habitual diet is postulated to be the most determinative factor in shaping the human gut microbiome. A "Westernised diet" has been identified as a causative link between nutrient and gut microbiota signalling, inflammation, and metabolic disease. Dysbiosis also characterises functional disorders of the gastrointestinal track like Irritable Bowel Syndrome and Inflammatory Bowel Disease. The microbiota of the gastrointestinal tract of older subjects differs from that of young healthy adults. A key aspect in future research is to identify the exact molecular mechanisms that link dysbiosis and disease in order to develop therapeutic strategies based on microbiota manipulation or administration of microbial products.

Investigation of the association of gut microbiota to phenotypes of health or disease is a major feature of contemporary microbiome research (Cho and Blaser 2012). Culture-independent techniques and massive parallel next-generation sequencing have generated data that enabled the profiling of the gut microbiota of various healthy and disease cohorts ranging from infancy to the extremities of life span.

The phyla *Firmicutes* and *Bacteroidetes* are the most abundant in the healthy adult gut microbiota and around 90% of the bacterial groups found in the intestine typically belong to these phyla, whereas *Actinobacteria*, *Proteobacteria*, and *Verrucomicrobia* comprise up to around 2%, 1%, and 0.1%, respectively, of the intestinal microbiota composition (Hold *et al.* 2002; Wang *et al.* 2003; Eckburg *et al.* 2005; Rajilic-Stojanovic *et al.* 2009; Tap *et al.* 2009; Qin *et al.* 2010). Dominant bacterial groups in the gut microbiota belong to the genera *Faecalibacterium, Ruminococcus, Eubacterium, Dorea, Bacteroides, Alistipes*, and *Bifidobacterium* (Tap *et al.* 2009). Eckburg *et al.* (2005) also identified substantial representation of butyrate-producing Firmicutes belonging mostly to clostridia (clusters IV, XIVa) and *Bacteroides thetaiotaomicron* in the intestinal microbiota, both subgroups known for exerting beneficial effects upon the host (Wrzosek *et al.* 2013).

The gut microbiota composition as described above is based on studies on the intestinal microbiota of Westernised urban cohorts. There are exceptions to this general picture of the composition of the gut microbiota. A recent study on the gut microbiota of healthy hunter-gatherers in Hadza of Tanzania revealed the absence of bifidobacteria and other major compositional differences compared to a European urban cohort (Schnorr *et al.* 2014). Strong selective pressure during coevolution of commensal gut bacteria and the mammalian host (Ley *et al.* 2009) apparently resulted in the establishment of symbiotic relationships that allow the host to extract the maximum energy from the available habitual

FIGURE 3.1. Overview of the factors affecting the composition of the gut microbiota throughout life (C-section: Caesarean section).

diet sources. It seems that the gut microbiota of the Hadza cohort is adapted to a certain diet which is significantly different to the Westernised urban diet.

Host genetics appears to play a significant role in shaping the microbiome profile and potentially explains some of the interindividual variations in predominant species, subgroups, and phylotypes, those belonging to a particular phylum (Eckburg *et al.* 2005; Benson *et al.* 2010). However, a more determinative role in the composition of the gut microbiota has been attributed to habitual diet compared to host-genetics (Zhang *et al.* 2010). Recently, bacterial groups have been identified that are stably either absent or abundant over time in most individuals (based on a cohort of 1,000 Western adults) without being affected by short-term diet intervention and which correlated to health status such as overweight and ageing (Lahti *et al.* 2014).

It is becoming accepted that there is a functional redundancy in the gut microbiota (Mahowald *et al.* 2009; Lozupone *et al.* 2012) which explains how variant bacterial groups can result in the same functional core of the human gut microbiome (Turnbaugh *et al.* 2009; Qin *et al.* 2010). Lessons from microbial ecology support the notion that bacterial community structure is explained by the repertoire of functional genes rather than by species composition (Burke *et al.* 2011).

The gut microbiome shifts that occur throughout the lifespan and the factors driving these changes are reviewed here, including the early colonisation events of the gastrointestinal track (GIT) and the changes observed during ageing. The important role of diet in modulating the intestinal microbiota is also discussed, summarising the evidence indicating that diet can underpin metabolic disorders and inflammation via the microbiota. Furthermore, we review the studies linking gut microbiota and exercise. Finally, dysbiosis associated with antibiotic intake and functional disorders of the gastrointestinal track are reviewed. Figure 3.1 provides an overview of the factors affecting the composition of the gut microbiota throughout life.

3.2. ESTABLISHMENT OF THE GUT MICROBIOTA

Several studies of germ-free (GF) mice demonstrate the importance of the gut microbiota in the development of gut physiology and the priming of the innate and adaptive immune systems (Sekirov *et al.* 2010; Sommer and Backhead 2013). Disrupted colonisation patterns of the GIT in early life may have an immediate or future impact on health

(Salminen *et al.* 2004; Penders *et al.* 2007, 2013; Cahenzli *et al.* 2013; El Aidy *et al.* 2013; Ardeshir *et al.* 2014).

In utero, the fetus has generally been considered to be devoid of bacteria although some recent studies suggest otherwise. Bacteria such as *Escherichia coli, Enterococcus faecium*, and *Streptococcus epidermidis* have been isolated from the meconium of healthy term infants (Jimenez *et al.* 2008). Bacteria have also been isolated from the amniotic fluid of preterm neonates which could however be indicative of a hidden infection and preterm delivery risk (DiGiulio *et al.* 2008). In a first study on the placental microbiome, Aagaard *et al.* (2014) conducted 16S rRNA amplicon sequencing and shot-gun metagenomics on placental samples from 320 individuals. Comparing the findings to other body site microbiota data they reported the presence of a distinct community comprised of nonpathogenic *Firmicutes, Tenericutes, Proteobacteria, Bacteroidetes*, and *Fusobacteria*. The placental microbiota was found to be similar to the microbiota of the oral cavity. Importantly, an association between placental microbiota and infection during the first pregnancy months and risk of preterm birth was observed. Despite data suggestive of prenatal presence of bacteria in the uterus environment, the role of these bacteria in the colonisation of the GIT remains largely unclear and debateable (Matamoros *et al.* 2013).

Colonisation of the human GIT is a dynamic and complex event and begins with facultative anaerobes that prime the niche for the strict anaerobes that follow and ultimately prevail in the healthy adult microbiota. After birth the first bacteria that colonise the infant gut are enterococci, enterobacteria (*E. coli, E. faecium, Enterococcus faecalis*), and some staphylococci, and when these organisms deplete the gut of oxygen, anaerobes such as bifidobacteria, lactobacilli, clostridia, and *Bacteroidetes* emerge (Palmer *et al.*, 2007; Alderberth and Wold 2009). In the repopulation of a GF murine model with normal murine gut microbiota, fermentation metabolites and cross-feeding between early colonising bacteria and successors facilitated the establishment of the latter (El Aidy *et al.* 2013). The colonisation event is regarded as leading to a dynamic eubiotic balance between the commensal bacteria and the host inhibiting the growth of pathobionts, a phenomenon called colonization resistance (Lawley *et al.* 2012).

It has been observed that in the first days of life the microbiota not only in the gut but in all body niches (skin, oral mucosa, nasopharyngeal cavity) is similar and reflects the mode of delivery (Dominguez-Bello *et al.* 2010). Samples taken from new-born babies within 24 hours of birth showed that vaginally-born neonates harboured a microbiota compositionally similar to the maternal vaginal microbiota with *Lactobacillus, Prevotella, Atopobium*, and *Sneathia* spp. as prevalent taxa, whereas in neonates delivered by Caesarean section (C-section) the microbiota of various body niches including the gut resembled that of maternal skin (Dominguez-Bello *et al.* 2010). A few days after birth bifidobacteria are among the most dominant taxa in the gut of naturally delivered babies (Biasucci *et al.* 2008) but they seem to be absent or in low abundance in the gut of new-borns delivered by C-section (Aires *et al.* 2011).

Jost *et al.* (2012) found that from the fourth day of life till the end of the first month the gut microbiota of vaginally delivered breast-fed babies was dominated by bifidobacteria. It was also reported that *Bacteroides* species showed variant patterns of establishment that correlated to low counts of *Bifidobacterium* species, that clostridia of the clusters IV and XIV were undetectable, and that the presence of *Lactobacillus* was not stable (Jost *et al.* 2012). Other studies also confirm that *Bifidobacterium* and *Lactobacillus* dominate the infant gut (Turroni *et al.* 2012). Aspects of the colonisation of the human gut by bifidobacteria have been comprehensively reviewed in Ventura *et al.* (2012).

Interestingly, gestation time seems to affect the colonisation patterns of the GIT. Vaginally delivered, preterm babies seem to have an intestinal microbiota of low phylogenetic diversity compared to full-term infants (Arboleya *et al.* 2012). When compared to full-term babies, preterm children were found to harbor more facultative and fewer strict anaerobes, and these differences could be attributed to exposure to intensive medical treatment, hospitalisation, and insufficient organ function (Arboleya *et al.* 2012).

The gut microbiota of preterm infants seems to be a risk factor for the development of necrotising enterocolitis (NEC)/late onset sepsis (LOS). Both culture- and nonculture-based analysis of stool samples from infants born around week 27 of pregnancy showed that those diagnosed with NEC and LOS had higher fecal abundance of enterobacteria and coagulase negative staphylococci (CONS), respectively (Stewart *et al.* 2012). As observed in another study on the gut microbiota of preterm babies (Mshvildadze *et al.* 2010), meconium was not sterile. Dysbiosis rather than the presence of pathobionts, for example CONS, was postulated to be significant in the onset of LOS in preterm patients (Mai *et al.* 2013). The group also observed a low abundance of bifidobacteria in stool samples of LOS patients. Interestingly, it has been recently suggested that irrespective of mode of birth, feeding type, or antibiotic treatment, the colonisation of the GIT of preterm children that reside in care units follows a certain pattern from members of the *Bacillus* class through to γ-*Proteobacteria* and onwards to Clostridia, in a pattern mainly influenced by gestation stage of birth (La Rosa *et al.* 2014).

It has been controversially proposed that during the final pregnancy stages maternal symbiotic gut bacteria could translocate to the mammary gland (Fernandez *et al.* 2013). Studies identified *Streptococcus, Staphylococcus* and *Corynebacterium*, and lactic acid bacteria and bifidobacteria as dominant bacteria in the breast-milk microbiota which, however, seems to change with lactation towards a community composition more rich in *Veillonella, Leptotrichia*, and *Prevotella* (Hunt *et al.* 2010; Cabrera-Rubio *et al.* 2012; Soto *et al.* 2014). The same strains of bifidobacteria, lactobacilli, and staphylococci have been isolated from both the feces of new-borns and the maternal breast-milk of their mothers (Martin *et al.* 2012) demonstrating that the two bacterial communities are somehow interrelated.

Many studies have confirmed that the intestinal microbiota of breast-fed, new-born children harbors more bifidobacteria than formula-fed, for example, Harmsen *et al.* (2000), Martin *et al.* (2008), Roger *et al.* (2010), and Bezirtzoglou *et al.* (2011). Others failed to report significant differences in the bifidobacteria population in infant stool samples between breast-fed and formula-fed babies, but an abundance of *E. coli* and *Clostridium difficile* was higher in formula-fed subjects (Penders *et al.* 2005).

Breast milk contains human milk oligosaccharides (HMO). New-born babies are not able to fully digest those oligosaccharides but infant-type bifidobacteria like *Bifidobacterium longum* subsp. *infantis* are genetically endowed with the ability to efficiently metabolise various short chain HMOs, an ability not detected in adult-type bifidobacteria like *B. longum* subsp. *longum* (LoCassio *et al.* 2010; Sela and Mills 2010; Garrido *et al.* 2011). A variety of pathways for HMO metabolism are distributed among the infant-type bifidobacteria revealing the selective pressure breast feeding has put on certain bacterial groups (Zivkovic *et al.* 2011). In all cases, research agrees that after weaning, the introduction of solid food marks clear changes in the microbiota although preweaning colonisation patterns are still detectable (Fallani *et al.* 2011; Koening *et al.* 2011).

One of the most accepted causes of dysbiosis is antibiotic treatment; regardless of age, antimicrobial therapy has impacts on the nature of the gut microbiota. Analysis of fecal microbiota of infants subjected to antibiotic treatment during the first days after birth showed that the microbiota had limited numbers of bifidobacteria and high abundance of enterobacteria and proteobacteria (Brunser *et al.* 2006; Tanaka *et al.* 2009; Persaud *et al.* 2014). In early life, antibiotics can disturb normal colonisation patterns which in turn can affect mammalian immunity and metabolism in later life (reviewed in Rautava *et al.* 2012). Cho *et al.* (2012) observed that infant mice treated with antibiotics developed a gut microbiota that was correlated to increased body fat. Cox *et al.* (2014) showed that in weaning mice, antibiotic treatment resulted in an altered gut microbiota capable of inducing obesity, a permanent trait of the microbiota that could be transmitted through faecal transfer to GF mice. The group postulated that the microbiota-related obesity was a result of the altered gut microbiota composition due to antibiotic treatment. This has profound implications for human weight management at population level, because pediatric antibiotic usage is very frequent.

Studies suggest that the microbiota begins to stabilise to a more adult-like profile from year one through two-and-a-half (Palmer *et al.* 2007; Koening *et al.* 2011). More recent research shows that the microbiota undergoes changes until age 36 months and that an enterotype establishment can be observed between nine months and three years of age (Yatsunenko *et al.* 2012; Bergstrom *et al.* 2014).

3.3. SHAPING FACTORS OF GUT MICROBIOTA COMPOSITION

3.3.1. Diet

Habitual diet plays a significant role in the programming of the composition of the intestinal microbiota (Scott *et al.* 2013; Simoes *et al.* 2013). The intestinal microbiota rapidly responds to diet variations which is a characteristic of the gut microbiota of many mammalian groups (Muegge *et al.* 2011; David *et al.* 2014). Most of the dietary nutrient absorption occurs in the duodenum and small intestine whereas undigested molecules like complex polysaccharides reach the large intestine (Wong and Jenkins 2007). In the colon, a phylogenetically diverse and metabolically active bacterial community salvages energy from the undigested food particles which is subsequently used for the maintenance of the microbiota and by the host (Blaut and Clavel 2007).

Fermentation of undigested polysaccharides, dietary and endogenous proteinaceous substrates, and polyphenols result in various end-products with tremendous impact on colonic health and homeostasis (Cummings and Macfarlane 1997; Nyangale et al. 2012, Windey et al. 2012). These fermentation compounds are predominantly the short chain fatty acids (SCFA) propionate, acetate, and butyrate, and gasses like H_2, H_2S, and CO_2 as well as potentially toxic phenolic and indolic compounds, ammonia, amines, and sulphur molecules (reviewed in Macfarlane and Macfarlane 2012).

Different nondigestible carbohydrates like resistant starch (RS), non-starch polysaccharides (NSP), and prebiotics can selectively enhance the growth of certain bacterial groups in the GIT (Martinez et al. 2010; Walker et al. 2011). Modulated microbial fermentation can yield beneficial end-products for the host. One such fermentation product is butyrate that is extensively studied for the beneficial effect it exerts on the host including directly upon colonic epithelial cells (Hamer et al. 2008). *Clostridium* clusters XIVa and IV, with representatives like *Eubacterium/Roseburia* species and *F. prausnitzii* respectively, are major saccharolytic players in butyrate generation in the colon (Louis and Flint 2009) and are potentially prominent regulators of innate immunity through the production of SCFAs (Atarashi et al. 2011, 2013; Smith et al. 2013).

A statistical model was generated by Faith et al. (2011) that allowed for up to 60% accuracy in the prediction of shifts in species abundance as a response to host diet. The authors isolated representatives of ten bacterial species from the human colon, sequenced their genomes, and introduced them in gnotobiotic mice (those with a defined microbiota). They subsequently conducted detailed measurements on species abundance and gene expression in response to host diet, from which the statistical model was developed. Animal models are valuable tools in studying shifts in the microbiome but there is a need for more extensive human studies to better simulate the complex conditions occurring naturally in the GIT (Walker et al. 2010; Muegge et al. 2011). Importantly, a recent study by Korpela et al. (2014) developed models to predict the gut microbiota responses to diet intervention based on the intestinal microbiota composition, predominantly on the Firmicutes abundance, of certain obese individuals.

A summary of culture-independent studies correlating habitual diet to gut microbiota is presented in Table 3.1. From metagenomic studies on human fecal data sets from different countries Arumugam et al. (2011) observed that collectively, the grouping of the data formed three

TABLE 3.1. Culture-Independent Studies Associating Gut Microbiota with Diet.

Cohort	Analysis Method	Diet	Main Findings	Reference
29 healthy adults	qPCR, PCR-DGGE	Habitual vegetarians, omnivores	↓ *Clostridium* cluster IV ↑ *Bacteroides* spp. (nonsignificant)	Liszt et al. (2009)
46 healthy adults	qPCR	Habitual vegetarians, omnivores	→ *Clostridium* cluster XIVa ↓ *Roseburia/E. rectale* group	Kabeerdoss et al. (2012)
6 obese adults with type 2 diabetes	16S rRNA sequencing, biochemical analysis	Strict vegetarian diet/1 month	→ Firmicutes/Bacteroides ratio ↓ No effect on enterotypes ↓ Pathobionts, Enterobacteria ↑ *Bacteroides fragilis*, *Clostridium* cluster XIVa, IV species ↓ Body weight, blood triglycerides, cholesterol	Kim et al. (2013)
14 healthy Mossi ethnic group children (rural community)	16S rRNA sequencing	• Habitual rural diet: low fat/protein/sugar, high fiber	↑ Bacteroidetes, unique presence of *Xylanibacter*, *Prevotella*, *Butyrivibrio*, *Treponema*	De Filippo et al. (2010)
15 healthy Europeans (urban community)	16S rRNA sequencing	• Habitual urban diet: high fat/protein/sugar, low fiber	↑ Firmicutes, ↑ Enterobacteria	
98 healthy subjects	16S rRNA sequencing	• Habitual fat-rich diet • Habitual fiber-rich diet	• ↑ *Bacteroides*, *Alistipes*, *Parabacteroides* • ↑ *Prevotella*, *Paraprevotella*, *Catenibacterium*	Wu et al. (2011)
10 healthy *Bacteroides* enterotype		Diet intervention: • high fat/protein • carbohydrates	• *Bacteroides* enterotype correlated • *Prevotella* enterotype related • No permanent enterotype switch	

(continued)

TABLE 3.1 (continued). Culture-Independent Studies Associating Gut Microbiota with Diet.

Cohort	Analysis Method	Diet	Main Findings	Reference
14 overweight subjects	16S rRNA sequencing, qPCR	• Resistant starch diet intervention • Reduced carbohydrates diet intervention	• ↑ *Roseburia*/*E. rectale* related Firmicutes • ↓ *Roseburia*/*E. rectale* related Firmicutes	Walker *et al.* (2011)
18 lean subjects, 33 mammalian species	16S rRNA sequencing, qPCR	Proteins, insoluble fibre	• Similar enzymatic activity and adaptation to diet across species	Muegge *et al.* (2011)
178 older subject: community, long-stay	16S rRNA sequencing	• Community: moderate fat/high fiber, diverse diet • Long-stay: high fat/protein/sugar, low fiber, less diverse diet	• ↑ Phylogenetic diversity • ↑ Firmicutes, *Roseburia, Coprococcuss* • ↓ Phylogenetic diversity, • ↑ Bacterodetes, *Parabacteroides*, • ↓ *Eubacterium, Anaerotruncus, Coprobacillus*	Claesson *et al.* (2012)
20 monozygotic twins	16S rRNA sequencing, DGGE	Habitual diet	*Bacteroides* spp. and bifidobacteria most affected by habitual diet	Simoes *et al.* (2013)
40 professional athletes	16S rRNA sequencing, biochemical	Strict habitual diet, high protein	• ↑ *Akkermansia* sp., • ↓ Inflammatory markers in blood	Clarke *et al.* (2014)
10 healthy individuals	16S rRNA sequencing, qPCR, RNA-sequencing, biochemical	Short-term intervention: • Animal-based diet • Plant-based diet	• ↑ Bile-tolerant bacteria: *Alistipes, Bilophila, Bacteroides* • ↓ Plant fiber metabolising Firmicutes • ↑ *Roseburia*/*Eubacterium*, Ruminococci	David *et al.* (2014)

clusters called "enterotypes". Enterotype number 1 was *Bacteroides*-dominated and positively corelated with *Parabacteroides, Lactobacillus, Clostridiales*, and *Alkaliphilus* species, and was characterised by enrichment in metabolism of a broad spectrum of carbohydrate utilization and proteolytic activity. In enterotype 2, *Prevotella* was predominant and co-occurred with *Desulfovibrio* probably in a synergistic consortium for mucin degradation. The *Ruminococcus* enterotype correlated with *Akkermansia* abundance which is a mucin-degrading bacterium. The "enterotype" classification was based on the high abundances of certain genera and this grouping should be regarded as metabolic variations in the way gut microbiota utilises available nutrients.

Consumption of a vegetarian diet leads to changes in the microbiota of the GIT. Low abundance of *Clostridium* cluster IV and higher but not significant abundance of *Bacteroides* was observed by qPCR analysis of fecal samples of vegetarians compared to omnivores (Liszt *et al.* 2008). Another culture-independent study reported low abundance of *Clostridium* cluster XIVa and the *Roseburia/E. rectale* group in the fecal samples of vegetarians compared to omnivores (Kabeerdoss *et al.* 2012). Diet intervention with a strictly vegetarian diet in obese individuals for one month led to changes in the intestinal microbiota composition of patients without any impact on enterotypes and improved inflammatory markers in blood (Kim *et al.* 2013). In that study, the diet intervention reduced the *Firmicutes/Bacteroidetes* ratio and the proportion of enterobacteria in the intestinal microbiota, and increased the abundances of *Clostridium* cluster XIVa and IV.

Based on diet interventions and the habitual diet of 98 individuals, Wu *et al.* (2011) reported that colonic communities clustered according to the aforementioned enterotypes 1 and 2 and that number 3 demonstrated a tendency to fuse with enterotype 1. The *Bacteroides*-dominated cluster co-occurred with *Alistipes* and *Parabacteroides* and associated with an animal protein and saturated fat-based habitual diet. In contrast, the *Prevotella*-cluster correlated with *Paraprevotella* and *Catenibacterium* and was related to fibre intake. Importantly, enterotypes as monitored by Wu *et al.* (2011) were driven by habitual diet and overall, remained stable during the diet intervention, although changes in the microbiome did occur as a response to the new diet. The group also concluded that the two clusters mirrored a "Westernised" and a more "agrarian" kind of diet respectively.

Similar conclusions were drawn when high-through put sequencing was conducted on 16S rDNA amplicons from fecal samples of children of two geographically distinct areas: urban in Florence, Italy, and rural in Boulpon, Burkina Faso (De Filippo *et al.* 2010). The rural diet was almost vegetarian and enriched in starch, fiber, and plant polysaccharides and was low in animal fat and protein. The fecal microbiota of subjects residing in rural areas was enriched in bacteria enabling maximum energy harvest from fiber and high rates of SCFAs production. Genera like *Xylanibacter, Prevotella, Butyrivibrio*, and *Treponema* were characteristic of the intestinal microbiota of people living in rural areas and indicative of the potential to harvest energy from complex polysaccharides. By contrast, the urban diet sustained a much less diverse microbiota.

Taken together, the results of the two studies above reveal the impact on gut microbiota of two kinds of diet: one "rural" enriched in fiber and low in fat and the other more "Westernised", rich in animal protein and saturated fat. The global convergence of dietary habits towards a westernised high fat/low fiber diet is an increasing phenomenon in modern society and is accompanied by the escalating prevalence of inflammatory and metabolic disease. The interplay between Westernised dietary habits, commensal bacteria, and inflammation are comprehensively reviewed by Thorburn *et al.* (2014).

3.3.2. Exercise

No extensive studies have been dedicated so far to the exploration of the effect of exercise on the human microbiota. There are, however, some indications that exercise affects the gut environment based on a murine model. Matsumoto *et al.* (2008) observed that rats that regularly exercise on the wheel-run had an intestinal microbiota distinct from that of microbiota of sedentary animals and notably the former had higher colonic butyrate concentration compared to sedentary littermates. Choi *et al.* (2013) observed exercise-induced changes in the intestinal microbiota of healthy mice. Another study showed that exercise can lead to shifts in the gut microbiota of obese mice (Evans *et al.* 2013). Recently, Clarke *et al.* (2014) studied the associations of multiple factors including exercise with the intestinal microbiota of the members of a professional rugby team. The study showed that athletes had a distinct fecal microbiota and lower inflammatory status in blood compared to controls. Athletes and low body mass index (BMI) controls had higher abundance of *Akkermansia* in the fecal samples compared to high BMI controls. The group correlated the gut microbiota changes not only with regular intense exercise but with the high protein diet regime the athletes consumed. However, it was impossible to rule out the effect of potential confounding factors. The existing evidence indicate that gut-microbiota is responsive to exercise but more research including intervention studies is necessary to elucidate the causality and mechanisms of such responses.

3.3.3. Antibiotics

Antibiotic treatment can reduce the compositional diversity and the metabolic potential of the gut microbiome, with an effect that attenuates with treatment cessation (Perez-Cobas *et al.* 2013). Jakobsson *et al.* (2010) studied subjects receiving clarithromycin and metronidazole treatment and monitored microbiota over a period of 4 years. After a week of treatment, distinct changes in the composition of the intestinal microbiota of all subjects were observed. Interestingly, the group also observed that although the general trend was gut microbiota recovery in individuals after completion of treatment, some cases failed to fully recover the pretreatment microbiota. The same pattern of interindividual responses to antibiotics, recovery after cessation of treatment, but some cases of reduced to absent restoration of taxa in the intestinal microbiota, was reported in other studies (Jernberg *et al.* 2007; Dethlefsen *et al.* 2008; Dethlefsen and Relman 2011). Interindividual responses reflected the variations of the healthy status microbiota among individuals. The fecal metabolome associated with the microbiota seemed to be affected. Streptomycin was shown to exert multiple effects on murine intestinal metabolic activity (Caetano *et al.* 2011) implying a potential effect on the microbiota involved in certain metabolic pathways.

Low diversity of the gut microbiota makes the intestinal environment susceptible to entero-pathogens. In a murine model after 2 days of treatment with vancomycin/streptomycin, a dose-dependent effect was apparent upon the composition of the gut microbiota (Sekirov *et al.*, 2008). The group observed that although total bacterial numbers were not affected, the composition of the enterica microbiota was altered making the animals vulnerable to *Salmonella* serovar *Typhimurium* infection.

Another common example of infection susceptibility and breach in the colonisation resistance phenomenon is that of *Clostridium difficile* infection (CDI). *C. difficile* is a common nosocomial pathogen causing antibiotic-associated diarrhoea (AAD) and affecting mostly hospitalised older patients. Epidemiology and risk factors of CDI were reviewed by Ananthakrishnan (2011). Cephalosporins, macrolides, clindamycin, and predominantly fluoroquinolones are some of the associated infection-risk antibiotics (Pepin *et al.* 2005). The successful treatment of CDI in a mouse model with a mixture of bacteria isolated from healthy subjects (Lawley *et al.* 2012), and in humans with a stool substitute (Petrof *et al.* 2013) showed that a phylogenetically diverse mix of fecal bacteria is sufficient to displace *C. difficile* from the gut more effectively than any known antibiotic.

Overall, antibiotic administration results in mostly short-term disturbances in the microbiota and these disturbances tend to revert with the end of antibiotic use. The dominant taxa in the gut microbiota tend to recover soon after short-term administration ends showing that balancing ecological mechanisms drive the microbiota back to its pretreatment status whereas long-term use may result in permanent alterations in the intestinal microbiota (De La Cochetiere *et al.* 2005) and increase risk of AAD.

3.3.4. Extra-Intestinal Disorders: Metabolic Syndrome and Related Obesity, Type 2 Diabetes, and Liver Disease

Metabolic syndrome is mainly characterised by insulin resistance and visceral obesity as well as raised blood pressure, atherogenic dyslipidaemia, and proinflammatory status (Grundy *et al.* 2004). The syndrome is a risk factor for the interrelated development of cardiovascular disease (CDV) and type 2 diabetes (T2D) (Huang 2009). Dysbiosis seems to play a role in this syndrome. The intermediate step is the interplay between the gut microbiota, nutrient absorption, fat storage, and host metabolism (Musso *et al.* 2011). Metabolic syndrome also affects hepatic physiology with the potential for the development of the nonalcoholic fatty liver disease (NAFLD) and the more severe nonalcoholic steatohepatitis (NASH) (Abu-Shanab and Quigley 2010).

3.3.4.1. Obesity and Type 2 Diabetes

There is accumulating evidence that commensal bacteria play a significant role in modulating fat storage. When lean GF mice were conventionalised with normal gut microbiota they gained body fat although food intake was not increased (Backhed *et al.* 2004). Cani *et al.* (2007, 2008) used a murine model to propose a sequence of events leading from high-fat diet to gut microbiota-controlled metabolic disorders. High-fat feeding was correlated to increased intestinal permeability and high lipopolysaccharides (LPS) plasma concentrations leading to metabolic endotoxemia, inflammation, and eventually metabolic conditions. Obesity and diabetes reflecting the metabolic deregulation were attenuated in a mouse strain lacking the main LPS-receptor CD14, demonstrating the importance of LPS in the initiation of the metabolic deregulation (Cani *et al.* 2007). In humans, plasma LPS levels increase after increase of fat intake (Erridge *et al.* 2007) and translocation of bacterial components across the gut barrier seem to occur in obese and diabetic subjects (Gummesson *et al.* 2011; Hawkesworth *et al.* 2013).

Importantly, LPS interacts with the endocannabinoid system (eCB) which emerges as an important player in the metabolic deregulation in obesity (Lambert and Muccioli 2007). The endocannabinoid system is implicated in gut barrier integrity as blocking the G protein-coupled receptor of eCB (CB_1) resulting in reduced plasma LPS and an increase in tight-junction proteins in murine model (Muccioli *et al.* 2010).

Ley *et al.* (2005) showed that the establishment of the diet-induced obesity phenotype in mice was accompanied by low intestinal microbiota diversity and altered relative abundances of the major taxa with increased abundance of *Firmicutes* and reduction of the abundance of phylum *Bacteroidetes* to 50% of that of lean mice. The group also demonstrated that the potential to gain energy from nutrients was improved in this "obese microbiome": more enzymes involved in starch and carbohydrate metabolism, more fermentation-end products like butyrate and reduced energy load in feces compared to lean littermates (Turnbaugh *et al.* 2006). Importantly, the increased energy uptake trait could be transmitted to GF mice by fecal transplantation (Turnbaugh *et al.* 2006). Studying obesity-associated shifts in gut microbiota in humans, Ley *et al.* (2006) obtained results analogous to data from the mouse studies. Accordingly, in obese humans *Bacteroidetes* relative abundance was lower compared to that of *Firmicutes*. However, studies by other groups in humans have been contradictory and report different changes in the ratios of the *Firmicutes* and *Bacteroidetes* (Duncan *et al.* 2008; Schwiertz *et al.* 2010).

As for obesity, research on the intestinal microbiota of patients with T2D reveals significant alterations compared to healthy controls. The "diabetic gut microbiota" had high proportions of *Bacteroidetes-Prevotella* group whereas the abundance of the butyrate-producer group *C. coccoides-E. rectale* was reduced (Larsen *et al.* 2010). A metagenomic analysis on the faecal microbiome of Chinese T2D patients showed decreased butyrate biosynthesis and metabolism of vitamins, enrichment in metabolism of xenobiotics, branched chain amino acids (BCAA), and methane, and sulphate reduction, to mention some (Qin *et al.*, 2012). The group also reported opportunistic pathogens, the mucin-degrader *Akkermansia municiphila*, and sulphur reducers like *Desulfovibrio* sp. to be higher in abundance in T2D faecal microbiota. A metagenome-wide study on a cohort of 145 healthy and with risk for diabetes European women also revealed composition and function alterations in the gut microbiota of the at-risk subjects (Karlsson *et al.* 2013). The group developed a diabetes risk predicting tool which, however, failed to predict diabetes when applied on the aforementioned Chinese cohort suggesting that the metagenomic markers for T2D between the two cohorts were cohort-specific and could not be generalised.

3.3.4.2. Nonalcoholic Fatty Liver Disease and Nonalcoholic Steatohepatitis

A range of hepatic conditions related to obesity are grouped under the term NAFLD including the inflammatory steatosis NASH (Brunt and Tiniakos 2010). In obesity accompanied by comorbidity the NAFLD prevalence is 95% and prevalence for NASH is 25%, whereas in diabetic patients NASH may reach 63% prevalence (Bajaj *et al.* 2012). The gut microbiota of patients with either NAFLD or NASH has not yet been extensively studied.

Mouzaki *et al.* (2013) conducted one of the first culture-independent studies on 50 adults with steatosis, nonalcoholic steatohepatitis, and healthy controls. The group observed a diet-independent reduced relative abundance of *Bacteroidetes* in the stool of NASH patients. A previous study on the intestinal microbiota of obese, healthy children and children with NASH showed that obese and NASH microbiomes could cluster in a *Prevotella*-enriched enterotype diversifying from the healthy controls (Zhu *et al.* 2012). *Bacteroides* abundance was higher in obese and NASH subjects compared to healthy controls, *Actinobacteria* had relatively lower abundance in nonhealthy subjects, and *Proteobacteria* abundance increased significantly from healthy to obese individuals with higher scores in NASH subjects (Zhu *et al.* 2012). The study also associated the elevated ethanol measurements in NASH patient blood samples to a gut microbiota enriched in ethanol-producing bacteria like *E. coli*.

Data on the gut microbiota of NAFLD patients is still limited. A clear link between gut microbiota, diet, and metabolic disorders exists. However, these two studies on NAFLD and intestinal microbiota yielded diet-independent results implying a causal role of the gut microbiota in the pathogenesis of the condition. Serino *et al.* (2012) investigating metabolic disorders and gut microbiota in a murine model showed that irrespective of diet and host genetics, gut microbiota and the naturally occurring metabolic variations of the intestinal microbiome among individuals predisposed the animals for metabolic conditions. Future work is expected to further elucidate both the mechanisms through which diet controls gut microbiota-driven host metabolism disorders and the existence of a diet-independent role of the gut microbiota in the progression of metabolic conditions.

3.4. FUNCTIONAL GASTROINTESTINAL DISORDERS: INFLAMMATORY BOWEL DISEASE AND IRRITABLE BOWEL SYNDROME

Functional gastrointestinal disorders (FGIDs) are a broad group of commonly occurring gastrointestinal conditions with irritable bowel syndrome (IBS) being the most prevalent (Talley 2008). Using Rome III criteria, similar symptoms in IBS and inflammatory bowel disease (IBD) could be identified (Bryant *et al.* 2011).

3.4.1. Inflammatory Bowel Disease

The two major forms of IBD are ulcerative colitis (UC) and Crohn's disease (CD), and both are influenced by host genetics and deregulated immunological responses to the intestinal microbiota (Abraham and Cho 2009). Culture-independent studies show abnormalities in the microbiome of individuals with IBD. Frank *et al.* (2007) compared the microbiota in biopsies of UC and CD to those of healthy controls. Although the abundances of the predominant taxa were not altered, significant variations in the relative abundances of subgroups distinguished the healthy from the nonhealthy microbiota, and health-promoting butyrate-producers belonging to *Bacteroidetes* and *Lachnospiraceae* were significantly reduced in abundance in the IBD samples. Conversely, Proteobacteria and *Bacillus* species were higher in proportional abundance.

Manichanh *et al.* (2006) applied metagenomics to the fecal microbiota of CD patients and observed reduced phylogenetic diversity. Patient fecal samples demonstrated a low abundance of the butyrate-producer groups *Clostridium leptum* and *Clostridium coccoides* and were enriched in Gram-negative species and the family *Porphyromonadaceae*. *F. prausnitzii* is a major butyrate-producer in the gut and the dysbiosis and symptomology characterized by the significant reduction of this Firmicutes in CD could be reversed by oral administration of this organism in a murine model (Sokol *et al.* 2008).

Another bacterium implicated in IBD is *Bacteroides fragilis*. Transcriptional analysis of colonic mucosal biopsies showed that the phylum *Bacteroidetes* was the most active based on rRNA gene libraries and *B. fragilis* was dominant in CD based on proportion of sequences in rRNA library/percentage of sequences in rRNA gene library (AIR) (Rehman *et al.* 2010). Previously, Swidsinski *et al.* (2005) had demonstrated the prevalence of this species in gut microbiota of IBD patients and especially CD and they suggested the potential role of the high abundance of this bacterium in increased barrier permeability.

Risk factors for IBD have been identified, and host genetics and dysbiosis have been associated to the progression of the condition. However, the exact mechanisms leading to IBD remain elusive thus making the unravelling of IBD causation a paradigm for the understanding of the complex interplay between microbiota, host, and inflammation (Huttenhower *et al.* 2014). Furthermore, the complex cross-talk between the causative factors of the condition, the interindividual variations in gut microbiota, and the fact that no universal colitogenic bacteria have been identified indicate the need for more personalised IBD therapeutic strategies (Stephens and Round 2014).

3.4.2. Irritable Bowel Syndrome

A common FGID disorder is IBS, and unlike IBD, it is not associated with histopathological conditions and it is believed to be related at least in some cases to the dysregulation or disruption of the gut-brain axis (GBA) (Ringel and Maharshak 2013). The cross-talk between gut and brain and the association with intestinal homeostasis and progression of FGID have been recently comprehensively reviewed by Collins and Bercik (2009) and Collins (2014). IBS progression can be outlined as follows: extrinsically-induced dysbiosis, for example after severe infection, triggers mucosal immune activity in genetically predisposed subjects resulting in enteric nervous system deregulation caused by intestinal permeability (Collins *et al.* 2009; Simren *et al.* 2009; Steck *et al.* 2013).

TABLE 3.2. Diagnostic Criteria for IBS, Adapted from Rome III (http://www.theromefoundation.org/criteria/).

Criteria for Diagnosis of IBS	
• Intermittent abdominal pain and discomfort related to at least two of the conditions below: 1. Improvement with defecation 2. Alterations in the defecation habits 3. Alterations in stool form	• Frequency of symptoms: 3 days per month for the last 3 months • Symptom detection at least 6 months before diagnosis

The diagnostic criteria set by the Rome Foundation, Rome criteria III (http://www.theromefoundation.org/criteria/), are widely used for FGID diagnosis (Soares 2014); outlined in Table 3.2. Three subtypes of IBS are defined based on laxation habits: diarrhoea-predominant (IBS-D), constipation-predominant (IBS-C), and mixed (IBS-M) (Longstreth et al. 2006).

Using culture-based techniques and PCR-denaturing gradient gel electrophoresis (PCR-DGGE) analysis Matto et al. (2005) observed temporal instability of the gut microbiota of IBS patients but failed to associate shifts in bacterial groups with symptoms of the condition. Other culture-independent studies also showed temporal instability of the intestinal microbiota of IBS subjects, a lower abundance of *Lactobacillus* species in stool samples of patients with IBS-D, an increase in abundance of *Veillonella* species in IBS-C individuals, and a reduction in clostridia species abundance (Malinen et al. 2005; Kassinen et al. 2007).

Other culture-independent studies showed that in the fecal microbiota of IBS patients compared to controls, the *Firmicutes* to *Bacteroidetes* ratio was increased, *Actinobacteria* had lower abundance, and *Proteobacteria* were more abundant (Krogius-Kurikka et al. 2009; Rajilic-Stojanovic et al. 2011). Similar results were published by Jeffrey et al. (2012), that is, significant differences between gut microbiota of IBS patients and healthy controls and an increase in *Firmicutes* to *Bacteroidetes* ratio. Subgroups were formed based on the fecal microbiota of patients and importantly, one of these was characterised by normal-resembling fecal microbiota (Jeffrey et al. 2012). These patients had high scores for anxiety and depression, suggesting a subclass of IBS of nonmicrobial aetiology.

The fermentation potential of the fecal microbiota of IBS-C patients was altered: fluorescent *in situ* hybridisation (FISH) revealed that the abundance of H_2-consuming sulphur-reducing bacteria and butyrate producing *Roseburia/E. rectale* group were significantly higher and lower respectively compared to control subjects (Chassard *et al.* 2012). The altered fermentation activity of the intestinal microbiota in IBS patients was also indicated by abnormal hydrogen and methane breath measurements in patients (King *et al.* 1998). Interestingly, the commensal methanogenic *Archaed* with the production of H_2 may have a role in IBS pathogenesis (Triantafyllou *et al.* 2014) but studies to date yield contradictory results (Rajilic-Stojanovic *et al.* 2011; Kim *et al.* 2012).

Studies so far have not fully revealed the exact mechanism through which the microbiota influences the onset and progression of IBS. The fact that probiotics seem to ameliorate the symptoms of the dysbiotic condition of this multifaceted syndrome further promotes the idea of microbiota disturbances implicated in the disease (Santos and Whorwell 2014). IBS is an interesting field of research as the unravelling of the disease mechanisms will enable further insight in the GBA communication and how mediators from the intestinal microbiota can influence cognitive function in subjects with intestinal disorders (Bercik *et al.* 2011).

3.5. AGEING AND MICROBIOTA ALTERATIONS

There is growing scientific interest in the connection between low-grade inflammation observed in older age and the role of gut microbiota. As mentioned earlier, age is associated with increased diversity of the gut microbiota species. This can shift towards imbalance leading to potential negative effects on quality of life. According to this concept, accumulation of antigenic stimulation and stressors throughout life leads to declined immune responses and low immune adaptiveness resulting in inflamm-ageing (Franceschi *et al.* 2000). An overview of how the protective activity of the immune system (IS) is dysregulated with immuno-senescence, that is the ageing of the IS, is presented by De Martinis *et al.* (2005). The major driver of immune function signalling, known as Nuclear Factor κB (NF-κB), seems to be at the core of this process by being the major signalling pathway in the innate IS (Salminen *et al.* 2008).

It is now accepted that this low-grade inflammatory status with characteristic markers circulating in the plasma of old people is associated with the degenerative conditions that characterise old age (Howcroft *et al.* 2013). Among the most usual proinflammatory markers correlated with muscle tissue damage and sarcopenia, neuro-degeneration, and obesity are C-reactive protein (CRP) and tumour necrosis factor-α (TNF-α), IL-6, and IL-18 (Howcroft et al. 2013).

Antigenic stimuli associated with the gut microbiota composition shifts observed in old age may contribute to inflamm-ageing. The exact correlation and mechanisms that link the microbiome to ageing remains unclear and this presents a promising field for future research and new therapeutic strategies.

3.5.1. Changes in the Gut Microbiota in Old Age

Changes in the GIT physiology, in the functionality of the IS, in lifestyle and nutritional habits, and hospitalisation and medical treatment can impact the composition of the gut microbiota in old age and consequently health status (Cusack and O'Toole 2013).

The combined data from older mainly culture-dependent studies suggest that age is accompanied by an increase in the proportion of facultative anaerobes, with a simultaneous decline of beneficial anaerobes such as bifidobacteria and lactobacilli and importantly, an overall decline in species diversity in several bacterial groups (Woodmansey 2007). During inflammation, pathobionts, potentially harmful bacteria that are in low abundance in the healthy gut, increase in abundance displacing commensals (Pédron and Sansonetti 2008). The example of the proliferation of the pathogenic *Salmonella enterica* serotype *Typhimurium* in the gut shows how a pathogen can compete with commensals and prevail in the dysbiotic condition of the inflamed ecosystem and further promote inflammation (Winter et al. 2010). Furthermore, a striking correlation exists between the low diversity profile of the intestinal microbiota of older subjects and infection or nonsymptomatic colonization by the common nosocomial pathogen *C. difficile* (Rea et al. 2012).

Studies based on culture-independent techniques offered more comprehensive information about the shifts occurring in the gut microbiome of old subjects. Overall, studies do not yield homogenous results and there are variations based on the cohorts and the methods employed. However, it seems that compared to controls, the intestinal microbiota of older people is characterised by differences in the abundance of specific bacterial groups and the composition and species diversity of these groups is altered (Tiihonen et al. 2010). Variations in the reported composition and significant interindividual variability in the microbiota could be attributed to the fact that ageing alone is not enough to alter the generally stable microbiota (Brussow 2013) and the alterations should be viewed and explained accompanied by data on the cohort diet, physical activity, drug and antibiotic intake, and even geographical and socioeconomical profile.

When the fecal bacterial population of older community-dwellers, hospitalized, and hospitalized on antibiotic treatment were compared, Bartosch *et al.* (2004) noted that the basic difference in the fecal microbiota between healthy and hospitalized subjects was the significant reduction in the *Bacteroides/Prevotella* group. In the same study, bifidobacteria, *Desulfovibrio* spp., *Clostridium clostridioforme*, and *F. prausnitzii* also declined in abundance after hospitalisation. However, the relative abundance of the aforementioned bacteria remained stable as the total bacterial load in the stool of hospitalised subjects also decreased. Antibiotic treatment had an additional negative impact on the fecal microbiota. Antibiotics lowered the diversity in the stool microbiota and even eliminated certain bacterial groups, promoting the growth of opportunistic species and enterococci like *E. faecalis* (Bartosch *et al.* 2004).

Mueller *et al.* (2006) studied the gut microbiota composition of 230 healthy individuals from four European countries. A significant increase in Enterobacteria abundance in the stool samples of the older volunteers was observed irrespective of country of origin whereas for most of the other dominant bacterial groups like *F. prausnitzii*, bifidobacteria, *Bacteroides/Prevotella* no generalised conclusions could be made that would unify the results from the stool samples of all individuals tested. However, the *Bacteroides/Prevotella* group was more abundant in male subjects compared to female. Zwielehner *et al.* (2009) compared the fecal microbiota of long-term residential elderly to healthy adults and reported an overall decline in diversity of the dominant phyla and significantly low representation of *Clostridium* cluster IV and bifidobacteria along with interindividual variations in *Bacteroides*. Comparing the fecal microbiota of 161 Irish subjects aged greater than 65 years and nine younger controls identified significant differences, and remarkable inter-individual variability was reported for *Bacteroidetes* and *Firmicutes* abundances (Claesson *et al.* 2011).

When fecal microbiota of healthy adults (around 30 years), older adults (around 70 years), and centenarian subjects was analysed, the intestinal microbiota of centenarians was distinct from the other subject categories (Biagi *et al.* 2010). With *Firmicutes* and *Bacteroidetes* remaining the major phyla throughout life, certain rearrangements were observed in the microbiota of centenarians with extensive remodelling in the population of *Clostridium* cluster XIVa (Biagi *et al.* 2010).

Data from the ELDERMET project demonstrated a clear correlation between diet, gut microbiota, and health status in a large cohort of older subjects (Claesson *et al.* 2012). The participants were stratified in community-dwelling and short-term and long-term hospitalized. The profiling of the intestinal microbiota was accompanied by data collection on habitual diet, inflammation status, cognitive function, and frailty. Overall, subjects frequenting or residing in long-stay care units had poor diet which correlated with higher frailty scores, higher inflammation indicators like IL-6, IL-8, TNFα, and CRP in the serum, and a distinct intestinal microbiota in comparison to community-dwellers of the same ethnogeographic region (Claesson *et al.* 2012). *Bacteroidetes*-related operational taxonomic units (OUTs) were more abundant in fecal samples of long-stay subjects whereas in community-residing subjects the phylum *Firmicutes* was at higher levels with dominant genera *Coprococcus* and *Roseburia* for the latter group of subjects, and *Parabacteroides, Eubacterium, Anaerotruncus, Lactonifactor*, and *Coprobacillus* for the former (Claesson *et al.* 2012).

The gut microbiota in older age is characterised by both compositional and functional alterations expressed in centenarians in the marked shift from a saccharolytic metabolism towards putrefaction with significant loss of genes involved in SCFA production (Rampelli *et al.* 2013). Given the importance of these metabolites in gut homeostasis (Smith *et al.* 2013), this shift could increase the risk for disease. Protein fermentation metabolites can also be detrimental for health by affecting gut homeostasis (Windey *et al.* 2012). However, in people reaching the hundredth decade of life, changes in the microbiome could be seen as an evolutionary advantage. In centenarians high phenylacetyglutamine (PAG) and p-cresol sulphate (PCS) secretion indicates shifts in the microbiome which could reflect an overall remodelling of structure and functionality in order for the human organism to survive in the extremities of life-span (Collino *et al.* 2013).

3.6. CONCLUDING REMARKS

There is accumulating evidence for the significance of gut microbiota in modulating health and disease. The gut microbiota is affected by the host life style, with a major factor being habitual diet. Intestinal microbiota profiling conducted on fecal samples of various cohorts ranging from the Hazda hunter-gatherers of Tanzania to subjects residing in Westernised urban areas has revealed how diet shapes the composition and function of the mammalian gut microbiome. Nutrient products generated by commensal bacterial fermentation in the colon affect fat storage, metabolism, and even the GBA. In Figure 3.2, a schematic overview of how extrinsic factors like diet and antibiotics influence normal gut microbiota leading to dysbiosis is presented. Interestingly, the gut microbiota is significantly associated with the rate of ageing of the human organism (Heinz and Mair 2014).

Thus, the gut microbiota is a promising target for novel therapeutics for a wide range of conditions and for the manipulation of the progression of ageing itself (Foxx-Orenstein *et al.* 2010; Cani and Delzenne 2011; Candela *et al.* 2014). Culturomics and phylogeny are indispens-

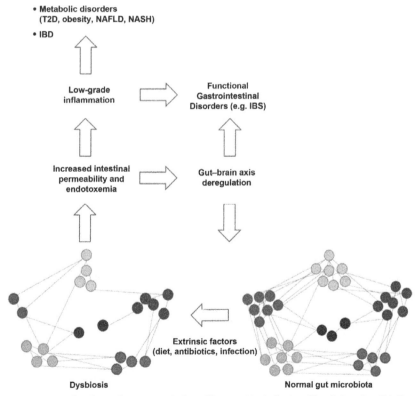

FIGURE 3.2. *A schematic representation of how extrinsic factors like diet and antibiotics influence normal gut microbiota leading to dysbiosis. The Wiggum plots at the bottom represent normal gut microbiota and gut microbiota in dysbiosis. Coloured circles represent correlated genera and their interactions are schematically indicated by lines.*

able for the development of novel microbial therapeutics as they offer tools for rationally designed studies of the full breadth of gut bacterial physiology and metabolism (Walker *et al.* 2014). Understanding gut bacterial physiology will also facilitate the development of *in silico* simulation models of host-microbe/microbe-microbe interactions (Thiele *et al.* 2013).

3.7. REFERENCES

Aagaard, K., Ma, J., Antony, K.M., Ganu, R., Petrosino, J., and Versalovic, J. 2014. The placenta harbours a unique microbiome. *Sci Transl Med, 6*(237), 237–65.

Abraham, C. and Cho, J.H. 2009. Inflammatory Bowel Disease. *N Engl J Med*, (361), 2066–2078.

Abu-Shanab, A. and Quigley, E.M.M. 2010. The role of the gut microbiota in nonalcoholic fatty liver disease. *Nat Rev Gastroenterol Hepatol, 7*(12), 691–701.

Adlerberth, I. and Wold, A.E. 2009. Establishment of the gut microbiota in Western infants. *Acta Paediatr, 98*(2), 229–38.

Aires, J., Thouverez, M., Allano, S., and Butel, M.J. 2011. Longitudinal analysis and genotyping of infant dominant bifidobacterial populations. *System Appl Microbiol, 34*(7), 536–41.

Ananthakrishnan, A.A. 2011. *Clostridium difficile* infection: Epidemiology, risk factors and management. *Nat Rev Gastroenterol Hepatol, 8*(1), 17–26.

Antunes, L.C.M., Han, J., Ferreira, R.B.R., Lolic, P., Borchers, C.H., and Finlay, B.B. 2011. Effect of antibiotic treatment on the intestinal metabolome. *Antimicrob Agents Chemother, 55*(4), 1494–1503.

Arboleya, S., Ang, L., Margolles, A., Yiyuan, L., Dongya, Z., Liang, X., *et al.* 2012. Deep 16S rRNA metagenomics and quantitative PCR analyses of premature infant fecal microbiota. *Anaerobe, 18*, 378–380.

Arboleya, S., Binetti, A., Salazar, N., Fernandez, N., Solís, G., Hernandez-Barranco, A., *et al.* 2012. Establishment and development of intestinal microbiota in preterm neonates. *FEMS Microbiol Ecol, 79*(3), 763–772.

Ardeshir, A., Narayan, N.R., Méndez-Lagares, G., Lu, D., Rauch, M., Huang, Y., *et al.* 2014. Breast-fed and bottle-fed infant rhesus macaques develop distinct gut microbiotas and immune systems. *Sci Transl Med, 6*(252), 252ra120.

Arumugam, M., Raes, J., Pelletier, E., Le Paslier, D., Yamada, T., Mende, D. R., *et al.* 2011. Enterotypes of the human gut microbiome. *Nature, 473*(7346), 174–180.

Atarashi, K., Tanoue, T., Oshima, K., Suda, W., Nagano, Y., Nishikawa, H., *et al.* 2013. Treg induction by a rationally selected mixture of Clostridia strains from the human microbiota. *Nature, 500*(7461), 232–6.

Atarashi, K., Tanoue, T., Shima, T., Imaoka, A., Kuwahara, T., Momose, Y., *et al.* 2011. Induction of colonic regulatory T cells by indigenous Clostridium species. *Science, 331*(6015), 337–341.

Backhed, F., Ding, H., Wang, T., Hooper, L.V., Koh, G.Y., Nagy, A., *et al.* 2004. The gut microbiota as an environmental factor that regulates fat storage. *Proc Natl Acad Sci USA, 101*(44), 15718–23.

Bajaj, J.S., Hylemon, P.B., and Younossi, Z. 2012. The intestinal microbiota and liver disease. *Am J Gastroenterol Suppl, 1*(1), 9–14.

Bartosch, S., Fite, A., MacFarlane, G.T., and McMurdo, M.E.T. 2004. Characterization of bacterial communities in faces from healthy elderly volunteers and hospitalized elderly patients by using Real-Time PCR and effects of antibiotics treatment on the fecal microbiota. *Appl Environ Microbiol, 70*(6), 3575–3581.

Benson, A.K., Kelly, S.A, Legge, R., Ma, F., Low, S.J., Kim, J., *et al.* 2010. Individuality in gut microbiota composition is a complex polygenic trait shaped by multiple environmental and host genetic factors. *Proc Natl Acad Sci USA, 107*(44), 18933–18938.

Bercik, P., Denou, E., Collins, J., Jackson, W., Lu, J., *et al.* 2011. The intestinal microbiota affect central levels of brain-derived neurotropic factor and behaviour in mice. *Gastroenterol, 141*(2), 599–609.

Bergstrom, A., Skov, T.H., Bahl, M.I., Roager, H.M., Christensen, L.B., Ejlerskov, K.T., *et al.* 2014. Establishment of intestinal microbiota during early life: A longitudinal, explorative study of a large cohort of Danish infants. *Appl Environ Microbiol, 80*(9), 2889–900.

Bezirtzoglou, E., Tsiotsias, A., and Welling, G.W. 2011. Microbiota profile in feces of breast- and formula-fed newborns by using fluorescence in situ hybridization (FISH). *Anaerobe, 17*(6), 478–82.

Biagi, E., Cnadel, M., Turroni, S., Garagnani, P., Franscechi, C., and Brigidi, P. 2013. Ageing and gut microbes: Perspectives for health maintenance and longevity. *Pharmacol Res, 69*, 11–20.

Biagi, E., Nylund, L., Candela, M., Ostan, R., Bucci, L., Pini, E., *et al.* 2010. Through ageing, and beyond: Gut microbiota and inflammatory status in seniors and centenarians. *PloS One, 5*(5), e10667.

Biasucci, G., Benenati, B., Morelli, L., Bessi, E., and Boehm, G. 2008. Cesarean delivery may affect the early biodiversity of intestinal bacteria. *J Nutr, 138*(9), 1796S–1800S.

Blaut, M. and Clavel, T. 2007. Metabolic diversity of the intestinal microbiota: Implications for health and disease. *J. Nutr, 137*(3), 751–755.

Brunser, O., Gotteland, M., Cruchet, S., Figueroa, G., Garrido, D., and Steenhout, P. 2006. Effect of a milk formula with prebiotics on the intestinal microbiota of infants after an antibiotic treatment. *Pediatr Res, 59*(3), 451–6.

Brunt, E.M. and Tiniakos, D.G. 2010. Histopathology of non-alcoholic fatty liver disease. *World J Gastroenterol, 16*(42), 5286–96.

Brussow, H. 2013. Microbiota and healthy ageing: Observational and nutritional intervention studies. *Microbial Biotechnology, 6*(4), 326–34.

Bryant, R.V, van Langenberg, D.R., Holtmann, G.J., and Andrews, J.M. 2011. Functional gastrointestinal disorders in inflammatory bowel disease: Impact on quality of life and psychological status. *J Gastroenterol Hepatol, 26*(5), 916–923.

Burke, C., Steinberg, P., Rusch, D., Kjelleberg, S., and Thomas, T. 2011. Bacterial community assembly based on functional genes rather than species. *Proc Natl Acad Sci USA, 108*(34), 14288–14293.

Cabrera-Rubio, R., Collado, M.C., Laitinen, K., Salminen, S., Isolauri, E., and Mira, A. 2012. The human milk microbiome changes over lactation and is shaped by maternal weight and mode of delivery. *Am J Clin Nutr, 96*(3), 544–51.

Cahenzli, J., Koller, Y., Wyss, M., Geuking, M.B., and McCoy, K.D. 2013. Intestinal microbial diversity during early-life colonization shapes long-term IgE levels. *Cell Host Microbe, 14*(5), 559–570.

Candela, M., Biagi, E., Brigidi, P., O'Toole, P.W., and de Vos, W.M. 2014. Maintenance of a healthy trajectory of the intestinal microbiome during aging: A dietary approach. *Mech Ageing Dev*, 136–137, 70–75.

Cani, P.C., Amar, J., Iglesias, M.A., Poggi, M., Knauf, C., Bastelica, D., et al. 2007. Metabolic endotoxemia initiates obesity and insulin resistance. *Diabetes, 56*(7), 1761–72.

Cani, P.C., Bibiloni, R., Knauf, C., Waget, A., Neyrinck, A.M., Delzenne, N.M., et al. 2008. Changes in the gut microbiota control metabolic endotoxemia-induced inflammation in high-fat diet-induced obesity and diabetes in mice. *Diabetes, 57*(6), 1470–1481.

Cani, P.D. and Delzenne, N.M. 2011. The gut microbiome as therapeutic target. *Pharmacol Therap, 130*(2), 202–12.

Chassard, C., Dapoigny, M., Scott, K.P., Crouzet, L., Del'homme, C., Marquet, P., et al. 2012. Functional dysbiosis within the gut microbiota of patients with constipated-irritable bowel syndrome. *Aliment Pharmacol Ther, 35*(7), 828–38.

Cho, I. and Blaser, M.J. 2012. The human microbiome: At the interface of health and disease. *Nat Rev Genet, 13*(4), 260–270.

Cho, I., Yamanishi, S., Cox, L., Methé, B.A, Zavadil, J., Li, K., et al. 2012. Antibiotics in early life alter the murine colonic microbiome and adiposity. *Nature, 488*(7413), 621–626.

Choi, J.J., Eum, S.Y., Rampersaud, E., Daunert, S., Abreu, M.T., and Toborek, M. 2013. Exercise attenuates PCB-induced changes in the mouse gut microbiome. *Environ Health Perspect, 121*(6), 725–730.

Clarke, S.F., Murphy, E.F., O'Sullivan, O., Lucey, A.J., Humphreys, M., Hogan, A., et al. 2014. Exercise and associated dietary extremes impact on gut microbial diversity. *Gut*, [Epub ahead of print].

Claesson, M.J., Cusack, S., O'Sullivan, O., Greene-Diniz, R., de Weerd, H., Flannery, E., et al. 2011. Composition, variability, and temporal stability of the intestinal microbiota of the elderly. *Proc Natl Acad Sci USA, 108* (Suppl 1), 4586–4591.

Claesson, M.J., Jeffrey, I.B., Conde, S., Power, S.E., O'Connor, E.M., Cusack, S., et al. 2012. Gut microbiota composition correlates with diet and health in elderly. *Nature, 488*, 178–184.

Collino, S., Montoliu, I., Martin, F.P.J., Scherer, M., Mari, D., Salvioli, S., et al. 2013. Metabolic signatures of extreme longevity in northern Italian centenarians reveal a complex remodelling of lipids, amino acids, and gut microbiota metabolism. *PloS One, 8*(3), e56564.

Collins, S.M. and Bercik, P. 2009. The relationship between intestinal microbiota and the central nervous system in normal gastrointestinal function and disease. *Gastroenterology, 136*(6), 2003–2014.

Collins, S., Verdu, E., Denou, E., and Bercik, P. 2009. The role of pathogenic microbes and commensal bacteria in irritable bowel syndrome. *Dig Dis, 27*(suppl 1), 85–89.

Cox, L.M., Yamanishi, S., Sohn, J., Alekseyenko, A.V., Leung, J.M., Cho, I., *et al.*, 2014. Altering the intestinal microbiota during a critical developmental window has lasting metabolic consequences. *Cell, 158*(4), 705–721.

Cummings, J.H. and Macfarlane, G.T. 1997. Role of intestinal bacteria in nutrient metabolism. *JPEN J Parenter Enteral Nutr, 21*(6), 357–365.

Cusack, S. and O'Toole, P.W. 2013. Diet, the gut microbiota and healthy ageing: How dietary modulation of the gut microbiota could transform the health of older populations. *Agro FOOD Ind Hi Tech, 24*(2), 54–57.

David, L.A, Maurice, C., Carmody, R.N., Gootenberg, D.B., Button, J.E., Wolfe, B.E., *et al.* 2014. Diet rapidly and reproducibly alters the human gut microbiome. *Nature, 505*(7484), 559–63.

De Filippo, C., Cavalieri, D., Di Paola, M., Ramazzotti, M., Poullet, J.B., Massart, S., *et al.* 2010. Impact of diet in shaping gut microbiota revealed by a comparative study in children from Europe and rural Africa. *Proc Nantl Acad Sci USA, 107*(33), 14691–14696.

De La Cochetiere, M.F., Durand, T., Lepage, P., Bourreille, A., Galmiche, J.P., and Dore, J. 2005. Resilience of the dominant human fecal microbiota upon short-course challenge. *J Clin Microbiol, 43*(11), 5588–5592.

De Martinis, M., Franceschi, C., Monti, D., and Ginaldi, L. 2005. Inflamm-ageing and lifelong antigenic load as major determinants of ageing rate and longevity. *FEBS Lett, 579*(10), 2035–2039.

Dethlefsen, L., Huse, S., Sogin, M.L., and Relman, D.A. 2008. The pervasive effect of an antibiotic on the human gut microbiota, as revealed by deep 16S rRNA sequencing. *PLoS Biol, 6*(11), e280

Dethlefsen, L. and Relman, D.A. 2011. Incomplete recovery and individualized responses of the human distal gut microbiota to repeated antibiotic perturbation. *Proc Nantl Acad Sci USA, 108*(suppl. 1), 4554–4561

DiGiulio, D.B., Romero, R., Amogan, H.P., Kusanovic, J.P., Bik, E.M., Gotsch, F., *et al.* 2008. Microbial prevalence, diversity and abundance in amniotic fluid during preterm labour: A molecular and culture-based investigation. *PloS One, 3*(8), e3056.

Dominguez-Bello, M.G., Costello, E.K., Contreras, M., Magris, M., Hidalgo, G., Fierer, N., *et al.* 2010. Delivery mode shapes the acquisition and structure of the initial microbiota across multiple body habitats in newborns. *Proc Nantl Acad Sci USA, 107*(26), 11971–5.

Duncan, S.H., Lobley, G.E., Holtrop, G., Ince, J., Johnstone, A.M., Louis, P., *et al.* 2008. Human colonic microbiota associated with diet, obesity and weight loss. *Int J Obes (Lond), 32*(11), 1720–1724.

Eckburg, P.B., Bik, E.M., Bernstein, C.N., Purdom, E., Dethlefsen, L., Sargent, M., *et al.* 2005. Diversity of the human intestinal microbial flora. *Science, 308*(5728), 1635–1638.

El Aidy, S., Derrien, M., Merrifield, C.A, Levenez, F., Dore, J., Boekschoten, M.V., et al. 2013. Gut bacteria-host metabolic interplay during conventionalisation of the mouse germfree colon. *ISME J, 7*(4), 743–755.

El Aidy, S., Hooiveld, G., Tremaroli, V., Backhed, F., and Kleerebezem, M. 2013. The gut microbiota and mucosal homeostasis: Colonized at birth or at adulthood, does it matter? *Gut Microbes, 4*(2), 118–124.

Erridge, C., Attina, T., Spickett, C.M., and Webb, D.J. 2007. A high-fat meal induces low-grade endotoxemia: Evidence of a novel mechanism of postprandial inflammation. *Am J Clin Nutr, 86*(5), 1286–1292.

Evans, C.C., LePard, K.J., Kwak, J.W., Stancukas, M.C., Laskowski, S., et al. 2014. Exercise prevents weight gain and alters the gut microbiota in a mouse model of high fat diet-induced obesity. *PLoS One, 9*(3), e92193.

Fallani, M., Amarri, S., Uusijarvi, A., Adam, R., Khanna, S., Aguilera, M., et al. 2011. Determinants of the human infant intestinal microbiota after the introduction of first complementary foods in infant samples from five European centres. *Microbiology, 157*(Pt 5), 1385–1392.

Fernandez, L., Langa, S., Martin, V., Maldonado, A., Jimenez, E., Martin, R., et al. 2013. The human milk microbiota: Origin and potential roles in health and disease. *Pharmacol Res, 69*(1), 1–10.

Foxx-Orenstein, A.E. and Chey, W.D. 2012. Manipulation of the gut microbiota as a novel treatment strategy for gastrointestinal disorders. *Am J Gastroenterol Suppl, 1*(1), 41–46.

Franceschi, C., Bonafe, M., Valensin, S., and Benedictis, G.D.E. 2000. Inflamm-ageing: An evolutionary perspective on immunosenescence. *Ann NY Acad. Sci*, (908), 244–254.

Frank, D.N., Amand, A.L.S.T., Feldman, R.A, Boedeker, E.C., Harpaz, N., and Pace, N.R. 2007. Molecular-phylogenetic characterization of microbial community imbalances in human inflammatory bowel diseases. *Proc Natl Acad Sci USA, 104*(34), 13780–13785.

Garrido, D., Kim, J.H., German, J.B., Raybould, H.E., and Mills, D.A. 2011. Oligosaccharide binding proteins form *Bifidobacterium longum* subsp. *infantis* reveal a preference for host glycans. *PLoS One, 6*(3), e17315.

Gummesson, A., Carlsson, L.M., Storlien, L.H., Backhed, F., Lundin, P., Lofgren L., et al. 2011. Intestinal permeability is associated with visceral adiposity in healthy women. *Obesity(Silver Spring), 19*(11), 2280–2282.

Grundy, S.M., Brewer, H.B. Jr., Cleeman, J.I., Smith, S.C., and Lenfant, C. 2004. Definition of metabolic syndrome: Report of the National Heart, Lung, and Blood Institute/American Heart Association conference on scientific issues related to definition. *Circulation, 109*(3), 433–438.

Hamer, H.M., Jonkers, D., Venema, K., Vanhoutvin, S., Troost, F.J., and Brummer, R.J. 2008. Review article: The role of butyrate on colonic function. *Aliment Pharmacol Ther, 27*(2), 104–119.

Harmsen, H.J., Wildeboer-Veloo, A.C., Raangs, G.C., Wagendorp, A.A., Klijn, N., Bindels, J.G., and Welling, G.W. 2000. Analysis of intestinal flora development in breast-fed and formula-fed infants by using molecular identification and detection methods. *J Pediatr Gastroenterol Nutr, 30*(1), 61–67.

Hawkesworth, S., Moore, S.E., Fulford, A.J.C., Barclay, G.R., Darboe, A.A., Mark, H., *et al.* 2013. Evidence for metabolic endotoxemia in obese and diabetic Gambian women. *Nutr Diabetes, 3,* e83.

Heinz, C. and Mair, W. 2014. You are what you host: Microbiome modulation of the aging process. *Cell, 156*(3), 408–411.

Hold, G.L., Pryde, S.E., Russell., V.J., Furrie, E., and Flint, H.J. 2002. Assessment of microbial diversity in human colonic samples by 16S rRNA sequence analysis. *FEMS Microbiol Ecol, 39*(1), 33–39.

Howcroft, T.K., Campisi, J., Louis, G.B., Smith, M.T., Wise, B., Wyss-Coray, T., *et al.* 2013. The role of inflammation in age-related disease. *Aging, 5*(1), 84–93.

Huang, P.L. 2009. A comprehensive definition for metabolic syndrome. *Dis Model Mech, 2*(5–6), 231–237.

Hunt, K.M., Foster, J.A., Forney, L.J., Schutte, U.M.E., Beck, D.L., Williams, J.E., *et al.* 2010. The human milk microbiome: A potential influence on mammary health and bacterial colonization of infant gut. *FASEB J, 206.5.*

Huttenhower, C., Kostic, A.D., and Xavier, R.J. 2014. Inflammatory bowel disease as a model for translating the microbiome. *Immunity, 40*(6), 843–854.

Jakobsson, H.E., Jernberg, C., Andersson, A.F., Sjolund-Karlsson, M., Jansson, J.K., and Engstrand, L. 2010. Short-term antibiotic treatment has differing long-term impacts on the human throat and gut microbiome. *PloS One, 5*(3), e9836.

Jeffrey, I.B., O'Toole, P.W., Ohman, L., Claesson, M.J., Deane, J., Quigley, E.M., *et al.* 2012. An irritable bowel syndrome subtype defined by specie-species alterations in faecal microbiota. *Gut, 61*(7), 997–1006.

Jernberg, C., Lofmark, S., Edlund, C., and Jansson, J.K. 2007. Long-term ecological impacts of antibiotic administration on the human intestinal microbiota. *ISME J, 1*(1), 56–66.

Jimnez, E., Marin, M.L., Martin, R., Odriozola, J.M., Olivares, M., Xaus, J., *et al.* 2008. Is meconium from healthy newborns actually sterile? *Res. Microbiol, 159*(3), 187–193.

Jost, T., Lacroix, C., Braegger, C.P., and Chassard, C. 2012. New insights in gut microbiota establishment in healthy breast fed neonates. *PLOS One, 7*(8), e44595.

Kabeerdoss, J., Devi, R.S., Mary, R.R., and Ramakrishna, B.S. 2012. Faecal microbiota composition in vegetarians: Comparison with omnivores in a cohort of young women in southern India. *Br J Nutr, 108*(6), 953–957.

Karlsson, F.H., Tremaroli, V., Nookaew, I., Bergstrom, G., Behre, C.J., Fagerberg, B., *et al.* 2013. Gut metagenome in European women with normal, impaired and diabetic glucose control. *Nature, 498*(7452), 99–103.

Kassinen, A., Krogius-Kurikka, L., Makivuokko, H., Rinttila, T., Paulin, L., Corander, J., *et al.* 2007. The fecal microbiota of irritable bowel syndrome patients differs significantly from that of healthy subjects. *Gastroenterology, 133*(1), 24–33.

Kim, G., Deepinder, F., Morales, W., Hwang, L., Weitsman, S., Chang, C., *et al.* 2012. Methanobrevibacter smithii is the predominant methanogen in patients with constipation-predominant IBS and methane on breath. *Dig Dis Sci, 57*(12), 3213–3218.

Kim, M.S., Hwang, S.S., Park, E.J., and Bae, J.W. 2013. Strict vegetarian diet improves the risk factors associated with metabolic diseases by modulating gut microbiota and reducing intestinal inflammation. *Environ Microbiol Rep, 5*(5), 765–75.

King, T.S., Elia, M., and Hunter, J.O. 1998. Abnormal colonic fermentation in irritable bowel syndrome. *Lancet, 352*(9135), 1187–1189.

Koenig, J.E., Spor, A., Scalfone, N., Fricker, A.D., Stombaugh, J., Knight, R., et al. 2011. Succession of microbial consortia in the developing infant gut microbiome. *Proc Nat Acad Sci USA, 108* (Suppl 1), 4578–4585.

Korpela, K., Flint, H.J., Johnstone, A.M., Lappi, J., Poutanen, K., Dewulf, E., et al. 2014. Gut microbiota signatures predict host and microbiota responses to dietary interventions in obese individuals. *PloS One, 9*(6), e90702.

Krogius-Kurikka, L., Lyra, A., Malinen, E., Aarnikunnas, J., Tuimala, J., Paulin, L., et al. 2009. Microbial community analysis reveals high level phylogenetic alterations in the overall gastrointestinal microbiota of diarrhoea-predominant irritable bowel syndrome sufferers. *BMC Gastroenterol, 9*(1), 95.

Lahti, L., Salojarvi, J., Salonen, A., Scheffer, M., and de Vos, W.M. 2014. Tipping elements in the human intestinal ecosystem. *Nat Commun, 5*, 4344.

Lambert, D.M. and Muccioli, G.G. 2007. Endocannabinoids and related N-acylethanolamines in the control of appetite and energy metabolism: Emergence of new molecular players. *Curr Opin Clin Nutr Metab Care, 10*(6), 735–744.

La Rosa, P.S., Warner, B.B., Zhou, Y., Weinstock, G.M., Sodergren, Hall-Moore, C.M., et al. 2014. Patterned progression of bacterial populations in the premature infant gut. *Proc Nat Acad Sci USA, 111*(34), 12522–12527.

Larsen, N., Vogensen, F.K., van den Berg, F.W.J., Nielsen, D.S., Andreasen, A.S., Pedersen, B.K., et al. 2010. Gut microbiota in human adults with type 2 diabetes differs from non-diabetic adults. *PloS One, 5*(2), e9085.

Lawley, T.D., Clare, S., Walker, A.W., Stares, M.D., Connor, T.R., Raisen, C., et al. 2012. Targeted restoration of the intestinal microbiota with a simple, defined bacteriotherapy resolves relapsing *Clostridium difficile* disease in mice. *PLoS Pathog, 8*(10), e1002995.

Lawley, T. D., Walker, A. W. (2013). Intestinal colonization resistance. *Immunology, 138*(1), 1–11.

Ley, R.E., Backhed, F., Turnbaugh, P., Lozupone, C.A, Knight, R.D., and Gordon, J.I. 2005. Obesity alters gut microbial ecology. *Proc Nat Acad Sci USA, 102*(31), 11070–11075.

Ley, R.E., Lozupone, C.A., Hamady, M., Knight, R., and Jeffrey, I. 2009. Worlds within worlds: Evolution of the vertebrate gut microbiota. *Nat Rev Microbiol, 6*(10), 776–788.

Ley, R.E., Turnbaugh, P.J., Klein, S., and Gordon, J.I. 2006. Human gut microbes associated with obesity. *Nature, 444*, 1022–23.

Liszt, K., Zwielehner, J., Handschur, M., Hippe, B., Thaler, R., and Haslberger, A.G. 2009. Characterization of bacteria, clostridia and Bacteroides in faeces of vegetarians using qPCR and PCR-DGGE fingerprinting. *Ann Nutr Metab, 54*(4), 253–257.

LoCasio, R.G., Desai, P., Sela, D.., Weimer, B., and Mils, D.A. 2010. Broad conservation of milk utilization genes in *Bifidobacterium longum* subsp. *infantis* as revealed by comparative genomic hybridization. *Appl Environ Microbiol, 76*(22), 7373–7381.

Longstreth, G.F., Thompson, W.G., Chey, W.D., Houghton, L.A., Mearin, F., and Spiller, R.C. 2006. Functional bowel disorders. *Gastroenterology, 130*(5), 1480–1491.

Louis, P. and Flint, H.J. 2009. Diversity, metabolism and microbial ecology of butyrate-producing bacteria from the human large intestine. *FEMS Microbiol Lett, 294*(1), 1–8.

Lozupone, C.A., Stombaugh, J.I., Gordon, J.I., Jansson, J.K., and Knight, R. 2012. Diversity, stability, and resilience of the human gut microbiota. *Nature, 489*(7415), 220–230.

Macfarlane, G.T. and Macfarlane, S. 2012. Bacteria, colonic fermentation, and gastrointestinal health. *J AOAC Int, 95*(1), 50–60.

Mahowald, M.A., Rey, F.E., Seedorf, H., Turnbaugh, P.J., Fulton, R.S., Wollam, A., et al. 2009. Caharacterizing a model human gut microbiota composed of members of its two dominant bacterial phyla. *Proc Natl Acad Sci USA, 106*(14), 5859–5864.

Mai, V., Torrazza, R.M., Ukhanova, M., Wang, X., Sun, Y., Li, N., et al. 2013. Distortions in development of intestinal microbiota associated with late onset sepsis in preterm infants. *PLOS One, 8*(1), e52876.

Malinen, E., Rinttila, T., Kajander, K., Matto, J., Kassinen, A., Krogius, L., et al. 2005. Analysis of the fecal microbiota of irritable bowel syndrome patients and healthy controls with real-time PCR. *Am J Gastroenterol, 100*(2), 378–382.

Manichanh, C., Rigottier-Gois, L., Bonnaud, E., Gloux, K., Pelletier, E., Frangeul, L., et al. 2006. Reduced diversity of faecal microbiota in Crohn's disease revealed by a metagenomic approach. *Gut, 55*(2), 205–11.

Martin, V., Maldonado-Barragan, A., Moles, L., Rodriguez-Banos, M., Campo, R.D., et al. 2012. Sharing of bacterial strains between breast milk and infant feces. *J Hum Lact, 28*(1), 36–44.

Martin, R., Jimenez, E., Heilig, H., Fernandez, L., Marin, M., Zoetendal, G., et al. 2008. Isolation of bifidobacteria from breast milk and assessment of the bifidobacterial population by PCR-denaturing gradient gel electrophoresis and quantitative real-time PCR. *Appl Environ Microbiol, 75*(4), 965–969.

Martinez, I., Kim, J., Duffy, P.R., Schlegel, V.L., and Walter, J. 2010. Resistant starches types 2 and 4 have differential effects on the composition of the fecal microbiota in human subjects. *PloS One, 5*(11), e15046.

Matamoros, S., Gras-Leguen, C., Le Vacon, F., Potel, G., and de La Cochetiere, M.F. 2013. Development of intestinal microbiota in infants and its impact on health. *Trends Microbiol, 21*(4), 167–173.

Matsumoto, M., Inoue, R., Tsukahara, T., Ushida, K., Chiji, H., Matsubara, N., et al. 2008. Voluntary running exercise alters microbiota composition and increases n-butyrate concentration in the rat cecum. *Biosci Biotechnol Biochem, 72*(2), 572–576.

Matto, J., Maunuksela, L., Kajander, K., Palva, A., Korpela, R., Kassinen, A., et al. 2005. Composition and temporal stability of gastrointestinal microbiota in irritable bowel syndrome—A longitudinal study in IBS and control subjects. *FEMS Immunol Med Microbiol, 43*(2), 213–222.

Mouzaki, M., Comelli, E.M., Arendt, B.M., Bonengel, J., Fung, S.K., Fischer, S.E., *et al.* 2013. Intestinal microbiota in patients with nonalcoholic fatty liver disease. *Hepatology,* 58(1), 120–127.

Mshvildadze, M., Neu, J., Shuster, J., Theriaque, D., Li, N., and Mai, V. 2010. Intestinal microbial ecology in premature infants assessed with non-culture-based techniques. *J Pediatr,* 156(1), 20–5.

Muccioli, G.G., Naslain, D., Backhed, F., Reigstad, C.S., Lambert, D M., Delzenne, N.M., *et al.* 2010. The endocannabinoid system links gut microbiota to adipogenesis. *Mol Syst Biol,* 6, 392.

Muegge, B.D., Kuczynski, J., Knights, D., Clemente, J.C., Gonzalez, A., *et al.* 2011. Diet drives convergence in gut microbiome functions across mammalian phylogeny and within humans. *Science,* 332(6032), 970–974.

Mueller, S., Saunier, K., Hanisch, C., Norin, E., Alm, L., Midtvedt, T., *et al.* 2006. Differences in fecal microbiota in different European study populations in relation to age, gender, and country: A cross-sectional study. *Appl Environ Microbiol,* 72(2), 1027–33.

Musso, G., Gambino, R., and Cassader, M. 2011. Interactions between gut microbiota and host metabolism predisposing to obesity and diabetes. *Annu Review Med,* 62, 361–380.

Nyangale, E., Mottram, D.S., and Gibson, G.R. 2012. Gut microbial activity, implications for health and disease: The potential role of metabolic analysis. *J Proteome Res,* 11(12), 5573–5585.

O'Toole, P.W. and Claesson, M.J. 2010. Gut microbiota: Changes throughout the lifespan from infancy to elderly. *Intl Dairy J,* 20(4), 281–291.

Palmer, C., Bik, E.M., DiGiulio, D.B., Relman, D.A., and Brown, P.O. 2007. Development of the Human Infant Intestinal Microbiota. *PLoS Biol,* 5(7), e177.

Pedron, T. and Sansonetti, P. 2008. Commensals, bacterial pathogens and intestinal inflammation: An intriguing ménage á trois. *Cell Host Microbe,* 3(6), 344–347.

Penders, J., Gerhold, K., Stobberingh, E.E., Thijs, C., Zimmermann, K., Lau, S., *et al.* 2013. Establishment of the intestinal microbiota and its role for atopic dermatitis in early childhood. *J Allergy Clin Immunol,* 132(3), 601–607.

Penders, J., Thijs, C., van den Brandt, P.A, Kummeling, I., Snijders, B., Stelma, F., *et al.* 2007. Gut microbiota composition and development of atopic manifestations in infancy: The KOALA Birth Cohort Study. *Gut,* 56(5), 661–667.

Penders, J, Thijs, C., Vink, C., Stelma, F.F., Snijders, B., Kummeling, I., *et al.* 2006. Factors influencing the composition of the intestinal microbiota in early infancy. *Paediatrics,* 118(2), 511–521.

Penders, J, Vink, C., Driessen, C., London, N., Thijs, C., and Stobberingh, E.E. 2005. Quantification of Bifidobacterium spp., Escherichia coli and *Clostridium difficile* in faecal samples of breast- fed and formula-fed infants by real time PCR. *FEMS Microbiol Lett,* 243(1), 141–147.

Pepin,J., Saheb, N., Coulombe, M.A., Alary, M.E., Corriveau, M.P., Authier, S., *et al.* 2005. Emergence of fluoroquinolones as the predominant risk factor for *Clostridium difficile*-associated diarrhea: A cohort study during an epidemic in Quebec. *Clin Infect Dis,* 41(9), 1254–60.

Perez-Cobas, A.E., Gosalbes, M.J., Friedrichs, A., Knecht, H., Artacho, A., Eismann,

K., et al. 2013. Gut microbiota disturbance during antibiotic therapy: A multi-omic approach. *Gut, 62*(11), 1591–601.

Persaud, R., Azad, M.B., Konya, T., Guttman, D.S., Chari, R.S., Sears, M.R., et al. 2014. Impact of perinatal antibiotic exposure on the infant gut microbiota at one year of age. *Allergy Asthma Clin Imunol, 10*(Suppl 1), A31.

Petrof, E.O., Gloor, G.B., Vanner, S.J., Weese, S.J., Carter, D., Daigneault, M C., et al. 2013. Stool substitute transplant therapy for the eradication of *Clostridium difficile* infection: "RePOOPulating" the gut. *Microbiome, 1*(1), 3.

Qin, J., Li, Y., Cai, Z., Li, S., Zhu, J., Zhang, F., et al. 2012. A metagenome-wide association study of gut microbiota in type 2 diabetes. *Nature, 490*(7418), 55–60.

Qin, J., Li, R., Raes, J., Arumugam, M., Burgdorf, K.S., Manichanh, C., et al. 2010. A human gut microbial gene catalogue established by metagenomic sequencing. *Nature, 464*(7285), 59–65.

Rajilic-Stojanovic, M., Biagi, E., Heilig, H.G., Kajander, K., Kekkonen, R.A., Tims, S., et al. 2011. Global and deep molecular analysis of microbiota signatures in fecal samples from patients with irritable bowel syndrome. *Gastroenterology, 14*(5), 1792–1801.

Rajilic-Stojanovic, M., Heilig, H.G., Molenaar, D., Kajander, K., Surakka, A., Smidt, H., et al. 2009. Development and application of the human intestinal tract chip, a phylogenetic microarray: Analysis of universally conserved phylotypes in the abundant microbiota of young and elderly adults. *Environ Microbiol, 11*(7), 1736–1751.

Rampelli, S., Candela, M., Turroni, S., Biagi, E., Collino, S., Franceschi, C., et al. 2013. Functional metagenomic profiling of intestinal microbiome in extreme ageing. *Aging, 5*(12), 902–912.

Rautava, S., Luoto, R., Salminen, S., and Isolauri, E. 2012. Microbial contact during pregnancy, intestinal colonization and human disease. *Nat Rev Gastroenterol Hepatol, 9*(10), 565–576.

Rea, M.C., O'Sullivan, O., Shanahan, F., O'Toole, P.W., Stanton, C., Ross, R.P., and Hill, C. 2012. *Clostridium difficile* carriage in elderly subjects and associated changes in the intestinal microbiota. *J Clin Microbiol, 50*(3), 867–875.

Rehman, A., Lepage, P., Nolte, A., Hellmig, S., Schreiber, S., and Ott, S.J. 2010. Transcriptional activity of the dominant gut mucosal microbiota in chronic inflammatory bowel disease patients. *J Med Microbiol, 59*(Pt 9), 1114–1122.

Roger, L.C., Costabile, A., Holland, D.T., Hoyles, L., and McCartney, A.L. 2010. Examination of faecal Bifidobacterium populations in breast—and formula—fed infants during the first 18 months of life. *Microbiology, 156*(11), 3329–3341.

Salminen, S., Gibson, G.R., McCartney, A.L., and Isolauri, E. 2004. Influence of mode of delivery on gut microbiota composition in seven year old children. *Gut, 53*(9), 1388–1389.

Salminen, A., Huuskonen, J., Ojala, J., Kauppinen, A., Kaarniranta, K., and Suuronen, T. 2008. Activation of innate immunity system during aging: NF-kB signalling is the molecular culprit of inflamm-aging. *Ageing Res Rev, 7*(2), 83–105.

Santos, A.R. and Whorwell, P.J. 2014. Irritable bowel syndrome: The problem and the problem of treating it—Is there a role for probiotics? *Proc Nutr Soc, 26*, 1–7.

Sekirov, I., Russell, S.L., Antunes, L.C.M., and Finlay, B.B. 2010. Gut microbiota in health and disease. *Physiol Rev,* 90(3), 859–904.

Schwiertz, A., Taras, D., Schafer, K., Beijer, S., Bos, N.A, Donus, C., *et al.* 2010. Microbiota and SCFA in lean and overweight healthy subjects. *Obesity,* 18(1), 190–195.

Schnorr, S.L., Candela, M., Rampelli, S., Centanni, M., Consolandi, C., Basaglia, G., *et al.* 2014. Gut microbiome of the Hadza hunter-gatherers. *Nat Commun,* 5, 3654.

Scott, K.P., Gratz, S.W., Sheridan, P.O., Flint, H.J., and Duncan, S.H. 2013. The influence of diet on the gut microbiota. *Pharmacol Res,* 69(1), 52–60

Sela, D.A and Mills, D.A. 2010. Nursing our microbiota: Molecular linkages between bifidobacteria and milk oligosaccharides. *Trends Microbiol,* 18(7), 298–307.

Serino, M., Luche, E., Gres, S., Baylac, A., Bergé, M., Cenac, C., *et al.* 2012. Metabolic adaptation to a high-fat diet is associated with a change in the gut microbiota. *Gut,* 61(4), 543–553.

Simoes, C.D., Maukonen, J., Kaprio, J., Rissanen, A., Pietilainen, K.H., and Saarela, M. 2013. Habitual dietary intake is associated with stool microbiota composition in monozygotic twins. *J Nutr,* 143(4), 417–423.

Simren, M., Barbara, G., Flint, H.J., Spiegel, B.M.R., Spiller, R.C., Vanner, S., *et al.* 2013. Intestinal microbiota in functional bowel disorders: A Rome foundation report. *Gut,* 62(1), 159–176.

Smith, P.M., Howitt, M.R., Panikov, N., Michaud, M., Gallini, C.A., Bohlooly, Y.M., *et al.* 2013. The microbial metabolites, short chain fatty acids, regulate colonic Treg cell homeostasis. *Science,* 341(6145), 569–573.

Soares, R.L. 2014. Irritable bowel syndrome: A clinical review. *World J Gastroenterol,* 20(34), 12144–12160.

Sokol, H., Pigneur, B., Watterlot, L., Lakhdari, O., Bermudez-Humaran, L.G., Gratadoux, J.J., *et al.* 2008. Faecalibacterium prausnitzii is an anti-inflammatory commensal bacterium identified by gut microbiota analysis of Crohn disease patients. *Proc Natl Acad Sci USA,* 105(43), 16731–6.

Sommer F. and Backhed, F. 2013. The gut microbiota-masters of host development and physiology. *Nat Rev Microbiol,* 11(4), 227–238.

Soto, A., Martin, V., Jimenez, E., Mader, I., Rodriguez, J.M., and Fernandez, L. 2014. Lactobacilli and bifidobacteria in human breast milk: Influence of antibiotherapy and other host and clinical factors. *J Pediatr Gastroenterol Nutr,* 59(1), 78–88.

Steck, N., Mueller, K., Schemann, M., and Haller, D. 2013. Republished: Bacterial proteases in IBD and IBS. *Postgrad Med,* 89(1047), 25–33.

Stephens, W.Z. and Round, J.L. 2014. IgA targets the troublemakers. *Cell Host Microbe.* 16(3), 265–267.

Stewart, C.J., Marrs, E.C.L., Magorrian, S., Nelson, A., Lanyon, C., Perry, J.D., *et al.* 2012. The preterm gut microbiota: Changes associated with necrotizing enterocolitis and infection. *Acta Paediatr,* 101(11), 1121–1127.

Swidsinski, A., Ladhoff, A., Pernthaler, A., Swidsinski, S., Loening-Baucke, V., Orther, M., *et al.* 2002. Mucosal flora in inflammatory bowel disease. *Gastroenterol,* 122(1), 44–54

Talley, N.J. 2008. Functional gastrointestinal disorders as a public health problem. *Neurogastroenterol Motil, 20* (Suppl 1), 121–129.

Tanaka, S., Kobayashi, T., Songjinda, P., Tateyama, A., Tsubouchi, M., Kiyohara, C., *et al.* 2009. Influence of antibiotic exposure in the early postnatal period on the development of intestinal microbiota. *FEMS Immunol Med Microbiol, 56*(1), 80–87.

Tap, J., Mondot, S., Levenez, F., Pelletier, E., Caron, C., Furet, J.P., *et al.* 2009. Towards the human intestinal microbiota phylogenetic core. *Environl Microbiol, 11*(10), 2574–2584.

Thiele, I., Heinken, A., and Fleming, R.M.T. (013. A systems biology approach to studying the role of microbes in human health. *Curr Opin Biotechnol, 24*(1), 4–12.

Thorburn, A.N., Macia, L., and Mackey, C.R. 2014. Diet, metabolites, and "Western-Lifestyle" inflammatory diseases. *Immunity, 40*(6), 833–842.

Tiihonen, K., Ouwehand, A.C., and Rautonen, N. 2010. Human intestinal microbiota and healthy ageing. *Ageing Res Rev, 9*(2), 107–116.

Triantafyllou, K., Chang, C., and Pimentel, M. 2014. Methanogens, methane and gastrointestinal motility. *J Neurogastroenterol Motil, 20*(1), 31–40.

Turnbaugh, P.J., Hamady, M., Yatsunenko, T., Cantarel, B.L., Duncan, A., Ley, R.E., *et al.* 2009. A core gut microbiome in obese and lean twins. *Nature, 457*(7228), 480–484.

Turnbaugh, P.J., Ley, R.E., Mahowald, M.A, Magrini, V., Mardis, E.R., and Gordon, J.I. 2006. An obesity-associated gut microbiome with increased capacity for energy harvest. *Nature, 444*(7122), 1027–1031.

Turroni, F., Peano, C., Pass, D.A., Foroni, E., Severgnini, M., Claesson, M. J., *et al.* 2012. Diversity of bifidobacteria within the infant gut microbiota. *PloS One, 7*(5), e36957.

Ventura, M., Turroni, F., Motherway, M.O., MacSarry, J., and van Sinderen, D. 2012. Host-microbiome interactions that facilitate gut colonization by commensal bifidobacteria. *Trends Microbol, 20*(10), 467–476.

Walker, A.W., Duncan, S.H., Louis, P., and Flint, H.J. 2014. Phylogeny, culturing, and metagenomics of the human gut microbiota. *Trends Microbiol, 22*(5), 267–274.

Walker, A.W., Ince, J., Duncan, S.H., Webster, L.M., Holtrop, G., Ze, X., *et al.* 2011. Dominant and diet-responsive groups of bacteria within the human colonic microbiota. *ISME J., 5*(2), 220–230.

Wang, X., Heazlewood, S.P., Krause, D.O., and Florin, T.H. 2003. Molecular characterization of the microbial species that colonize human ileal and colonic mucosa by using 16S rDNA sequence analysis. *J Appl Microbiol, 95*(3), 508–520.

Windey, K., De Preter, V., and Verbeke, K. 2012. Relevance of protein fermentation to gut health. *Mol Nutr Food Res, 56*(1), 184–96.

Winter, S.E., Thiennimitr, P., Winter, M.G., Butler, B.P., Huseby, D.L., Crawford, R.W., *et al.* 2010. Gut inflammation provides a respiratory electron acceptor for Salmonella. *Nature, 467*(7314), 426–429.

Wong, J.M. and Jenkins, D.J. 2007. Carbohydrate digestibility and metabolic effects. *J Nutr, 137*(11 Suppl), 2539S–2546S.

Woodmansey, E.J. 2007. Intestinal bacteria and ageing. *J Appl Microbiol, 102*(5), 1178–1186.

Wrzosek, L., Miquel, S., Noordine, M.L., Bouet, S., Chevalier-Curt, M.J., Robert, V., et al. 2013. *Bacteroides thetaiotaomicron* and *Faecalibacterium prausnitzii* influence the production of mucus glycans and the development of goblet cells in the colonic epithelium of a gnotobiotic model rodent. *BMC Biol, 11*, 61.

Wu, G.D., Chen, J., Hoffmann, C., Bittinger, K., Chen, Y.Y., Keilbaugh, S.A, et al. 2011. Linking long-term dietary patterns with gut microbial enterotypes. *Science, 334*(6052), 105–108.

Yatsunenko, T., Rey, F.E., Manary, M.J., Trehan, I., Dominguez-Bello, M.G., Contreras, M., et al. 2012. Human gut microbiome viewed across age and geography. *Nature, 486*(7402), 222–227.

Zhang, C., Zhang, M., Wang, S., Han, R., Cao, Y., Hua, W., et al. 2010. Interactions between gut microbiota, host genetics and diet relevant to development of metabolic syndromes in mice. *ISME J, 4*(2), 232–241

Zhu, L., Baker, S.S., Gill, C., Liu, W., Alkhouri, R., et al. 2013. Characterization of gut microbiomes in nonalcoholic steatohepatitis (NASH) patients: A connection between endogenous alcohol and NASH. *Hepatology, 57*(2), 601–9.

Zivkovic, A.M., German, J.B., Lebrilla, C.B., and Mills, D.A. 2011. Human milk glycobiome and its impact on the infant gastrointestinal microbiota. *Proc Natl Acad Sci USA, 108* (Suppl 1), 4653–4658.

Zwielehner, J., Liszt, K., Handschur, M., Lassl, C., Lapin, A., and Haslberger, A.G. 2009. Combined PCR-DGGE fingerprinting and quantitative-PCR indicates shifts in fecal population sizes and diversity of Bacteroides, bifidobacteria and *Clostridium* cluster IV in institutionalized elderly. *Exp Gerontol, 44*(6–7), 440–446.

CHAPTER 4

Microbial Biochemical Processes Critical to Human Health

VICKY DE PRETER and KRISTIN VERBEKE

4.1. INTRODUCTION

THE microbial ecosystem inhabiting the large intestine has been increasingly recognized as a metabolically active organ that profoundly affects human physiology and health. Within the gastrointestinal tract, both the concentration and diversity of the bacteria gradually increase towards the more distal parts. With up to 10^{14} predominantly anaerobic microorganisms belonging to more than 1,000 different species, the colon is the most densely colonized intestinal organ. The *Firmicutes* and *Bacteroidetes* phyla constitute the vast majority of dominant gut microbiota, followed by the *Actinobacteria* and *Proteobacteria* phyla (Arumugam *et al.* 2011). The genetic potential encoded within the resident microbial population exceeds by 100-fold that of the human genome and provides metabolic traits that complement those encoded within our own genome (Qin *et al.* 2010). As a consequence, this symbiotic relationship between the gut microbiota and the host is considered more and more to be critical to health. With the emergence of technical advancements, such as high throughput sequencing techniques and "omic" methodologies, our knowledge of the composition of the microbiota and their metabolic capacities has increased tremendously. In this chapter, several biochemical processes critical to health will be discussed.

4.2. METABOLIC FUNCTION OF THE COLONIC MICROBIOTA

The colonic microbiota plays a major role in salvaging energy from the diet through the fermentation of substrates that escaped digestion in the upper GI tract. Colonic fermentation is an anaerobic process in which carbohydrates, proteins, but also other dietary derived (polyphenols) and host-derived residues (host-produced enzymes, mucus, etc.) are metabolized to a variety of end products. These fermentation reactions enable the bacteria to maintain their cellular functions and to obtain energy for growth. Several factors may influence the activity of the colonic microbiota such as the availability of substrates (amount and nature), intestinal transit time, gut luminal pH, age, presence of antimicrobial compounds, and immunological interactions (Gibson *et al.* 1995). Changes in dietary composition and in digestive processes in the proximal intestinal tract may result in both qualitative and quantitative changes in the supply of nondigestible carbohydrates and proteins to the colon. Small intestinal digestion is affected by a variety of parameters including the macronutrient composition of the food (e.g., the presence of fat), the food matrix, the viscosity of the meals, or the type of preparation (such as raw versus cooked food) (Hooda *et al.* 2011).

Fermentation of the above mentioned residues by the microbiota results in the production of a wide variety of metabolites such as short-chain fatty acids (SCFA), protein metabolites (phenolic compounds, nitrogen- and sulfur-containing compounds), and gases that are in close contact with the epithelial cells and might impact on their cellular functions. After absorption, these metabolites can also influence host systemic functions. As an example, the interest in the fermentation products of polyphenols has considerably increased due to the demonstration of their antioxidant effects and their role in the prevention of several chronic diseases, certain cancers, or Type 2 diabetes (Arts *et al.* 2005; Scalbert *et al.* 2005).

4.3. CARBOHYDRATE METABOLISM

4.3.1. Production of Short-Chain Fatty Acids

Using direct instillation of ^{14}C-labelled glucose, acetate, and lactate in the cecum of germ-free (GF) and conventional rats, Bond and Levitt showed already in 1975 that the colonic bacteria anaerobically metabolize glucose to SCFAs which are consequently absorbed and oxidized by the host (Bond and Levitt 1976). The authors suggested that this mechanism might reduce the osmotic load in the colon due to nonabsorbed carbohydrate, preventing osmotic diarrhea and resulting in the production of normal formed stool.

The pathways of bacterial SCFA production from complex carbohydrates have mainly been elucidated within the field of ruminant nutrition (Baldwin and Allison 1983) and were confirmed in the human colonic ecosystem using incubations with radiolabelled substrates (Miller and Wolin 1996). The first step involves the hydrolysis of polymers to monosaccharides by bacterial enzymes such as amylases, cellulases, and xylosidases. Hexose monosaccharides are primarily fermented to pyruvate via the Embden-Myerhof pathway generating 2 ATP and 2 NADH per hexose. Also, pentoses are fermented to pyruvate via transketolase and transaldolase yielding 1.67 mol ATP/mol pentose. Pyruvate undergoes several reactions with production of acetate or propionate as depicted in Figure 4.1. Butyrate is not directly formed from pyruvate but arises from butyryl-CoA, either using phosphotransbutyrylase and

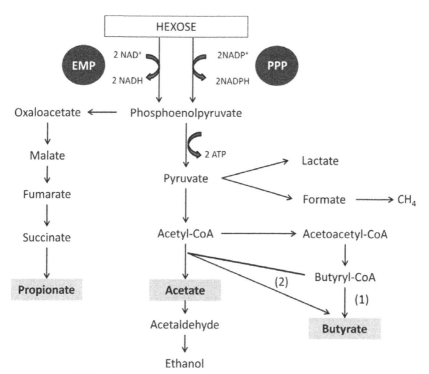

FIGURE 4.1. Schematic representation of the reactions involved in the bacterial fermentation of carbohydrates: (1) production of butyrate via phosphotransbutyrylase and butyrate kinase, (2) production of butyrate via butyryl-CoA:acetate CoA-transferase.

butyrate kinase or by transfer of the acetyl-CoA moiety from butyryl-CoA onto acetate employing a CoA-transferase (Louis et al. 2007). A screen of 38 butyrate-producing bacteria from the human colon for both pathways revealed that the CoA-transferase route was the predominant route, whereas butyrate kinase was only found in a few isolates (Louis et al. 2004).

4.3.2. Factors That Affect the Production of Short-Chain Fatty Acids

The amount and type of SCFA produced from indigestible carbohydrates is determined by a number of factors including the degree of polymerization (DP), degree of substitution (linear versus branched backbones), the type of glycosidic bond, and the type of the monosaccharide building blocks, accessibility (association with other molecules), and solubility (Flint et al. 2008). For instance, fructans with a higher DP are more resistant to bacterial degradation and are fermented in the more distal parts of the colon (van de Wiele et al. 2007). Similarly, arabinoxylan oligosaccharides, consisting of a linear backbone of xylose substituted with arabinose moieties, are fermented in in different regions of gastrointestinal tract depending on their DP and DS (Sanchez et al. 2009). Several in vitro fermentation studies compared the patterns of SCFA produced from incubation of different carbohydrate sources with human fecal inocula. Starch consistently produces the highest proportions of butyrate (Wang and Gibson 1993; Brouns et al. 2002). Oligofructose, lactose, and lactulose favor higher proportions of acetate (Wang and Gibson 1993) whereas methylcellulose, arabinogalactan, arabinoxylan, guar gum, and wheat dextrins are major contributors to propionate (Pylkas et al. 2005; Hobden et al. 2013; Noack et al. 2013; Van den Abbeele et al. 2013). In a study comparing the SCFA production from different cereal grains, rice resulted in the highest production of acetate whereas the highest levels of butyrate were obtained from fermentation of rye (Yang et al. 2013). In addition, the molar SCFA ratios produced depended on the BMI of the fecal donors. Propionate production was significantly correlated with BMI even after correction for age, gender, and race ($p = 0.0069$) (Yang et al. 2013). This differential fermentation by lean and obese inocula was attributed to a different microbiota composition with obese microbiota being characterized by higher levels of *Bacteroides* (Schwiertz et al. 2010).

4.4. PHYSIOLOGICAL EFFECTS OF SHORT-CHAIN FATTY ACIDS

4.4.1. Luminal Effects of Short-Chain Fatty Acids

SCFA are the major anions within the large intestine and are responsible for the drop in the pH from the ileum to the proximal colon. Upon progression to the distal colon, SCFA concentrations decline and the pH increases inversely (Cummings *et al.* 1987). Lower colonic pH might prevent the overgrowth of acid-sensitive pathogenic bacteria such as *Salmonella* and *E. coli* (Cherrington *et al.* 1991), is less favorable for *Bacteroides*, and more favorable to many Gram-positive species (Duncan *et al.* 2009). In *in vitro* fermentation experiments with human fecal bacteria, known butyrate producing bacteria such as *Roseburia* spp and *Faecalibacterium prausnitzii* were stimulated at pH 5.5, corresponding with high levels of butyrate whereas propionate production was higher at pH 6.5 and correlated with increased levels of *Bacteroides* (Walker *et al.* 2005). In addition, a more acidic environment results in a decreased activity of various microbial enzymes such as β-glucuronidase and β-glucosidase (Ballongue *et al.* 1997), the bile acid 7-α-dehydroxylase (Midtvedt *et al.* 1968), or proteases (Macfarlane *et al.* 1988).

4.4.2. Short-Chain Fatty Acids as Energy Substrates

SCFA are transported across the apical membrane of the colonocytes by different mechanisms. Passive diffusion of the undissociated form most likely does not play a prominent role in SCFA absorption as only a small fraction of SCFA is protonated in the intestinal lumen. With a pKa of ~4.8 for SCFA, greater than 99% of SCFA is ionized at physiological pH.

Active transport of dissociated SCFA are mediated by specific transporters and/or carrier-mediated anion transport. The involvement of the monocarboxylate transporter (MCT-1) was shown in *in vitro* experiments using known inhibitors of MCT-1 and antisense technology that suppressed the expression of MCT-1 (Ritzhaupt *et al.* 1998; Hadjiagapiou *et al.* 2000). MCT-1 is the predominant isoform of the monocarboxylate transporters localized to the apical membrane of the enterocytes with increasing abundance along the human intestinal tract, while the MCT-4 and MCT-5 isoforms are localized to the basolateral membrane (Gill *et al.* 2005).

In addition, SLC5A8, known as a tumour-suppression gene that is down regulated in human colon cancer, codes for a sodium-dependent transporter for SCFA, later renamed as SMCT-1 (Miyauchi et al. 2004). The transporter is expressed in the small intestine and the colon with the strongest expression in the distal parts of the large intestine (Takebe et al. 2005). A recent study measuring uptake of radiolabelled butyrate in intestinal epithelial cells (IEC-6) concluded that uptake of butyrate mainly involves MCT-1 with a small contribution of SMCT-1.

Using luminal membrane vesicles, the existence of $SCFA/HCO_3^-$ anion exchange that was independent of the Cl^-/HCO_3^- and Na^+/H^+ exchange systems was convincingly shown (Harig et al. 1996). However, the molecular identity of this apical membrane transporter is not known (Vidyasagar et al. 2005).

Within the epithelial cells, SCFA and in particular butyrate are considered the major energy sources for normal colonocytes. Isolated human colonocytes are able to oxidize each of the three major SCFA with no major differences in the apparent V_{max} (maximum rate of metabolism) (Clausen and Mortensen 1995). However, the energy contribution as estimated from production rates of ATP, were similar for acetate, propionate, and glucose but were significantly higher for butyrate, confirming the importance of butyrate as an energy substrate. Colonocytes from GF mice are in an energy-deprived state and exhibit a decreased expression of key enzymes of the mitochondrial fatty acid metabolism, ultimately leading to autophagy (Donohoe et al. 2011). Oxidation of butyrate accounts for up to 70% of the oxygen consumption in colonocytes (Roediger 1982) and in humans, SCFA oxidation has been estimated to provide up to 10% of the daily energy requirements (Bergman 1990). In colonic epithelial tumoral cells, butyrate uptake is reduced due to a reduction in the expression of MCT-1 and SMCT-1 which is associated with an increase in the rate of glucose uptake and activity of glycolysis. In those cells, glucose becomes the primary energy source (Goncalves et al. 2013).

The fraction of the SCFA that is not consumed by the colonocytes is transported across the basolateral membrane and reaches the liver via the portal blood stream. In the liver, acetate is used as a substrate for lipogenesis and cholesterol synthesis whereas propionate is a precursor for gluconeogenesis. Infusion of the stable isotope labelled substrates $[1-^{13}C]$ acetate and $[2-^{13}C]$ propionate into the caecum of mice showed incorporation of 0.7% of acetate in cholesterol and 2.8% in palmitate while 62% of propionate was used in whole body glucose production (den Besten *et al.* 2013). Comparison of SCFA levels in portal, arterial, and hepatic vein blood showed that acetate was released in the splanchnic area in significant amounts (17.3 [7.2] µmol/kg.h) whereas virtually all propionate and butyrate was scavenged by the liver (Bloemen *et al.* 2009). After consumption of ^{13}C-labelled barley by healthy humans, ^{13}C-SCFA were retrieved in plasma and urine in a molar ratio of acetate:propionate:butyrate of 98:1:1 reflecting the considerable differential absorption and metabolism in colonic epithelial cells and liver (Verbeke *et al.* 2010).

4.5. SHORT-CHAIN FATTY ACIDS AS SIGNALING MOLECULES

4.5.1. G-Protein Coupled Receptors

SCFA are considered to be one of the chemical signals that pass information from the luminal microbiota to the host through receptors on the intestinal epithelial cells (Kaji *et al.* 2014). Recently, the G-protein coupled receptors GPR43 and GPR41 were deorphanized as receptors for SCFA using a ligand fishing strategy (Brown *et al.* 2003) and were later renamed as free fatty acid (FFA) receptor 2 and 3, respectively. Both receptors are tandemly encoded at a single chromosomal locus and display 43% amino acid identity in humans. They are activated by similar ligands but with differing specificity. FFA2 has been reported to have a preference for acetate and propionate whereas pentanoate, butyrate, and propionate are the most potent agonists for FFA3 (Brown *et al.* 2003). Both receptors are widely expressed throughout the body. Within the colon, immunohistochemical studies indicated a particularly strong expression in PYY- and GLP-1 producing enteroendocrine L-cells suggesting that activation of FFA2 and FFA3 by SCFA may regulate secretion of PYY and/or GLP-1 (Karaki *et al.* 2008; Tazoe *et al.* 2009). Direct infusion of SCFA in the colon of rats induced the release of PYY, explaining the anorexigenic effects observed (Roelofsen *et al.* 2010). Increasing luminal SCFA by administration of fructooligosaccharides to rats for 4 weeks induced L cell proliferation and increased the density of L cells expressing FFA2 and GLP-1 (Kaji *et al.* 2011). Since PYY, GLP-1 as well as other hormones like GIP are key modulators in the energy homeostasis and glucose metabolism, the potential role of FFA2 and FFA3 in the prevention of obesity was investigated using mice with a null mutation. FFA2$^{-/-}$ mice were found protected from diet-induced obesity most likely due to increases in resting energy expenditure and higher fecal energy excretion (Bjursell et al. 2011). In addition, SCFA-triggered secretion of GLP-1 is reduced in mice lacking FFA2 but not in FFA3 knockout mice (Tolhurst et al. 2012). Accordingly, body weight gain and adiposity were similar in FFA3$^{-/-}$ mice compared to wild type animals although the knockouts showed mild hyperphagia on a high-fat diet (Lin et al. 2012).

The niacin receptor GPR109a, also known as the hydroxycarboxylic acid receptor 2 or HCA2, is another receptor for SCFA, first identified as G-protein coupled receptor induced by IFN-γ in macrophages (Schaub et al. 2001). Subsequent studies indicated that GPR109A is also expressed abundantly on the luminal membrane of mouse and human colonic epithelial cells (Cresci et al. 2010). GPR109a recognizes butyrate with low affinity requiring millimolar concentrations of butyrate to activate the receptor. In colon cancer in humans, the expression of GPR109a is silenced. However, re-expression of GPR109a in colon cancer cells induces apoptosis in the presence of its ligands butyrate and nicotinate (Thangaraju et al. 2009). Besides, GPR109a is also highly expressed in adipocytes and to a lesser extent in immune cells (Tan et al. 2014).

4.5.2. Inhibition of Histone Deacetylase

Acetylation of histones and DNA methylation are the major epigenetic modifications of DNA that modulate gene expression. Acetylation of the lysine residues within histones facilitates the access of transcription factors to the promotor regions, resulting in gene activation. Histone deacetylases are a class of enzymes that remove the acetyl group from the ε-N-acetyl lysine amino acid on a histone which allows the histones to wrap the DNA more tightly. Mammalian cells contain 18 different HDACs which are grouped in four classes (I to IV) (Santini et al. 2007). The SCFA butyrate and propionate, but not acetate, are inhibitors of the HDACs, with butyrate being a more potent inhibitor than propionate, and induce hyperacetylation of the histones which can in this way affect gene expression. Depending on the cell-type, inhibition of HDACs can suppress cell proliferation, induce differentiation, and cause cell cycle blockade (Wu et al. 2012). The exact mechanism by which SCFA inhibit HDAC is not known. SCFA may enter the cell via a transporter and directly act as a noncompetitive inhibitor of HDAC (Cousens et al. 1979). Davie (2003) proposed that two molecules of butyrate could occupy the hydrophobic pocket of the HDAC in a similar way as trichostatin A (TSA), whose interaction with HDAC has been characterized by X-ray diffraction.

Interestingly, activation of FFA3 by butyrate in a FFA3-expressing Chinese Hamster Ovary (CHO) cell line reduced the sustained elevation of histone acetylation, and prevented the antiproliferative and pro-apoptotic effects of butyrate, indicating that butyrate displays dual effects mediated by both HDAC inhibition and FFA3 activation (Wu *et al.* 2012).

4.6. THE ROLE OF SHORT-CHAIN FATTY ACIDS IN THE IMMUNE RESPONSE

SCFA exhibit anti-inflammatory effects by regulating the release of cytokines. The exact mechanism seems to depend on the cell type. In particular, butyrate has been shown to decrease the activity of nuclear factor-kappa B (NF-κB), a transcription factor involved in the regulation of many cytokine genes, including TNF-α. In LPS-stimulated PBMC cells, butyrate significantly reduced the mRNA levels of TNF-α, TNF-β, IL-1β, and IL-6. This effect seems not mediated by inhibitory HDAC activity as the HDAC inhibitor TSA did not affect NF-κB translocation to the nucleus (Segain *et al.* 2000). A recent study in intestinal macrophages, the most abundant immune cell type in the lamina propria, confirmed the downregulation of proinflammatory mediators such as NO, IL-6, and IL-12 by butyrate. However, the effects were found independent of NF-κB downregulation or activation of G-protein coupled receptors. Several lines of evidence suggested that butyrate behaves as an HDAC inhibitor in these cells (Chang *et al.* 2014).

In particular, FFA2 is widely expressed in different types of immune cells, suggesting a broad role of SCFA in the regulation of immune responses. A recent study in mice investigated the role of SCFA and their FFA2 and FFA3 receptors in regulating the response to immunological challenges. Mice deficient in FFA2 or FFA3 could not mount a normal Th1 response to TNBS and failed to induce an acute inflammatory response to clear a *Citrobacter rodentium* infection (Kim *et al.* 2013). Activation of FFA2 and FFA3 on intestinal epithelial cells with SCFA led to mitogen-activated protein kinase signaling and rapid production of chemokines and cytokines, mediating protective immunity. Treatment of FFA2 knockout mice with dextran sulfate sodium resulted in significantly worse colonic inflammation than in wild-type mice. Supplementation of the drinking water with acetate improved the disease indices in wild-type mice, but had no effect on FFA2 knockout mice (Masui *et al.* 2013).

Recent work has shown that SCFA induce regulatory T (T_{reg}) cells in the colon. Regulatory T cells express the transcription factor Foxp3 and control intestinal inflammation by limiting proliferation of $CD4^+$ T cells (Smith *et al.* 2013). Singh *et al.* (2014) showed that GPR109a signaling by butyrate promoted anti-inflammatory properties in colonic macrophages and dendritic cells and enabled them to induce differentiation of T_{reg} cells and IL-10-producing T cells. Also, FFA2 seems involved as propionate enhanced colonic T_{reg} in $FFA^{+/+}$ but not $FFa2^{-/-}$ mice (Smith *et al.* 2013). In mice, oral administration of propionate and butyrate, but not acetate, facilitated the extra-thymic *de novo* generation of anti-inflammatory T_{reg}. In contrast, rectal administration of acetate and propionate, but not butyrate, promoted accumulation of T_{reg} cells, suggesting that butyrate promotes *de novo* generation but not colonic accumulation of T_{reg} cells, whereas acetate has an opposite activity and propionate is capable of both (Arpaia *et al.* 2013).

4.7. FERMENTATION OF PROTEINS

Bacterial fermentation of proteins generates metabolites including branched-chain fatty acids, aromatic amino acid metabolites, nitrogen-containing compounds like ammonia and amines, and sulfur-containing compounds. Several of these metabolites are presumed to be detrimental for health. However, the toxic potential of these compounds has mainly been derived from *in vitro* studies or animal studies whereas little evidence for adverse effects of protein fermentation metabolites has emerged from human studies (Windey *et al.* 2012). Higher concentrations of protein fermentation products are found in the distal colon compared to the proximal colon, indicating that the distal colon is the major site for proteolysis (Hughes *et al.* 2000). The amount of protein available for bacterial fermentation in the colon depends on the protein content of available sources and their digestibility. Several sources of protein, such as undigested dietary protein, desquamated epithelial cells, digestive secretions such as enzymes, mucins, free amino acids, and peptides, enter the colon (Gibson *et al.* 1976; Blachier *et al.* 2007). The origin of the protein source, as well as the type of food processing may influence the digestibility of proteins. Animal-derived proteins (dairy and animal proteins) are generally more digestible (94–99%)

than proteins derived from pinto and kidney beans and lentils (72–84%) (Gilbert *et al.* 2011). The impact of cooking on the digestibility of egg protein was nicely demonstrated by quantifying the protein fermentation metabolites in urine after a meal of cooked versus raw eggs (Evenepoel *et al.* 1999) with cooked egg protein being clearly more digestible than raw egg protein. In addition, the ratio of available carbohydrate to nitrogen is an important determinant that governs saccharolytic versus proteolytic fermentation. Indeed, by increasing the amount of fermentable carbohydrate reaching the colon in the form of resistant starch (Birkett *et al.* 1996) or prebiotic oligosaccharides (Swanson *et al.* 2002; De Preter *et al.* 2007; Wutzke *et al.* 2010; Damen *et al.* 2012; Lecerf *et al.* 2012), the production of protein fermentation metabolites can be reduced. The following paragraphs describe the production and the metabolic fate as well as the impact on the health of several protein fermentation metabolites in more detail.

4.7.1. Branched-Chain Fatty Acids

Branched-chain fatty acids (BCFA) are solely formed from the fermentation of proteins and amino acids. Isobutyrate, 2-methylbutyrate, and isovalerate are produced by bacterial fermentation of valine, isoleucine, and leucine, respectively. Relatively little is known on the effects of BCFA on colonic epithelial cells and BCFA are generally considered as "neutral" to health. Therefore, BCFA are most often quantified as markers of proteolysis in the colon rather than as markers of health.

4.7.2. Aromatic Amino Acid Metabolites

Bacterial species that ferment aromatic amino acids mainly belong to the *Clostridium* and *Bacteroides* genera, but also *Lactobacillus*, *Enterobacter*, *Bifidobacterium* (Chung *et al.*; 1975, Elsden *et al.*, 1976; Yokoyama and Carlson 1979; Cummings 1983; Russell 1983) are able to ferment amino acids. The major metabolites include phenolic (phenol and p-cresol) and indolic (indole and skatole) compounds (Figure 4.2). Besides, degradation of tyrosine results in the formation of intermediary metabolites like 4-ethylphenol, 4-hydroxyphenylpyruvate, 4-hydroxyphenyllactate, 4-hydroxyphenylpropionate, and 4-hydroxyphenylacetate (Hughes *et al.* 2000). Phenylalanine bacterial metabolism leads to similar derivates, such as phenylpyruvate, phenyllactate, phenylpropionate, and phenylacetate. Besides indole and skatole, tryp-

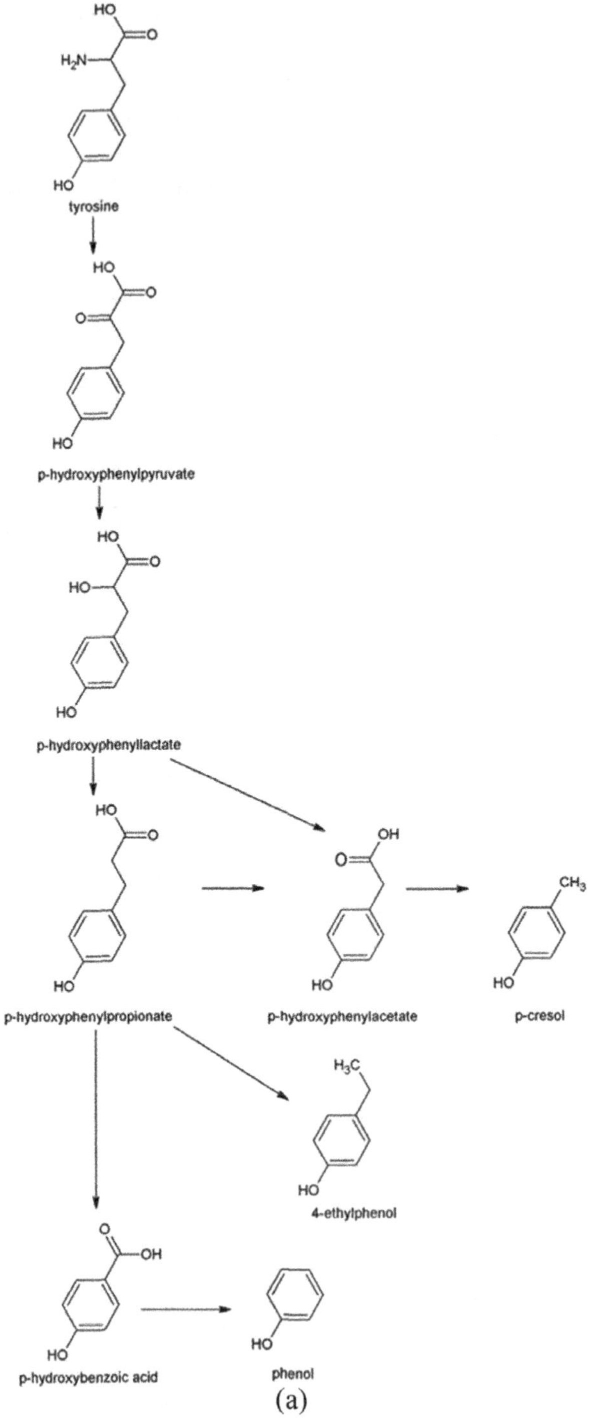

FIGURE 4.2(a). Colonic microbial metabolism of the aromatic amino acids (a) tyrosine.

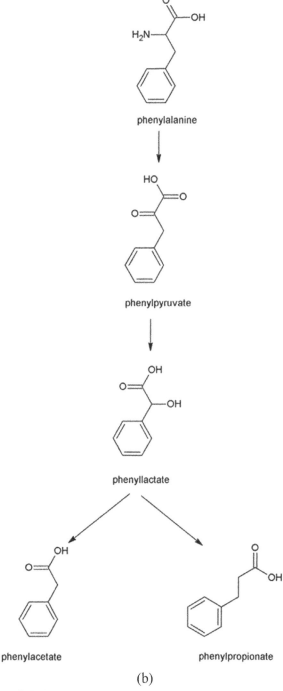

FIGURE 4.2(b). Colonic microbial metabolism of the aromatic amino acids (b) phenylalanine.

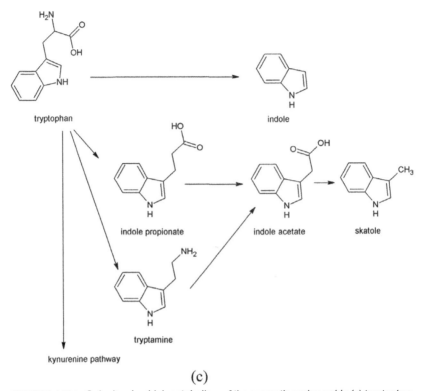

FIGURE 4.2(c). Colonic microbial metabolism of the aromatic amino acids (c) tryptophan.

tophan metabolism also generates indole acetate and indole propionate (Figure 4.2).

After production, phenolic compounds are largely absorbed from the colonic lumen, detoxified in the colonic mucosa and the liver by sulfate and, to a lesser extent, glucuronide conjugation, and finally excreted in urine (Hughes *et al.* 2000). Only a minor fraction of the phenolic compounds is recovered in fecal samples. Indoles are absorbed and metabolized to indoxyl sulfate, also called indican, in the liver (Cummings 1983). In patients with active and quiescent UC, the phenol conjugation capacity of the colonic mucosa is impaired which may pose a metabolic burden on the epithelial cells (Ramakrishna *et al.* 1991).

The toxic potential of the aromatic amino acid metabolites has been evaluated in *in vitro* incubation experiments with epithelial cells (phenol and p-cresol) and bacteria (indole). In primary human colonic epithelial cells, viability was decreased after exposure to 1.25 mM phenol, a physiologically relevant concentration. In contrast, higher concentrations of phenol (20 mM) were required to reduce the viability in HT-29 cells (Pedersen *et al.* 2002). After exposure to p-cresol (10–50 µg/ml), permeability of endothelial cells was significantly decreased which may increase the accessibility of different agents to the colon (Cerini *et al.* 2004). Indole, but not phenol and p-cresol, at concentrations of 20 µg/ml and 100 µg/ml appeared to be toxic for lactic acid bacteria. However, the time required to induce the toxic effect was strain-dependent (Nowak and Libudzisz 2006).

In patients with chronic kidney disease, bacterial metabolites such as phenols and indoles may contribute to uremic toxicity. As protein assimilation is impaired in uremia, increased amounts of protein escape digestion and absorption in the small intestine resulting in a decreased ratio of available carbohydrate to nitrogen (Bammens *et al.* 2003). This may favor the proliferation of proteolytic species with increased generation of phenolic compounds. Within the plasma, phenolic compounds are highly protein-bound and accumulate when renal function deteriorates. In addition, the high protein binding hampers efficient removal using dialysis techniques (Evenepoel *et al.* 2009). Furthermore, p-cresyl- and indoxyl sulfate may be involved in the pathogenesis of cardiovascular disease associated with chronic kidney disease by inducing endothelial dysfunction (Dou *et al.* 2004). Free serum p-cresol levels are independently associated with overall mortality (Bammens *et al.* 2006) and cardiovascular disease (Meijers *et al.* 2008).

4.7.3. Sulfur-Containing Metabolites

Hydrogen sulfide (H_2S) results from bacterial fermentation of dietary and mucinous sulfur containing amino acids such as methionine, cystine, cysteine, and taurine (Florin *et al.* 1991; Roediger *et al.* 1997) and from the reduction of sulfate by sulfate-reducing bacteria (SRB) like *Desulfomonas* spp. and *Desulfovibrio* spp. Luminal concentrations in the colon range between 1 mM and 2.4 mM (Macfarlane *et al.* 1992; Blachier *et al.* 2010), with an estimated production of 1.5–16 mmol/day from dietary inorganic sulfate and 3.8 mmol/day from protein derived sulfur (Blachier *et al.* 2007). Fecal matter has a large capacity to bind H_2S, which renders it inactive and only a minor proportion (60 µmol/L) of the sulfide in the colonic lumen is suggested to be free in the form of the sulfide ion (Jorgensen and Mortensen 2001). The consumption of sulfur-containing amino acids fluctuates with protein intake. In a randomized cross-over trial in healthy men, an increased intake of dietary protein correlated positively with an increase in fecal sulfide concentrations, indicating a dose-dependent relationship (Magee *et al.* 2000).

It has been postulated that, in millimolar concentrations, H_2S exerts a dose-dependent toxicity. A crucial mechanism involves the inhibition of the cellular respiration due to a noncompetitive inhibition of the oxygen binding to mitochondrial cytochrome-c-oxidase, which is the final step in the production of cellular adenosine triphosphate (ATP) (Cooper and Brown 2008). Lower cellular levels of ATP activate ATP-sensitive potassium channels, which account for most of the biological effects of H_2S. At concentrations of about 30 µM, respiration of intact cells is half inhibited (Leschelle *et al.* 2005). Besides, H_2S affects different cellular pathways (cell cycle, inflammation, cell adhesion) at concentrations similar to those found in the colon (Attene-Ramos *et al.* 2006). Two different enzymes have been proposed for the endogenous detoxification of sulfide in the colonic epithelial cells. Thiol S-methyltransferase (TMT) methylates H_2S to methanethiol (Roediger *et al.* 2000), whereas thiosulfate sulfurtransferase (TST, also known as rhodanese) oxidizes H_2S to thiosulfate that is further converted to the less toxic thiocyanate in the presence of cyanide (Levitt *et al.* 1999; De Preter *et al.* 2012). Using human tissue homogenates, it was shown that TST is the principal enzyme in H_2S detoxification (Picton *et al.* 2002).

In colonocytes, H_2S impedes the oxidation of butyrate, which is their principal energy source (Roediger *et al.* 1993) inducing an energy-deficient state that ultimately results in reduced absorption of sodium, reduced secretion of mucin, and a shorter life of the colonocytes (Roediger *et al.* 1997). H_2S has been implicated in the pathogenesis of UC (Roediger *et al.* 1993; Medani *et al.* 2011) as increased fecal levels of H_2S, a reduced detoxification capacity, and a higher metabolic activity of SRB along with impaired butyrate oxidation have been reported in UC as compared to healthy subjects (Florin *et al.* 1990; Pitcher *et al.* 1996; Levine *et al.* 1998; De Preter *et al.* 2012). However, the evidence linking H_2S to UC is not yet conclusive.

A potential role for H_2S in carcinogenesis has been suggested based on genotoxicity assays in CHO and human HT-29 colonic epithelial cells that showed significant genomic DNA damage at H_2S concentration as low as 250 µM, thus at levels below those occurring in the colonic lumen (Attene-Ramos *et al.* 2006, 2007). Furthermore, sulfide was shown to modulate the expression of genes in cell-cycle progression, inflammation, and DNA repair response (Attene-Ramos *et al.* 2010).

In contrast to the previous reports on harmful effects, hydrogen sulfide has emerged as an endogenous gaseous-signaling molecule. It is endogenously produced in micromolar concentrations from cysteine by the action of cystathionine gamma-lyase (CSE) and cystathionine beta-synthase (CBS). At these low concentrations, H_2S would be involved in neuromodulation of chloride secretion, in controlling ileum contractility, and in nociception from the large intestine (Blachier *et al.* 2010; Farrugia *et al.* 2014).

4.7.4. Nitrogen Containing Compounds

4.7.4.1. Ammonia

Ammonia is produced via bacterial deamination of amino acids and to a lesser extent through urea hydrolysis catalyzed by bacterial urease activity (Blachier *et al.* 2007). Up to 3.5–4 g of NH_3 is released every day in the gut (Visek 1978) resulting in luminal concentrations between 2 and 44 mM in healthy subjects (Mouille *et al.* 2004). The concentration of NH_3 increases progressively from the proximal to the distal colon, which is considered the major site of protein and amino acid fermentation (Macfarlane *et al.* 1992). In humans, increasing amounts of dietary protein resulted in enhanced fecal ammonia levels (Geypens *et al.* 1997). Ammonia concentrations as low as 5–10 mM have been shown to alter the metabolism of intestinal cells, thereby affecting DNA synthesis and decreasing the lifespan of the cells (Visek 1978). In addition, ammonia is able to inhibit mitochondrial oxygen consumption in a dose-dependent manner (Andriamihaja *et al.* 2010). Furthermore, high millimolar concentrations of ammonia inhibit SCFA oxidation in colonic epithelial cells (Cremin *et al.* 2003).

NH_3 can be used by the bacteria for their own metabolism and protein synthesis. Alternatively, it is absorbed by the colonocytes, converted in the liver to urea, and excreted in urine. In subjects with liver cirrhosis and portal hypertension, ammonia is able to bypass the liver into the systemic circulation. After uptake in the astrocytes of the brain, ammonia is converted to glutamine that exerts an osmotic effect and contributes to the development of hepatic encephalopathy (Tranah *et al.* 2013). In patients with acute liver failure, arterial ammonia levels exceeding 100 µmol/l predicted the development of severe hepatic encephalopathy with 70% accuracy (Bernal *et al.* 2007).

4.7.4.2. Amines

Amines are produced mainly by decarboxylation of amino acids and, to a lesser extent, by transamination and dealkylation reactions. For instance, decarboxylation of glycine results in the production of methylamine, and N-dealkylation of choline results in the production of dimethylamine. Amines are also produced by the degradation of polyamines (Macfarlane *et al.* 1994). Different amines are produced in the colon such as (di-)methylamine, tyramine, and phenylethylamine, as well as polyamines such as spermine, spermidine, cadaverine, and putrescine.

Normally, amines produced by colonic bacteria are rapidly absorbed from the colon and detoxified by monoamine and diamine oxidases in the gut mucosa and liver (to NH_3 and CO_2). However, some amines are also excreted in urine, including the heterocyclic products of putrescine and cadaverine oxidative deamination (i.e., pyrrolidine and piperidine). In addition, amines play a role in the formation of N-nitrosamines by condensation of a secondary amine with nitrite in an acidic environment or at neutral pH, when catalyzed by bacterial enzymes (Tricker 1997).

Trimethylamine (TMA) is generated through microbial degradation of dietary phosphatidylcholine/choline and L-carnitine, which are found in significant quantities in dietary red meats, eggs, saltwater fish, and dairy products. Following intestinal absorption, TMA is converted in the liver into trimethylamine-N-oxide (TMAO) by the enzyme flavin mono-oxygenase (FMO), which is then released into the circulation (Wang *et al.* 2011). In a metabolomics study profiling the plasma of patients undergoing elective cardiac evaluation, TMAO was identified and confirmed as a predictor of cardiovascular disease (CVD). Increased aortic atherosclerotic plaques after dietary supplementation of atherosclerosis-prone mice (C57BL/6J *Apoe*$^{-/-}$) with choline or TMAO suggested that a proatherosclerotic mechanism underlied the association between TMAO and CVD risk. Choline and TMAO promoted the up-regulation of multiple macrophage scavenger receptors linked to atherosclerosis and enhanced endogenous macrophage foam cell formation. Suppression of the intestinal microbiota with broad spectrum antibiotics significantly inhibited this effect. Recently, it was demonstrated that also metabolism of L-carnitine, a trimethylamine derivative abundantly found in red meat, by the intestinal microbiota produces TMAO and accelerates atherosclerosis in mice (Koeth *et al.* 2013). In a human cohort study, omnivorous subjects produced more TMAO than long-term (>1 year) vegans or vegetarians following an oral challenge with d3-(methyl)-carnitine. In addition, vegans and omnivores had significantly higher post-challenge plasma levels of d3-(methyl) L-carnitine which might be due to a lower microbial metabolism of carnitine prior to absorption. In a 3-year follow-up study of 4,007 patients undergoing elective coronary angiography, increased plasma levels of TMAO were associated with an increased risk of incident major adverse cardiovascular events (hazard ratio for highest versus lowest TMAO quartile, 2.54; 95% confidence interval, 1.96 to 3.28; $P < 0.001$) (Tang et al. 2013).

Besides, the bacterial metabolism of choline and the intestinal production of TMA also links the microbiota to nonalcoholic fatty liver disease (NAFLD). When a mouse strain susceptible to NAFLD was fed a high fat diet, choline was increasingly metabolized to methylamines resulting in high urinary excretion of dimethylamine (DMA) and TMA and correspondingly low levels of serum phosphatidylcholine (Dumas et al. 2006). Conversion of choline into methylamines by the gut microbiota reduces the bioavailability of choline and results in the inability to synthesize phosphatidylcholine with subsequent accumulation of triglycerides in the liver (Vance 2008). This mechanism mimics choline-deficient diets, which have been consistently associated with hepatic steatosis (Corbin and Zeisel 2012).

Finally, amines play a role as N-nitrosation precursors resulting in the formation of potentially carcinogenic N-nitroso compounds (NOC). NOC are formed endogenously in the gastrointestinal tract via nitrosation of primary and secondary amines by nitrite sources (Kuhnle et al. 2007). Heme iron, abundantly present in red meat, has a catalytic effect on the endogenous formation of NOC (Cross et al. 2003). Several secondary NOC are known to be carcinogenic and to alkylate DNA in gastrointestinal tissues leading to base pair transitions (GC-to-AT transitions) in genes mutated in colorectal cancer such as ras (Hall et al. 1991; Cross et al. 2003; Joosen et al. 2009). In a fully controlled study with subjects remaining in a metabolic suite, a strong correlation was found between red meat intake (0, 60, 120, 240, or 420 g/d) and total fecal output of NOC (R^2 value of 0.96) (Bingham et al. 2002). In contrast, an intervention for 7 days with red meat (300 g/day) did not affect fecal NOC levels, but significantly increased fecal water genotoxicity (Hebels et al. 2012). An equivalent amount of white meat intake had no effect on fecal NOC levels (Bingham et al. 2002), suggesting a role for heme, more specifically iron, in this phenomenon.

4.8. MICROBIAL METABOLISM OF POLYPHENOLS

4.8.1. Degradation of Polyphenols By the Intestinal Microbiota

Polyphenols are natural compounds, occurring in plants, fruits, and vegetables (Manach *et al.* 2004), with highly diverse chemical structures characterized by hydroxylated phenyl moieties. Many polyphenols occur in plants as glycosides with different sugar units and acylated sugars at different positions of the polyphenol backbone, although other modifications such as esterification or polymerization are also commonly found (Perez-Jimenez *et al.* 2010). Polyphenols are generally classified according to the chemical structures of the aglycones (Tsao 2010). Phenolic acids are nonflavanoid compounds that can be further divided into benzoic acids and cinnamic acids. Flavanoids have a general diphenylpropane (C6-C3-C6) backbone and consist of different subgroups (anthocyanins, flavan-3-ols, flavones, flavanones, flavonols, isoflavones, neoflavonoids, and chalcones). Polyphenolic amides have N-containing functional substituents with capsaicinoids (in chilli peppers) and avenanthramides (in oats) as the major subgroups. Finally, several nonflavanoid polyphenols, including resveratrol and lignans have been found in foods. The average human consumption of polyphenols amounts to about 1 g/day. Main sources are fruits, tea, and coffee, and to a lesser extent, vegetables, cereals and legume seeds, onions, broccoli, and red wine (Rastmanesh 2011). Phenol-Explorer (www.phenol-explorer.eu) is the first comprehensive web-based database on polyphenol content in foods and contains representative mean content values for more than 500 polyphenols in over 500 foods (Neveu *et al.* 2010; Rothwell *et al.* 2013).

Depending on the degree of structural complexity and polymerization, some polyphenols (aglycones, monomers to trimers of flavonols, and some intact glycosides [anthocyanins]) may be absorbed in the small intestine to a limited extent (5–10%), most likely by passive diffusion. During intestinal absorption and passage in the liver, polyphenols are extensively glucuronidated and/or sulfated (phase II metabolism), whereas phase I metabolism (oxidation/reduction reactions) appears to be of minor importance (Lampe 2009). The majority of dietary polyphenols (90–95%) reach the colon and become available for fermentation by the microbiota (Walle 2004; Manach *et al.* 2005; Cardona *et al.* 2013). Different bacterial species (e.g., *Escherichia coli, Bifidobacterium* spp., *Lactobacillus* spp., *Bacteroides* spp., *Eubacterium* spp.)

catalyzing the metabolism of phenolics have been identified, together with the catabolic pathways implicated (Kutschera *et al.* 2011; Cardona *et al.* 2013).

In the colon, dietary polyphenols and phase II metabolites, extruded via the bile throughout the enterohepatic circulation, are degraded to aglycones and oligomers by microbial glycosidases and esterases (Figure 4.3) (Cardona *et al.* 2013). In addition, the microbiota performs transformations of the polyphenolic core, for example, ring fission by *Eubacterium ramulus*, (Clavel *et al.* 2006) and extensive breakdown of the polyphenolic structure into relatively simple aromatic carboxylic acids or phenolic acids (Aura 2008). These metabolites are absorbable and may be responsible for the bioactivity of the polyphenol-rich food, rather than the original molecules present in the food. During and after absorption through the colonic wall, the majority of these microbial polyphenolic metabolites undergo phase I/II metabolism and are detected as such in urine and serum (Bolca *et al.* 2010).

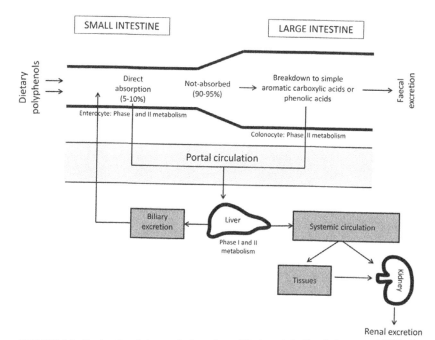

FIGURE 4.3. Routes for dietary polyphenols and their metabolites in humans. Within the host, dietary polyphenols and their microbial metabolites successively undergo intestinal and liver Phase I and II metabolism, biliary secretion, absorption in the systemic circulation, interaction with organs, and excretion in the urine (Cardona et al. 2013).

4.8.2. Impact of Polyphenols on Health

A wide range of protective activities including antioxidant, antimicrobial, anti-inflammatory, anticarcinogenic, antidiabetic, and antiadipogenic effects have been reported, suggesting an association between the consumption of polyphenol-rich foods and a reduced risk of several chronic diseases. However, studies evaluating the potential health benefits of polyphenolic intake have been hampered by the fact that the phenolic compounds appear in the systemic circulation as phase II metabolites and not as the parent compounds.

4.8.3. Modulation of the Colonic Microbial Environment

Only a limited number of studies have investigated the influence of polyphenolic consumption on the composition and activity of the gut microbial community. Nevertheless, evidence from *in vitro*, animal and human studies indicates that intake of polyphenols influences gut microbial populations. Therefore, the concept of polyphenols as potential prebiotic candidates could be considered as a newly emerging concept (Etxeberria *et al.* 2013). In a recent placebo controlled trial in nine healthy subjects, consumption of red wine significantly increased the fecal concentration of *Bifidobacterium* and *Enterococcus*. The increase in bifidobacteria correlated with increases in microbial metabolites derived from wine anthocyanins (Boto-Ordonez *et al.* 2014). Similar results on the fecal levels of *Bifidobacterium* and *Enterococcus* were observed in a randomized, crossover, controlled intervention study in 10 healthy male volunteers after consuming red wine (Queipo-Ortuno *et al.* 2012). Two rat studies showed an impact of a polyphenol-rich diet on particular gut bacterial populations. Rats treated with red wine polyphenols had significantly lower levels of *Clostridium* spp. and higher levels of *Lactobacillus* spp. (Dolara *et al.* 2005). Rats given a tannin-rich diet had significantly lower levels of species of the *Clostridium leptum* cluster, while the *Bacteroides* group significantly increased (Smith *et al.* 2005). In an *in vitro* mixed culture model of human intestinal microbiota, fermentation of polyphenols (chlorogenic acid, caffeic acid, rutin, and quercetin) stimulated proliferation of bifidobacteria and resulted in a decrease in the ratio of *Firmicutes:Bacteroidetes* (Parkar *et al.* 2013). In addition, all polyphenols increased generation of SCFAs (Parkar *et al.* 2013). Upon incubation of pure cultures of *Bifidobacteria* with either fermented media isolated from the incubations, the pure test polyphenols, or the biotransformation products detected in the fermentations, growth stimulation was observed only with fermented polyphenol media and the pure biotransformation products suggesting that the ability to modify the gut microbial balance is mediated by biotransformation products, rather than the original plant compounds. The mechanism for the reduced *Firmicutes:Bacteroidetes* ratio has been attributed to the fact that *Firmicutes* possess smaller genomes and a disproportionately smaller number of glycan-degrading enzymes compared to *Bacteroidetes*. Glycans are the product of glycosidic cleavage of polyphenolic compounds and are necessary for survival of the intestinal microbiota. In addition, *Firmicutes* are repressed to a larger extent than the *Bacteroidetes* by the antimicrobial properties of the phenolic compounds (Rastmanesh 2011). As obesity has been associated with increased relative ratios of *Firmicutes:Bacteroidetes*, supplementation with polyphenols with high bioavailability in obese individuals with a higher *Firmicutes:Bacteroides* community ratio phenotype has been proposed as a strategy to induce weight loss (Rastmanesh 2011).

4.8.4. Antioxidant Activity

Polyphenols are capable of slowing or preventing the oxidation of other molecules. Their efficiency as antioxidants depends on the chemical structure. For example, the antioxidant efficiency of flavonoids is directly correlated to their degree of hydroxylation and decreases with the presence of a sugar moiety (Cotelle 2001). Due to the antioxidant properties, polyphenols exhibit antioxidant activity. Resveratrol, a phytoalexin produced by grapes, was previously found to protect lipids from peroxidative degradation and inhibited the uptake of oxidized low-density lipoprotein–cholesterol (LDL) in the vascular wall (Nigdikar *et al.* 1998; Das and Das 2010).

4.8.5. Antimicrobial Activity

Several mechanisms for the antimicrobial activity of polyphenols have been proposed (Kemperman *et al.* 2010). Polyphenols have the ability to bind to bacterial cell membranes in a dose-dependent manner, thereby disturbing membrane function, resulting in reduced cell growth (Kemperman *et al.* 2010). Differences in antimicrobial activity of catechins were related to differences in binding affinity to the lipid bilayer of bacterial cell membranes (Sirk *et al.* 2009). Other proposed mechanisms include the formation of polyphenols—metal ions complexes which lead to an iron deficiency in the gut and affect sensitive bacterial populations (Smith *et al.* 2005). Treatment of *Candida albicans* with polyphenols significantly reduced biofilm formation due to a reduction in proteasomal activity, which may contribute to cellular metabolic and structural disruptions that expedite the inhibition of biofilm formation and maintenance by the yeast (Evensen and Braun 2009). Furthermore, the observed rise in bifidobacteria upon polyphenol administration may protect the gut mucosa from bacterial invasion. Through acidification of the lumen by SCFAs and through competition for nutrients, pathogens such as *Salmonella* are inhibited (Roy *et al.* 2006)

4.8.6. Anti-inflammatory Activity

The mechanisms of the anti-inflammatory activity of polyphenols has been demonstrated mainly in *in vitro* studies, especially in myeloid cells. The predominant observed effect was an inhibition of the NF-κB signaling and a down regulation of the expression and/or function of a number of inflammatory mediators including eicosanoids, NO, adhesion molecules, and cytokines (Santangelo *et al.* 2007; Gonzalez *et al.* 2011). However, the results observed in vitro cannot always be translated to the *in vivo* situation. The current results do not suggest that flavonoids and other polyphenols may produce a high magnitude effect at the concentrations that may be present in the serum or other body compartments (Gonzalez *et al.* 2011). A more profound characterization of polyphenol pharmacokinetics and structure-activity relations is necessary to increase our understanding of the anti-inflammatory effects of polyphenols.

4.9. CONCLUSION

The time that the large intestine was merely considered as an organ that prepared undigested food residues for defecation has definitively passed. Advances in the functional analysis of the intestinal microbiota have allowed us to unravel the metabolic capacities of this bacterial ecosystem and to appreciate its impact on the host physiology. In particular, short-chain fatty acids have emerged as metabolites that beneficially affect many cellular processes. However, the mechanisms of these pathways have mainly been elucidated in *in vitro* and animal studies with presently little evidence from human studies. In the past years, intestinal microbial metabolism has been associated with the development of diseases like NAFLD, cardiovascular disease, or kidney failure. Additional research is ongoing to evaluate a potential causal relationships and bi-directional relationships between the microbiota and host function. To fully understand the impact of the microbial activity on health, it is important to integrate metabolic signatures into a systems biology approach that correlates those data to the individual's metadata, microbiota composition datasets, and responses to interventions.

4.10. REFERENCES

Andriamihaja, M., Davila, A.M., Eklou-Lawson, M., Petit, N., Delpal, S., Allek, F., et al. 2010. Colon luminal content and epithelial cell morphology are markedly modified in rats fed with a high-protein diet. *Am. J Physiol Gastrointest. Liver Physiol 299*: G1030–G1037.

Arpaia, N., Campbell, C., Fan, X., Dikiy, S., van der Veeken, J., deRoos, P., et al. 2013. Metabolites produced by commensal bacteria promote peripheral regulatory T-cell generation. *Nature 504*: 451–455.

Arts, I.C. and Hollman, P.C. 2005. Polyphenols and disease risk in epidemiologic studies. *Am. J. Clin. Nutr. 81*: 317s–325s.

Arumugam, M., Raes, J., Pelletier, E., Le Paslier, D., Yamada, T., Mende, D.R., et al. 2011. Enterotypes of the human gut microbiome. *Nature 473*: 174–180.

Attene-Ramos, M.S., Nava, G.M., Muellner, M.G., Wagner, E.D., Plewa, M.J., and Gaskins, H.R. 2010. DNA damage and toxicogenomic analyses of hydrogen sulfide in human intestinal epithelial FHs 74 Int cells. *Environ. Mol. Mutagen. 51*: 304–314.

Attene-Ramos, M.S., Wagner, E.D., Gaskins, H.R., and Plewa, M.J. 2007. Hydrogen sulfide induces direct radical-associated DNA damage. *Mol. Cancer Res. 5*: 455–459.

Attene-Ramos, M.S., Wagner, E.D., Plewa, M.J., and Gaskins, H.R. 2006. Evidence that hydrogen sulfide is a genotoxic agent. *Mol. Cancer Res. 4*: 9–14.

Aura, A.M. 2008. Microbial metabolism of dietary phenolic compounds in the colon. *Phytochemistry Reviews 7*: 407–429.

Baldwin, R.L. and Allison, M.J. 1983. Rumen metabolism. *J. Anim. Sci. 57*, Suppl 2: 461–477.

Ballongue, J., Schumann, C., and Quignon, P. 1997. Effects of lactulose and lactitol on colonic microflora and enzymatic activity. *Scand. J. Gastroenterol. 32*: 41–44.

Bammens, B., Evenepoel, P., Keuleers, H., Verbeke, K., and Vanrenterghem, Y. 2006. Free serum concentrations of the protein-bound retention solute p-cresol predict mortality in hemodialysis patients. *Kidney Int. 69*: 1081–1087.

Bammens, B., Verbeke, K., Vanrenterghem, Y., and Evenepoel, P. 2003. Evidence for impaired assimilation of protein in chronic renal failure. *Kidney Int. 64*: 2196–2203.

Bergman, E.N. 1990. Energy contributions of volatile fatty-acids from the gastrointestinal-tract in various species. *Physiol. Rev. 70*: 567–590.

Bernal, W., Hall, C., Karvellas, C.J., Auzinger, G., Sizer, E., and Wendon, J. 2007. Arterial ammonia and clinical risk factors for encephalopathy and intracranial hypertension in acute liver failure. *Hepatology 46*: 1844–1852.

Bingham, S.A., Hughes, R., and Cross, A.J. 2002. Effect of white versus red meat on endogenous N-nitrosation in the human colon and further evidence of a dose response. *J. Nutr. 132*: 3522s–3525s.

Birkett, A., Muir, J., Phillips, J., Jones, G., and O'Dea, K. 1996. Resistant starch lowers fecal concentrations of ammonia and phenols in humans. *Am. J. Clin. Nutr. 63*: 766–772.

Bjursell, M., Admyre, T., Goransson, M., Marley, A.E., Smith, D.M., Oscarsson, J., and Bohlooly, Y.M. 2011. Improved glucose control and reduced body fat mass in free fatty acid receptor 2-deficient mice fed a high-fat diet. *Am. J. Physiol. Endocrinol. Metab. 300*: E211–220.

Blachier, F., Davila, A.M., Mimoun, S., Benetti, P.H., Atanasiu, C., Andriamihaja, M., et al. 2010. Luminal sulfide and large intestine mucosa: Friend or foe? *Amino Acids 39*: 335–347.

Blachier, F., Mariotti, F., Huneau, J.F., and Tome, D. 2007. Effects of amino acid-derived luminal metabolites on the colonic epithelium and physiopathological consequences. *Amino Acids 33*: 547–562.

Bloemen, J.G., Venema, K., de Poll, M.C.V., Damink, S.W.O., Buurman, W.A., and Dejong, C.H. 2009. Short chain fatty acids exchange across the gut and liver in humans measured at surgery. *Clin. Nutr. 28*: 657–661.

Bolca, S., Urpi-Sarda, M., Blondeel, P., Roche, N., Vanhaecke, L., Possemiers, S., et al. 2010. Disposition of soy isoflavones in normal human breast tissue. *Am. J. Clin. Nutr. 91*: 976–984.

Bond, J.H., Jr. and Levitt, M.D. 1976. Fate of soluble carbohydrate in the colon of rats and man. *J. Clin. Invest. 57*: 1158–1164.

Boto-Ordonez, M., Urpi-Sarda, M., Queipo-Ortuno, M. I., Tulipani, S., Tinahones, F.J., and Andres-Lacueva, C. 2014. High levels of Bifidobacteria are associated with increased levels of anthocyanin microbial metabolites: A randomized clinical trial. *Food & function 5*: 1932–1938.

Brouns, F., Kettlitz, B., and Arrigoni, E. 2002. Resistant starch and "the butyrate revolution". *Trends in Food Science & Technology 13*: 251–261.

Brown, A.J., Goldsworthy, S.M., Barnes, A.A., Eilert, M.M., Tcheang, L., Daniels, D., et al. 2003. The orphan G protein-coupled receptors GPR41 and GPR43 are activated by propionate and other short chain carboxylic acids. *J. Biol. Chem. 278*: 11312–11319.

Cardona, F., Andrés-Lacueva, C., Tulipani, S., Tinahones, F.J., and Queipo-Ortuño, M.I. 2013. Benefits of polyphenols on gut microbiota and implications in human health. *J. Nutr. Biochem. 24*: 1415–1422.

Cerini, C., Dou, L., Anfosso, F., Sabatier, F., Moal, V., Glorieux, G., et al. 2004. P-cresol, a uremic retention solute, alters the endothelial barrier function *in vitro*. *Thromb. Haemost. 92*: 140–150.

Chang, P.V., Hao, L., Offermanns, S., and Medzhitov, R. 2014. The microbial metabolite butyrate regulates intestinal macrophage function via histone deacetylase inhibition. *Proc. Natl. Acad. Sci. USA. 111*: 2247–2252.

Cherrington, C.A., Hinton, M., Pearson, G.R., and Chopra, I. 1991. Short-chain organic acids at ph 5.0 kill Escherichia coli and Salmonella spp. without causing membrane perturbation. *J. Appl. Bacteriol. 70*: 161–165.

Chung, K.T., Anderson, G.M., and Fulk, G.E. 1975. Formation of indoleacetic-acid by intestinal anaerobes. *J. Bacteriol. 124*: 573–575.

Clausen, M.R. and Mortensen, P.B. 1995. Kinetic studies on colonocyte metabolism of short chain fatty acids and glucose in ulcerative colitis. *Gut 37*: 684–689.

Clavel, T., Borrmann, D., Braune, A., Dore, J., and Blaut, M. 2006. Occurrence and activity of human intestinal bacteria involved in the conversion of dietary lignans. *Anaerobe 12*: 140–147.

Cooper, C.E. and Brown, G.C. 2008. The inhibition of mitochondrial cytochrome oxidase by the gases carbon monoxide, nitric oxide, hydrogen cyanide and hydrogen sulfide: Chemical mechanism and physiological significance. *J. Bioenerg. Biomembr. 40*: 533–539.

Corbin, K.D. and Zeisel, S.H. 2012. Choline metabolism provides novel insights into nonalcoholic fatty liver disease and its progression. *Curr. Opin. Gastroenterol. 28*: 159–165.

Cotelle, N. 2001. Role of flavonoids in oxidative stress. *Curr. Top. Med. Chem. 1*: 569–590.

Cousens, L.S., Gallwitz, D., and Alberts, B.M. 1979. Different accessibilities in chromatin to histone acetylase. *J. Biol. Chem. 254*: 1716–1723.

Cremin, J.D., Jr., Fitch, M.D., and Fleming, S.E. 2003. Glucose alleviates ammonia-induced inhibition of short-chain fatty acid metabolism in rat colonic epithelial cells. *Am. J. Physiol. Gastrointest. Liver Physiol. 285*: G105–G114.

Cresci, G.A., Thangaraju, M., Mellinger, J.D., Liu, K., and Ganapathy, V. 2010. Colonic gene expression in conventional and germ-free mice with a focus on the butyrate receptor GPR109A and the butyrate transporter SLC5A8. *J. Gastrointest. Surg. 14*: 449–461.

Cross, A.J., Pollock, J.R., and Bingham, S.A. 2003. Haem, not protein or inorganic iron, is responsible for endogenous intestinal N-nitrosation arising from red meat. *Cancer Res. 63*: 2358–2360.

Cummings, J.H. 1983. Fermentation in the human large intestine: Evidence and implications for health. *Lancet 1*: 1206–1209.

Cummings, J.H., Pomare, E.W., Branch, W.J., Naylor, C.P.E., and Macfarlane, G.T. 1987. Short Chain Fatty-Acids in Human Large-Intestine, Portal, Hepatic and Venous-Blood. *Gut 28*: 1221–1227.

Damen, B., Cloetens, L., Broekaert, W.F., Francois, I., Lescroart, O., Trogh, I., et al. 2012. Consumption of breads containing in situ-produced arabinoxylan oligosaccharides alters gastrointestinal effects in healthy volunteers. *J. Nutr. 142*: 470–477.

Das, M. and Das, D.K. 2010. Resveratrol and cardiovascular health. *Mol. Aspects Med. 31*: 503–512.

Davie, J.R. 2003. Inhibition of histone deacetylase activity by butyrate. *J. Nutr. 133*: 2485s–2493s.

De Preter, V., Arijs, I., Windey, K., Vanhove, W., Vermeire, S., Schuit, F., et al. 2012. Decreased mucosal sulfide detoxification is related to an impaired butyrate oxidation in ulcerative colitis. *Inflamm. Bowel Dis. 18*: 2371–2380.

De Preter, V., Vanhoutte, T., Huys, G., Swings, J., De Vuyst, L., Rutgeerts, P. Et al. 2007. Effects of Lactobacillus casei Shirota, Bifidobacterium breve, and oligofructose-enriched inulin on colonic nitrogen-protein metabolism in healthy humans. *Am. J. Physiol. Gastrointest. Liver Physiol. 292*: G358–G368.

den Besten, G., Lange, K., Havinga, R., van Dijk, T.H., Gerding, A., van Eunen, K., et al. 2013. Gut-derived short-chain fatty acids are vividly assimilated into host carbohydrates and lipids. *Am. J. Physiol. Gastrointest. Liver Physiol. 305*: G900–910.

Dolara, P., Luceri, C., De Filippo, C., Femia, A. P., Giovannelli, L., Caderni, G., et al. 2005. Red wine polyphenols influence carcinogenesis, intestinal microflora, oxidative damage and gene expression profiles of colonic mucosa in F344 rats. *Mutat. Res. 591*: 237–246.

Donohoe, D.R., Garge, N., Zhang, X., Sun, W., O'Connell, T.M., Bunger, M. K. et al. 2011. The microbiome and butyrate regulate energy metabolism and autophagy in the mammalian colon. *Cell metabolism 13*: 517–526.

Dou, L., Bertrand, E., Cerini, C., Faure, V., Sampol, J., Vanholder, R., et al. 2004. The uremic solutes p-cresol and indoxyl sulfate inhibit endothelial proliferation and wound repair. *Kidney Int. 65*: 442–451.

Dumas, M.E., Barton, R.H., Toye, A., Cloarec, O., Blancher, C., Rothwell, A., et al. 2006. Metabolic profiling reveals a contribution of gut microbiota to fatty liver phenotype in insulin-resistant mice. *Proc. Natl. Acad. Sci. USA. 103*: 12511–12516.

Duncan, S.H., Louis, P., Thomson, J.M., and Flint, H.J. 2009. The role of pH in determining the species composition of the human colonic microbiota. *Environ. Microbiol. 11*: 2112–2122.

Elsden, S.R., Hilton, M.G., and Waller, J.M. 1976. The end products of the metabolism of aromatic amino acids by Clostridia. *Arch. Microbiol. 107*: 283–288.

Etxeberria, U., Fernandez-Quintela, A., Milagro, F. I., Aguirre, L., Martinez, J.A., and Portillo, M.P. 2013. Impact of polyphenols and polyphenol-rich dietary sources on gut microbiota composition. *J. Agric. Food Chem. 61*: 9517–9533.

Evenepoel, P., Claus, D., Geypens, B., Hiele, M., Geboes, K., Rutgeerts, P. Et al. 1999. Amount and fate of egg protein escaping assimilation in the small intestine of humans. *Am. J. Physiol. 277*: G935–G943.

Evenepoel, P., Meijers, B.K.I., Bammens, B.R.M., and Verbeke, K. 2009. Uremic toxins originating from colonic microbial metabolism. *Kidney Int. 76*: S12–S19.

Evensen, N.A. and Braun, P.C. 2009. The effects of tea polyphenols on Candida albicans: Inhibition of biofilm formation and proteasome inactivation. *Can. J. Microbiol. 55*: 1033–1039.

Farrugia, G. and Szurszewski, J.H. 2014. Carbon monoxide, hydrogen sulfide, and nitric oxide as signaling molecules in the gastrointestinal tract. *Gastroenterology 147*: 303–313.

Flint, H.J., Bayer, E.A., Rincon, M.T., Lamed, R., and White, B.A. 2008. Polysaccharide utilization by gut bacteria: Potential for new insights from genomic analysis. *Nat. Rev. Microbiol. 6*: 121–131.

Florin, T., Gibson, G.R., and Neale, G. 1990. A role for sulfate reducing bacteria in ulcerative colitis? *Gastroenterology 98*: A170.

Florin, T., Neale, G., Gibson, G.R., Christl, S.U., and Cummings, J H. 1991. Metabolism of dietary sulfate - absorption and excretion in humans. *Gut 32*: 766–773.

Geypens, B., Claus, D., Evenepoel, P., Hiele, M., Maes, B., Peeters, M., *et al.* 1997. Influence of dietary protein supplements on the formation of bacterial metabolites in the colon. *Gut 41*: 70–76.

Gibson, G.R. and Roberfroid, M.B. 1995. Dietary modulation of the human colonic microbiota—introducing the concept of prebiotics. *J. Nutr. 125*: 1401–1412.

Gibson, J.A., Sladen, G.E., and Dawson, A.M. 1976. Protein absorption and ammonia production: The effects of dietary protein and removal of the colon. *Br. J. Nutr. 35*: 61–65.

Gilbert, J.A., Bendsen, N.T., Tremblay, A., and Astrup, A. 2011. Effect of proteins from different sources on body composition. *Nutr. Metab. Cardiovasc. Dis. 21 Suppl 2*: B16–31.

Gill, R.K., Saksena, S., Alrefai, W.A., Sarwar, Z., Goldstein, J.L., Carroll, R.E., *et al.* 2005. Expression and membrane localization of MCT isoforms along the length of the human intestine. *Am. J. Physiol. Cell Physiol. 289*: C846–852.

Goncalves, P. and Martel, F. 2013. Butyrate and colorectal cancer: The role of butyrate transport. Curr. Drug Metab. 14: 994–1008.

Gonzalez-Vallinas, M., Gonzalez-Castejon, M., Rodriguez-Casado, A., and Ramirez de Molina, A. 2013. Dietary phytochemicals in cancer prevention and therapy: A complementary approach with promising perspectives. *Nutr. Rev. 71*: 585–599.

Gonzalez, R., Ballester, I., Lopez-Posadas, R., Suarez, M.D., Zarzuelo, A., Martinez-Augustin, O. *et al.* 2011. Effects of flavonoids and other polyphenols on inflammation. *Crit. Rev. Food Sci. Nutr. 51*: 331–362.

Hadjiagapiou, C., Schmidt, L., Dudeja, P.K., Layden, T.J., and Ramaswamy, K. 2000. Mechanism(s) of butyrate transport in Caco-2 cells: Role of monocarboxylate transporter 1. *Am. J. Physiol. Gastrointest. Liver Physiol. 279*: G775–780.

Hall, C.N., Badawi, A.F., O'Connor, P.J., and Saffhill, R. 1991. The detection of alkylation damage in the DNA of human gastrointestinal tissues. *Br. J. Cancer 64*: 59–63.

Harig, J.M., Ng, E.K., Dudeja, P.K., Brasitus, T.A., and Ramaswamy, K. 1996. Transport of n-butyrate into human colonic luminal membrane vesicles. *Am. J. Physiol. 271*: G415–422.

Hebels, D.G., Sveje, K.M., de Kok, M.C., van Herwijnen, M.H., Kuhnle, G.G., Engels, L.G., *et al.* 2012. Red meat intake-induced increases in fecal water genotoxicity correlate with pro-carcinogenic gene expression changes in the human colon. *Food Chem. Toxicol. 50*: 95–103.

Hobden, M.R., Martin-Morales, A., Guerin-Deremaux, L., Wils, D., Costabile, A., Walton, G.E., *et al.* 2013. In vitro fermentation of NUTRIOSE® FB06, a wheat dextrin soluble fibre, in a continuous culture human colonic model system. *PLoS One 8*: e77128.

Hooda, S., Metzler-Zebeli, B.U., Vasanthan, T., and Zijlstra, R.T. 2011. Effects of viscosity and fermentability of dietary fibre on nutrient digestibility and digesta characteristics in ileal-cannulated grower pigs. *Br. J. Nutr. 106*: 664–674.

Hughes, R., Magee, E., and Bingham, S. 2000. Protein degradation in the large intestine: Relevance to colorectal cancer. *Curr. Issues Intest. Microbiol. 1*: 51–58.

Joosen, A.M.C.P., Kuhnle, G.G.C., Aspinall, S.M., Barrow, T.M., Lecommandeur, E., Azqueta, A., et al. 2009. Effect of processed and red meat on endogenous nitrosation and DNA damage. *Carcinogenesis 30*: 1402–1407.

Jorgensen, J. and Mortensen, P.B. 2001. Hydrogen sulfide and colonic epithelial metabolism—Implications for ulcerative colitis. *Dig. Dis. Sci. 46*: 1722–1732.

Kaji, I., Karaki, S., and Kuwahara, A. 2014. Short-chain fatty acid receptor and its contribution to glucagon-like peptide-1 release. *Digestion 89*: 31–36.

Kaji, I., Karaki, S., Tanaka, R., and Kuwahara, A. 2011. Density distribution of free fatty acid receptor 2 (FFA2)-expressing and GLP-1-producing enteroendocrine L cells in human and rat lower intestine, and increased cell numbers after ingestion of fructo-oligosaccharide. *J. Mol. Histol. 42*: 27–38.

Karaki, S., Tazoe, H., Hayashi, H., Kashiwabara, H., Tooyama, K., Suzuki, Y., and Kuwahara, A. 2008. Expression of the short-chain fatty acid receptor, GPR43, in the human colon. *J. Mol. Histol. 39*: 135–142.

Kemperman, R.A., Bolca, S., Roger, L.C., and Vaughan, E.E. 2010. Novel approaches for analysing gut microbes and dietary polyphenols: Challenges and opportunities. *Microbiology 156*: 3224–3231.

Kim, M.H., Kang, S.G., Park, J.H., Yanagisawa, M., and Kim, C.H. 2013. Short-chain fatty acids activate GPR41 and GPR43 on intestinal epithelial cells to promote inflammatory responses in mice. *Gastroenterology 145*: 396–406.

Koeth, R.A., Wang, Z., Levison, B.S., Buffa, J.A., Org, E., Sheehy, B.T., et al. 2013. Intestinal microbiota metabolism of L-carnitine, a nutrient in red meat, promotes atherosclerosis. *Nat. Med. 19*: 576–585.

Kuhnle, G.G., Story, G.W., Reda, T., Mani, A.R., Moore, K.P., Lunn, J.C., et al. 2007. Diet-induced endogenous formation of nitroso compounds in the GI tract. *Free Radic. Biol. Med. 43*: 1040–1047.

Kutschera, M., Engst, W., Blaut, M., and Braune, A. 2011. Isolation of catechin-converting human intestinal bacteria. *J. Appl. Microbiol. 111*: 165–175.

Lampe, J.W. 2009. Interindividual differences in response to plant-based diets: Implications for cancer risk. *Am. J. Clin. Nutr. 89*: 1553s–1557s.

Lecerf, J.M., Depeint, F., Clerc, E., Dugenet, Y., Niamba, C.N., Rhazi, L., et al. 2012. Xylo-oligosaccharide (XOS) in combination with inulin modulates both the intestinal environment and immune status in healthy subjects, while XOS alone only shows prebiotic properties. *Br. J. Nutr. 108*: 1847–1858.

Leschelle, X., Goubern, M., Andriamihaja, M., Blottiere, H.M., Couplan, E., Gonzalez-Barroso, M.D., et al. 2005. Adaptative metabolic response of human colonic epithelial cells to the adverse effects of the luminal compound sulfide. *Biochim. Biophys. Acta 1725*: 201–212.

Levine, J., Ellis, C.J., Furne, J.K., Springfield, J., and Levitt, M.D. 1998. Fecal hydrogen sulfide production in ulcerative colitis. *Am. J. Gastroenterol. 93*: 83–87.

Levitt, M.D., Furne, J., Springfield, J., Suarez, F., and DeMaster, E. 1999. Detoxification of hydrogen sulfide and methanethiol in the cecal mucosa. *J. Clin. Invest. 104*: 1107–1114.

Lin, H.V., Frassetto, A., Kowalik, E.J., Jr., Nawrocki, A.R., Lu, M.M., Kosinski, J. R., et al. 2012. Butyrate and propionate protect against diet-induced obesity and regulate gut hormones via free fatty acid receptor 3-independent mechanisms. *PLoS One* 7: e35240.

Louis, P., Duncan, S.H., McCrae, S.I., Millar, J., Jackson, M.S., and Flint, H.J. 2004. Restricted distribution of the butyrate kinase pathway among butyrate-producing bacteria from the human colon. *J. Bacteriol.* 186: 2099–2106.

Louis, P., Scott, K.P., Duncan, S.H., and Flint, H.J. 2007. Understanding the effects of diet on bacterial metabolism in the large intestine. *J. Appl. Microbiol.* 102: 1197–1208.

Macfarlane, G.T., Allison, C., and Gibson, G.R. 1988. Effect of pH on protease activities in the large-intestine. *Lett. Appl. Microbiol.* 7: 161–164.

Macfarlane, G.T. and Gibson, G.R. 1994. Metabolic activities of the normal colonic flora. Human health—The contribution of microorganisms. S. A. W. Gibson. London, Springer-Verlag: 17–52.

Macfarlane, G.T., Gibson, G.R., and Cummings, J.H. 1992. Comparison of fermentation reactions in different regions of the human colon. *J. Appl. Bacteriol.* 72: 57–64.

Magee, E.A., Richardson, C.J., Hughes, R., and Cummings, J.H. 2000. Contribution of dietary protein to sulfide production in the large intestine: An *in vitro* and a controlled feeding study in humans. *Am. J. Clin. Nutr.* 72: 1488–1494.

Manach, C., Scalbert, A., Morand, C., Remesy, C., and Jimenez, L. 2004. Polyphenols: Food sources and bioavailability. *Am. J. Clin. Nutr.* 79: 727–747.

Manach, C., Williamson, G., Morand, C., Scalbert, A., and Remesy, C. 2005. Bioavailability and bioefficacy of polyphenols in humans. I. Review of 97 bioavailability studies. *Am. J. Clin. Nutr.* 81: 230s–242s.

Masui, R., Sasaki, M., Funaki, Y., Ogasawara, N., Mizuno, M., Iida, A., et al. 2013. G Protein-Coupled Receptor 43 moderates gut inflammation through cytokine regulation from mononuclear cells. *Inflamm. Bowel Dis.* 19: 2848–2856.

Medani, M., Collins, D., Docherty, N.G., Baird, A.W., O'Connell, P.R., and Winter, D.C. 2011. Emerging role of hydrogen sulfide in colonic physiology and pathophysiology. *Inflamm. Bowel Dis.* 17: 1620–1625.

Meijers, B.K., Bammens, B., De Moor, B., Verbeke, K., Vanrenterghem, Y., and Evenepoel, P. 2008. Free p-cresol is associated with cardiovascular disease in hemodialysis patients. *Kidney Int.* 73: 1174–1180.

Midtvedt, T. and Norman, A. 1968. Parameters in 7-alpha-dehydroxylation of bile acids by anaerobic lactobacilli. *Acta Pathol. Microbiol. Scand.* 72: 313–329.

Miller, T.L. and Wolin, M.J. 1996. Pathways of acetate, propionate, and butyrate formation by the human fecal microbial flora. *Appl. Environ. Microbiol.* 62: 1589–1592.

Miyauchi, S., Gopal, E., Fei, Y.J., and Ganapathy, V. 2004. Functional identification of SLC5A8, a tumor suppressor down-regulated in colon cancer, as a Na(+)-coupled transporter for short-chain fatty acids. *J. Biol. Chem.* 279: 13293–13296.

Mouille, B., Robert, V., and Blachier, F. 2004. Adaptative increase of ornithine production and decrease of ammonia metabolism in rat colonocytes after hyperproteic diet ingestion. *Am. J. Physiol. Gastrointest. Liver Physiol.* 287: G344–351.

Neveu, V., Perez-Jimenez, J., Vos, F., Crespy, V., du Chaffaut, L., Mennen, L., et al.

2010. Phenol-Explorer: An online comprehensive database on polyphenol contents in foods. *Database (Oxford) 2010*: bap024.

Nigdikar, S.V., Williams, N.R., Griffin, B.A., and Howard, A.N. 1998. Consumption of red wine polyphenols reduces the susceptibility of low-density lipoproteins to oxidation in vivo. *Am. J. Clin. Nutr. 68*: 258–265.

Noack, J., Timm, D., Hospattankar, A., and Slavin, J. 2013. Fermentation profiles of wheat dextrin, inulin and partially hydrolyzed guar gum using an in vitro digestion pretreatment and in vitro batch fermentation system model. *Nutrients 5*: 1500–1510.

Nowak, A. and Libudzisz, Z. 2006. Influence of phenol, p-cresol and indole on growth and survival of intestinal lactic acid bacteria. *Anaerobe 12*: 80–84.

Parkar, S.G., Trower, T.M., and Stevenson, D.E. 2013. Fecal microbial metabolism of polyphenols and its effects on human gut microbiota. *Anaerobe 23*: 12–19.

Perez-Jimenez, J., Neveu, V., Vos, F., and Scalbert, A. 2010. Systematic analysis of the content of 502 polyphenols in 452 foods and beverages: An application of the phenol-explorer database. *J. Agric. Food Chem. 58*: 4959–4969.

Picton, R., Eggo, M.C., Merrill, G.A., Langman, M.J.S., and Singh, S. 2002. Mucosal protection against sulphide: Importance of the enzyme rhodanese. *Gut 50*: 201–205.

Pitcher, M.C.L. and Cummings, J.H. 1996. Hydrogen sulphide: A bacterial toxin in ulcerative colitis? *Gut 39*: 1–4.

Pylkas, A.M., Juneja, L.R., and Slavin, J.L. 2005. Comparison of different fibers for in vitro production of short chain fatty acids by intestinal microflora. *J. Med. Food 8*: 113–116.

Qin, J.J., Li, R.Q., Raes, J., Arumugam, M., Burgdorf, K.S., Manichanh, C., et al. 2010. A human gut microbial gene catalogue established by metagenomic sequencing. *Nature 464*: 59-U70.

Queipo-Ortuno, M.I., Boto-Ordonez, M., Murri, M., Gomez-Zumaquero, J.M., Clemente-Postigo, M., Estruch, R., et al. 2012. Influence of red wine polyphenols and ethanol on the gut microbiota ecology and biochemical biomarkers. *Am. J. Clin. Nutr. 95*: 1323–1334.

Ramakrishna, B.S., Robertsthomson, I.C., Pannall, P.R., and Roediger, W.E.W. 1991. Impaired sulfation of phenol by the colonic mucosa in quiescent and active ulcerative colitis. *Gut 32*: 46–49.

Rastmanesh, R. 2011. High polyphenol, low probiotic diet for weight loss because of intestinal microbiota interaction. *Chem. Biol. Interact. 189*: 1–8.

Ritzhaupt, A., Wood, I.S., Ellis, A., Hosie, K.B., and Shirazi-Beechey, S.P. 1998. Identification and characterization of a monocarboxylate transporter (MCT1) in pig and human colon: Its potential to transport l-lactate as well as butyrate. *J. Physiol. 513*: 719–732.

Roediger, W.E. 1982. Utilization of nutrients by isolated epithelial-cells of the rat colon. *Gastroenterology 83*: 424–429.

Roediger, W.E. and Babidge, W.J. 2000. Thiol methyltransferase activity in inflammatory bowel disease. *Gut 47*: 206–210.

Roediger, W.E., Moore, J., and Babidge, W. 1997. Colonic sulfide in pathogenesis and treatment of ulcerative colitis. *Dig. Dis. Sci. 42*: 1571–1579.

Roediger, W.E.W., Duncan, A., Kapaniris, O and Millard, S. 1993. Reducing sulfur compounds of the colon impair colonocyte nutrition—implications for ulcerative colitis. *Gastroenterology 104*: 802–809.

Roelofsen, H., Priebe, M.G., and Vonk, R.J. 2010. The interaction of short-chain fatty acids with adipose tissue: Relevance for prevention of type 2 diabetes. *Benef. Microbes 1*: 433–437.

Rothwell, J.A., Pérez-Jiménez, J., Neveu, V., Medina-Ramon, A., M'Hiri, N., Garcia Lobato, P., *et al.* 2013. Phenol-Explorer 3.0: A major update of the Phenol-Explorer database to incorporate data on the effects of food processing on polyphenol content. Database, 10.1093/database/bat070.

Roy, C.C., Kien, C.L., Bouthillier, L., and Levy, E. 2006. Short-chain fatty acids: Ready for prime time? *Nutr. Clin. Pract. 21*: 351–366.

Russell, J.B. 1983. Fermentation of peptides by Bacteroides ruminicola B(1)4. *Appl. Environ. Microbiol. 45*: 1566–1574.

Sanchez, J.I., Marzorati, M., Grootaert, C., Baran, M., Van Craeyveld, V., Courtin, C.M., *et al.* 2009. Arabinoxylan-oligosaccharides (AXOS) affect the protein/carbohydrate fermentation balance and microbial population dynamics of the Simulator of Human Intestinal Microbial Ecosystem. *Microb. Biotechnol. 2*: 101–113.

Santangelo, C., Vari, R., Scazzocchio, B., Di Benedetto, R., Filesi, C., and Masella, R. 2007. Polyphenols, intracellular signalling and inflammation. *Ann. Ist. Super. Sanita 43*: 394–405.

Santini, V., Gozzini, A., and Ferrari, G. 2007. Histone deacetylase inhibitors: Molecular and biological activity as a premise to clinical application. *Curr. Drug Metab. 8*: 383–393.

Scalbert, A., Manach, C., Morand, C., Remesy, C., and Jimenez, L. 2005. Dietary polyphenols and the prevention of diseases. *Crit. Rev. Food Sci. Nutr. 45*: 287–306.

Schaub, A., Futterer, A., and Pfeffer, K. 2001. PUMA-G, an IFN-gamma-inducible gene in macrophages is a novel member of the seven transmembrane spanning receptor superfamily. *Eur. J. Immunol. 31*: 3714–3725.

Schwiertz, A., Taras, D., Schafer, K., Beijer, S., Bos, N.A., Donus, C., and Hardt, P.D. 2010. Microbiota and SCFA in lean and overweight healthy subjects. *Obesity 18*: 190–195.

Segain, J.P., Raingeard de la Bletiere, D., Bourreille, A., Leray, V., Gervois, N., Rosales, C., *et al.* 2000. Butyrate inhibits inflammatory responses through NF-κB inhibition: Implications for Crohn's disease. *Gut 47*: 397–403.

Singh, N., Gurav, A., Sivaprakasam, S., Brady, E., Padia, R., Shi, H., *et al.* 2014. Activation of Gpr109a, receptor for niacin and the commensal metabolite butyrate, suppresses colonic inflammation and carcinogenesis. *Immunity 40*: 128–139.

Sirk, T.W., Brown, E.F., Friedman, M., and Sum, A.K. 2009. Molecular binding of catechins to biomembranes: Relationship to biological activity. *J. Agric. Food Chem. 57*: 6720–6728.

Smith, A.H., Zoetendal, E., and Mackie, R.I. 2005. Bacterial mechanisms to overcome inhibitory effects of dietary tannins. *Microb. Ecol. 50*: 197–205.

Smith, P.M., Howitt, M.R., Panikov, N., Michaud, M., Gallini, C.A., Bohlooly, Y.M., *et* al. 2013. The microbial metabolites, short-chain fatty acids, regulate colonic T_{reg} cell homeostasis. *Science 341*: 569–573.

Swanson, K.S., Grieshop, C.M., Flickinger, E.A., Bauer, L.L., Wolf, B.W., Chow, J., *et al.* 2002. Fructooligosaccharides and Lactobacillus acidophilus modify bowel function and protein catabolites excreted by healthy humans. *J. Nutr. 132*: 3042–3050.

Takebe, K., Nio, J., Morimatsu, M., Karaki, S., Kuwahara, A., Kato, I., and Iwanaga, T. 2005. Histochemical demonstration of a Na(+)-coupled transporter for short-chain fatty acids (slc5a8) in the intestine and kidney of the mouse. *Biomed. Res. 26*: 213–221.

Tan, J., McKenzie, C., Potamitis, M., Thorburn, A.N., Mackay, C.R., and Macia, L. 2014. The role of short-chain fatty acids in health and disease. *Adv. Immunol. 121*: 91–119.

Tang, W.H., Wang, Z., Levison, B. ., Koeth, R.A., Britt, E.B., Fu, X., *et al.* 2013. Intestinal microbial metabolism of phosphatidylcholine and cardiovascular risk. *N. Engl. J. Med. 368*: 1575–1584.

Tazoe, H., Otomo, Y., Karaki, S., Kato, I., Fukami, Y., Terasaki, M., *et al.* 2009. Expression of short-chain fatty acid receptor GPR41 in the human colon. *Biomed. Res. 30*: 149–156.

Thangaraju, M., Cresci, G.A., Liu, K., Ananth, S., Gnanaprakasam, J.P., Browning, D.D., *et al.* 2009. GPR109A is a G-protein-coupled receptor for the bacterial fermentation product butyrate and functions as a tumor suppressor in colon. *Cancer Res. 69*: 2826–2832.

Tolhurst, G., Heffron, H., Lam, Y.S., Parker, H.E., Habib, A.M., Diakogiannaki, E., *et al.* 2012. Short-chain fatty acids stimulate glucagon-like peptide-1 secretion via the G-protein-coupled receptor FFAR2. Diabetes 61: 364–371.

Tranah, T.H., Vijay, G.K., Ryan, J.M., and Shawcross, D.L. 2013. Systemic inflammation and ammonia in hepatic encephalopathy. *Metab. Brain Dis. 28*: 1–5.

Tricker, A.R. 1997. N-nitroso compounds and man: Sources of exposure, endogenous formation and occurrence in body fluids. *Eur. J. Cancer Prev. 6*: 226–268.

Tsao, R. 2010. Chemistry and biochemistry of dietary polyphenols. *Nutrients 2*: 1231–1246.

van de Wiele, T., Boon, N., Possemiers, S., Jacobs, H., and Verstraete, W. 2007. Inulin-type fructans of longer degree of polymerization exert more pronounced *in vitro* prebiotic effects. *J. Appl. Microbiol. 102*: 452–460.

Van den Abbeele, P., Venema, K., van de Wiele, T., Verstraete, W., and Possemiers, S. 2013. Different human gut models reveal the distinct fermentation patterns of arabinoxylan versus inulin. *J. Agric. Food Chem. 61*: 9819–9827.

Vance, D.E. 2008. Role of phosphatidylcholine biosynthesis in the regulation of lipoprotein homeostasis. *Curr. Opin. Lipidol. 19*: 229–234.

Verbeke, K., Ferchaud-Roucher, V., Preston, T., Small, A.C., Henckaerts, L., Krempf, M., *et al.* 2010. Influence of the type of indigestible carbohydrate on plasma and urine short-chain fatty acid profiles in healthy human volunteers. *Eur. J. Clin. Nutr. 64*: 678–684.

Vidyasagar, S., Barmeyer, C., Geibel, J., Binder, H.J., and Rajendran, V.M. 2005. Role of short-chain fatty acids in colonic HCO(3) secretion. *Am. J. Physiol. Gastrointest. Liver Physiol. 288*: G1217–1226.

Visek, W.J. 1978. Diet and cell-growth modulation by ammonia. *Am. J. Clin. Nutr. 31*: S216–S220.

Walker, A.W., Duncan, S.H., McWilliam Leitch, E.C., Child, M.W., and Flint, H.J. 2005. pH and peptide supply can radically alter bacterial populations and short-chain fatty acid ratios within microbial communities from the human colon. *Appl. Environ. Microbiol. 71*: 3692–3700.

Walle, T. 2004. Absorption and metabolism of flavonoids. *Free Radic. Biol. Med. 36*: 829–837.

Wang, X. and Gibson, G.R. 1993. Effects of the in vitro fermentation of oligofructose and inulin by bacteria growing in the human large intestine. *J. Appl. Bacteriol. 75*: 373–380.

Wang, Z., Klipfell, E., Bennett, B.J., Koeth, R., Levison, B.S., Dugar, B., et al. 2011. Gut flora metabolism of phosphatidylcholine promotes cardiovascular disease. *Nature 472*: 57–63.

Windey, K., De Preter, V., and Verbeke, K. 2012. Relevance of protein fermentation to gut health. *Mol. Nutr. Food Res. 56*: 184–196.

Wu, J., Zhou, Z., Hu, Y., and Dong, S. 2012. Butyrate-induced GPR41 activation inhibits histone acetylation and cell growth. *J. Genet. Genomics 39*: 375–384.

Wutzke, K.D., Lotz, M., and Zipprich, C. 2010. The effect of pre- and probiotics on the colonic ammonia metabolism in humans as measured by lactose-[(1)(5)N(2)]ureide. *Eur. J. Clin. Nutr. 64*: 1215–1221.

Yang, J., Keshavarzian, A., and Rose, D.J. 2013. Impact of dietary fiber fermentation from cereal grains on metabolite production by the fecal microbiota from normal weight and obese individuals. *J. Med. Food 16*: 862–867.

Yokoyama, M.T. and Carlson, J.R. 1979. Microbial metabolites of tryptophan in the intestinal tract with special reference to skatole. *Am. J. Clin. Nutr. 32*: 173–178.

CHAPTER 5

The Gut Microbiome: Pathways to Brain, Stress, and Behavior

AADIL BHARWANI, JOHN BIENENSTOCK, and PAUL FORSYTHE

5.1. INTRODUCTION

THE scientific literature has recently witnessed an explosive growth in the field of research examining the communication pathways that bridge the central nervous system (CNS) and the intestines. Particularly, studies have paved the way to elucidate the modulatory role of the host intestinal microbiome and the corollary of such a relationship on neural functioning and behavior. The result is the advent of a recurring theme in the literature: the systemic role of the microbiota-gut-brain axis in the maintenance of homeostasis and contributions to well-being and mental health.

Recent advances have further extended the role of bidirectional communication along this axis to include influencing CNS development and, consequentially, establishing the foundations for normal cognition and behavior (for a detailed review, see Collins *et al.* 2012; Forsythe and Kunze 2013). Altered host-microbiome interactions during the developmental period have been implicated in autism and other neurodevelopmental disorders. Furthermore, such interactions are also capable of influencing mood and behavioral responses in the developed brain, as demonstrated upon modulation of the gut microbiome composition. Indeed, such studies have coincided with the growing recognition of the need for a multifaceted approach to understanding mood and psychiatric disorders, with particular emphasis on the complex interactions between genes and the environment. Acknowledgment of such a need has led to a greater appreciation for the complexity of the host-microbiome interactions, which are informed by a number of factors such as environmental effects on microbiome diversity, maternal factors, sex, and epigenetic mechanisms—many of which lie beyond the scope of this chapter (see Stilling *et al.* 2014).

This chapter emphasizes the theme of bidirectional communication along the microbiota-gut-brain axis and provides a general overview of its role in neural development, behavior, and the stress response (Figure 5.1). Additionally, evidence is examined for the implications of such a form of signaling on mood, depression, and neurodevelopmental disorders. Evidence is also put forward to highlight the roles of the immune

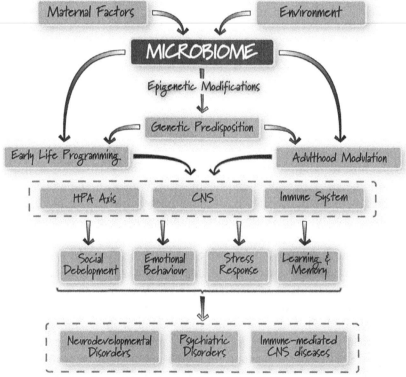

FIGURE 5.1. Broad overview of the complex relationship between various factors that inform host-microbiome interactions—some of which are discussed throughout Chapter 5 while others lie beyond its scope. The diversity and composition of the gut flora are heavily influenced by maternal factors, stressors, and other environmental changes during early life and throughout adulthood. These factors alter communication along the microbiota-gut-brain axis and can affect the development and/or regulation of the HPA axis, CNS architecture and physiology, and the immune system. The effects of such interkingdom interplay are reflected downstream in various aspects of the host's physiology and behavior, and can have potent implications on the appearance of pathologies and disorders.

system and the vagus nerve: two of the numerous pathways involved in the transmission of information from the gut bacteria to the brain and vice versa.

5.2. EVIDENCE OF MICROBIOTA–CNS INTERACTIONS

5.2.1. Behavior and Cognition

An oft-referenced and classical example of communication along the gut-brain axis involves the regulation of feeding behavior. Afferent projections from the GI tract, armed with various sensors at dendritic terminals in the gut, respond to signals that collectively control ingestivebehavior and satiety by influencing the activity of the paraventricular nucleus (PVN)—the "feeding center" in the hypothalamus. Further research however, has also revealed a role of the gut microbiota in appetitive behavior. In both humans and rodent models, the fermentation of dietary nondigestible carbohydrate compounds by intestinal bacteria induces the secretion of appetite-regulating peptides, leading to increased satiety and lowered energy intake (Cani *et al.* 2004, 2009). These compounds, referred to as "prebiotics", alter the balance of the microbiome and promote the growth of certain species such as bifidobacteria (Cani *et al.* 2007). Such observations have compelled the expansion of the gut-brain axis to include the commensal flora in the regulation of appetitive behavior, thus intricately linking the microbiome composition and food ingestion.

The influence of the microbiota however, is much more pervasive and expands beyond merely the regulation of satiety and feeding behavior. Several studies have demonstrated a temporal association between diet-induced shifts in the balance and composition of the intestinal microbiome and changes in behavior. Relative to a control group, rats fed a diet of fermentable carbohydrates exhibited increased levels of anxiety-like behavior and social aggression, as demonstrated during the light/dark emergence test and the social interaction test respectively (Hanstock *et al.* 2004). Implicating an active role for the intestinal bacteria was evidence suggesting that these behavioral changes occurred independently of changes in dopamine (DA) and serotonin (5-HT) concentrations in the CNS and instead, were associated with raised caecum concentrations of fermentation metabolites. Further evidence positing that shifts in the commensal population could bring about changes in behavior was put forward by a study of mice that were fed a beef-enriched diet.

Relative to those receiving standard rodent chow, this group exhibited a more diverse microbiome (Li *et al.* 2009). Moreover, in addition to reduced levels of anxiety-like behavior, these mice displayed improved performance on measures of temporary and long-term memory (working and reference memory respectively).

Another well-established instance of signaling between the CNS and bacterial species is sickness behavior: a term used to describe the adaptive changes in behavior and motivational state that occur as a result of infection by pathogenic organisms (Dantzer 2001). For example, an oral challenge with a common foodborne pathogen—*Campylobacter jejuni*—results in reduced exploration, which is indicative of increased anxiety-like behavior (Goehler *et al.* 2008). Such inflammation-induced behavioral changes are driven by CNS activity that coordinates the reorganization of the host's priorities in order to effectively cope with an infection. However, gut bacteria are also capable of exerting effects that, in addition to outlasting the duration of the illness, appear to be beyond the scope of sickness behavior altogether. In mice challenged with *Citrobacter rodentium*, a noninvasive enteric pathogen, exposure to a psychological stressor engendered nonspatial and working memory dysfunction—even after pathogen clearance (30 days) (Gareau *et al.* 2011). That these abnormalities occurred only in the presence of acute stress suggests that such changes were independent of symptoms encompassed by sickness behavior. Further prompting the role of the microbiota was that only the *C. rodentium* infection, not the acute stressor, altered the microbiome composition, and that pretreatment with *Lactobacillus*-containing probiotics prevented both infection-driven microbiome changes and stress-induced memory impairments.

Paralleling these observations are results from studies examining the effects of probiotics on cytokine-induced anxiety. Administration of *B. longum* abolished anxiety-like behavior induced by infection- or chemical colitis-associated GI inflammation (Bercik *et al.* 2010, 2011). These anxiolytic effects however, occurred without a corresponding reduction in proinflammatory cytokine levels, arguing against the hypothesis that the probiotics normalized behavior by merely suppressing inflammation. Such pathogen-induced anxiogenic effects have also been demonstrated in the absence of overt inflammation in a study of subclinical bacterial infection, therefore dissociating it from sickness behavior (Lyte *et al.* 1998).

Observations from studies employing transient and deliberate modifications of the commensal population elucidate the intricate effects of gut bacteria on behavior and implicate a casual relationship. In disruption studies, oral administration of nonabsorbable antimicrobials—neomycin, bacitracin, and pimaricin—to specific pathogen free (SPF) mice altered the composition of the rodent microbiome and induced modifications in behavior, such as increased exploration and decreased apprehension (Bercik *et al.* 2011). These behavioral changes were reversible and, along with the microbiota profile, began to normalize upon cessation of antimicrobial administration. Neither intraperitoneal administration of the antimicrobial drugs to SPF mice nor oral administration to GF mice induced changes in behavior or the microbiome profile, lending evidence for a causal relationship. Further compelling evidence has been provided by studies demonstrating the striking ability to adoptively transfer behavioral phenotypes via cecal bacteria (Bercik *et al.* 2011). Balb/c and NIH Swiss mice exhibit contrasting behavioral phenotypes—the former strain displays greater levels of anxiety-like behavior. The role of the microbiome in influencing host behavior was revealed by demonstrating that the colonization of GF Balb/c mice with NIH Swiss flora led to increased exploration, indicating lowered levels of anxiety. Conversely, transferring Balb/c gut bacteria into GF NIH Swiss mice led to greater levels of anxiety. Exposure studies have revealed that even a single bacterial species is often sufficient to alter behavior. Chronic administration of *Lactobacillus rhamnosus* (JB-1) engendered antidepressant-like effects on the forced swim test (FST), during which probiotic-fed mice spent less time immobile relative to broth-fed mice (Bravo *et al.* 2011). *Lactobacillus rhamnosus* (JB-1) ingestion also induced contrasting effects on conditioned and unconditioned components of anxiety behavior. Probiotic-fed mice exhibited an increased number of entries into the open arms of the elevated plus maze (EPM), which is suggestive of reduced unconditioned anxiety. In stark contrast, increased freezing behavior during the fear-conditioning paradigm was the corollary of greater levels of conditioned anxiety in the same group.

This behavioral phenotype was likely driven by region-dependent changes in the expression of GABA receptors, which have been implicated in regulating emotional behavior, including depression and anxiety. In previous studies, $GABA_{B\,(1)}$ KO mice exhibited anxiety-like behavior and markedly diminished responses to the anxiolytic effects of benzodiazepines (Cryan and Kaupmann 2005). In agreement with this, the anxiolytic profile of *Lactobacillus rhamnosus* (JB-1)-fed mice was accompanied by increased mRNA levels of the $GABA_{B1b}$ subunit in the prefrontal cortex (Bravo *et al.* 2011). In contrast, mRNA levels of this isoform were reduced in the amygdala and the hippocampus; this change was reflected in the reduced levels of despair exhibited by the probiotic group, consistent with the antidepressant effects observed upon administration of GABABR antagonists (Cryan and Kaupmann 2005).

Although microbial influence on behavior may appear to be limited to mammalian models, there is evidence to suggest that microbiome-host interactions are an evolutionarily conserved relationship. Examining this relationship in various systems through an evolutionary lens can yield insight into novel perspectives that may be obfuscated in complex systems. For instance, locust-congregation behavior is highly dependent on the gut microbial community, given its role in the production of aggregation pheromones (Dillon *et al.* 2000, 2002). Gut bacteria, often even a single species, can also influence fruit fly mating preferences (Sharon *et al.* 2011). Preferences in mating behavior, which were abolished in antibiotic-treated colonies, could be reestablished with bacteria obtained from fly media. Despite the two groups having diverged roughly around 782.7 million years ago (Hedges *et al.* 2006), such observations in insect models and rodent studies demonstrate a close relationship between the host and its microbiome.

5.2.2. Neurodevelopment

Observations stemming from studies of GF mice have well established the critical role of gut bacteria in contributing to the development of the mucosal immune system (Umesaki *et al.* 1995). However, their influence during the host's development appears to be even more widespread. Examining the behavior and brain structure of animals from GF model studies has revealed important clues about how the microbiome impacts the development of the brain and thus, modulates behavior and cognition.

Numerous studies have attempted to elucidate the neural correlates of altered behavior development in animals lacking a microbiome since birth. GF status appears to affect the physiology of the CNS, engendering a lower rate of DA, 5-HT, and norepinephrine (NE) turnover in in various regions of the brain (Crumeyrolle-Arias *et al.* 2014; Nishino *et al.* 2013). mRNA levels of brain-derived neurotrophic factor (BDNF) are significantly altered in the brains of GF mice, specifically in the hippocampus, amygdala, and the cingulate cortex—components of the memory, fear, and anxiety-regulating neural systems (Gareau et al. 2011; Heijtz et al. 2011; Lau and Pine 2008; Neufeld et al. 2011; Sudo et al. 2004). Moreover, such changes also appear to be sex-specific, with only male GF mice exhibiting decreased BDNF expression and increased 5-HT concentrations in the hippocampus, unlike their female counterparts (Clarke et al. 2013).

BDNF, which plays an important role in neurogenesis and neural differentiation, has also been implicated in numerous aspects of cognition and emotional behavior, including depression and anxiety (Martinowich and Lu 2007; Ren-Patterson et al. 2005). It is hardly surprising then that these animals display altered behavior during adulthood, including anxiolytic or anxiogenic behavior and baseline impairments in working and reference memory (Gareau et al. 2011; Heijtz et al. 2011; Neufeld et al. 2011). Moreover, that BDNF- and 5-HT-associated changes differ between sexes suggests an influence of the oestrous cycle hormones, which may engender certain sex-specific differences in the influence of the microbiome.

Similar changes in BDNF and associated behaviors have also been observed upon the adoptive transfer of behavioral phenotypes via the gut microbiota (Bercik et al. 2011). The adoptive transfer of the anxiogenic Balb/c phenotype into GF NIH Swiss mice was associated with a reduction in BDNF levels. The converse relationship also held true, with greater levels of BDNF and lower levels of anxiety being observed upon the transfer of the NIH Swiss mice microbiota into GF Balb/c mice. Albeit preliminary, these results demonstrate the changes in the neural circuitry and physiology that likely form the biological underpinnings of the microbiome's influence on cognition and behavior.

Understanding microbiota-CNS interactions may also have important implications for modeling the pathophysiology of Autism Spectrum Disorder (ASD) and other neurodevelopmental conditions. GI symptoms, often correlating with the severity of ASD, have been frequently reported in diagnosed patients and are hypothesized to be the corollary to changes in the activity and diversity of the microbiome (Adams et al. 2011; de Theije et al. 2011). Prenatal valproic acid (VPA) challenge and Maternal Immune Activation (MIA)—known risk factors for ASD development (reviewed by Hyman et al. 2005)—generate animal models that mimic the core symptoms encountered in ASD: communication deficits, social impairment, and repetitive and stereotypical behavior (Iwata et al. 2010; Malkova et al. 2012). Indeed, both these models also yield offspring with dysbiosis of the intestinal microbiome (Hsiao *et al.* 2013; de Theije *et al.* 2014).

The converse relationship holds true as well: GF mice exhibit various symptoms resembling neurodevelopmental disorders, including deficits in social cognition, changes in motivation and behavior, and a penchant for engaging in stereotypical and repetitive behavior (Desbonnet *et al.* 2014). These models also produce the GI symptoms reported in human ASD studies, including a compromised intestinal barrier that leads to increased epithelial permeability. Positing a causal role for the microbiome in neurodevelopmental disorders was the discovery that treatment of wild type mice with 4-ethylphenylsulfate (4EPS), whose serum levels are modulated by the gut microbiota and are elevated in the MIA model, induces anxiety-like behavior resembling that exhibited by MIA mice (Hsiao *et al.* 2013).

Further evidence is also provided by exposure and conventionalization studies. Human *Bacteroides fragilis* treatment normalized the serum metabolite profile of MIA mice and restored gut barrier integrity, while reversing microbiome dysbiosis and certain ASD-like behavioral abnormalities—communicative, repetitive, and anxiety-like behavior. Similarly, colonization of GF mice abolished certain social deficits and repetitive behavior (Desbonnet *et al.* 2014), thus implicating the microbiota-gut-brain axis in not merely CNS development, but also the pathogenesis of prevalent neurodevelopmental disorders.

5.2.3. Stress Response

The hypothalamic-pituitary-adrenal (HPA) axis is the neuroendocrine network responsible for eliciting and regulating responses to acute and chronic stress. There is plentiful evidence in the psychoneuroendocrinology literature suggesting that the programming of this system is sensitive to early-life influences such as stressors and, indeed, the gut microbiome (Meaney *et al.* 1993; Sudo *et al.* 2004). GF rodents exhibit a hyperactive HPA axis, as demonstrated by exaggerated ACTH and corticosterone levels at baseline (Neufeld *et al.* 2011) as well as in response to acute stress (Crumeyrolle-Arias *et al.* 2014; Sudo *et al.* 2004). This is a consequence of an observed decrease in the expression of cortical glucocorticoid receptors, given their role in the negative feedback loops that tightly regulate the stress response.

In contrast, association with even a single bacterial species—such as *B. infantis*, a common probiotic strain and the primary inhabitant of the neonate gut—appears to attenuate HPA axis activity and increase its sensitivity to feedback inhibition. These effects were replicated upon colonization of GF mice with SPF flora at 6 weeks of age. Interestingly, colonization at 14 weeks of age failed to normalize the stress response, suggesting that the effects of the commensal population on the HPA axis may be most influential during a period of plasticity in early development.

In addition to the bottom-up effects of the commensal flora on the programming and reactivity of the HPA axis, there exists a top-down influence of the CNS on the composition, balance, and activity of the intestinal microbiome. Maternal separation, an animal model for early life adversity, induces downstream changes in HPA axis activity that are retained into adulthood. Also associated with aberrant HPA axis responses are changes in the composition of the microbiome (O'Mahony *et al.* 2009), suggesting an influence of the stressor and its consequential neurochemical responses via signaling along the microbiota-gut-brain axis. The microbiome is also capable of responding to stressors that occur beyond the host's developmental period. Stress-associated release of NE and epinephrine are detected via Qsec sensor kinase, a quorum-sensing molecule to which bacteria respond by altering the expression of virulence genes, such as those associated with *Escherichia coli* O157:H7 (Clarke *et al.* 2006).

In conjunction with previously discussed results, these observations produce a complex model of interkingdom cross- signaling, in which bidirectional communication enables incessant interplay between the components of the microbiota-gut-brain axis. The discovery of such interactions, especially amidst the development of the stress axis, sheds light on a novel mechanism by which early life adversity may engender long-term downstream changes. Associated with early life stress in both human and animal models are changes in the programming of the stress response, emotional and social behavior, cognition and learning, and mental health conditions (Anisman *et al.* 1998; Blanchard *et al.* 2002; Frankenhuis and de Weerth 2013; Levine 2002; Weinburg and Tronick 1998). In accordance with the studies discussed above, many of these changes may be mediated via signaling along the microbiota-gut-brain axis.

The ability of the gut microbiome to influence the stress response and its neural correlates in developed, adult rodents has far reaching implications for enduring efforts in the field of medicine. There is emerging interest in the role of the microbiota-gut-brain axis in mental health, particularly depression (Dinan and Cryan, 2013). Early maternal separation, a valid rodent model for depression (Vetulani 2013), predictably increases immobility and decreases swimming behavior on the FST—changes that are suggestive of despair (Desbonnet *et al.* 2010). Albeit to different degrees, this behavior was attenuated by chronic treatment with citalopram, an antidepressant medication, or a probiotic preparation of *B. infantis*. Similar results were also demonstrated in one of the few human studies conducted. A double blind, placebo-controlled, randomized parallel group study examined the effects of a probiotic formulation containing *L. helvetica* and *B. longum* in a small group of healthy human volunteers (Messaoudi *et al.* 2011). Administration of the formulation for thirty days reduced scores on various measures of psychological distress, particularly depression and anger-hostility, as measured by the Hopkins Symptom Checklist (HSCL-90), and anxiety, as measured by the Hospital Anxiety and Depression Scale (HADS). Volunteers in the treatment group also exhibited reduced self-blame and greater problem solving on the Coping Checklist (CCL).

In a separate study using functional magnetic resonance imaging, the consumption of a fermented milk product with added probiotics by healthy women for 4 weeks resulted in changes in resting-state brain connectivity within a network that included affective and prefrontal regions (Tillisch *et al.* 2013). These changes were also associated with reduced activity of a widely distributed network—including frontal, prefrontal, and parahippocampal regions of the brain—in response to an emotional task. Albeit preliminary, such results provide promising evidence for the influence of gut bacteria on human health.

5.3. BIOLOGICAL UNDERPINNINGS OF NEURAL COMMUNICATION

5.3.1. Immune

Despite that nonpathological states are characterized by an intact intestinal epithelial barrier that mostly prevents microbial translocation, there exists overwhelming evidence of microbial influence on neurophysiology and behavior, thus implicating indirect routes of communication. Given its role in the processes that maintain and respond to changes in the microbiome, the immune system presents an important interface for gut-brain signaling. The recognition of surface molecular patterns and antigens by ubiquitous Pattern Recognition Receptors (PRRs)—localized on host cells, including those of immune, epithelial, and neuronal origin—enables the detection of the lumen-confined bacterial population (Murphy 2011; Rolls et al. 2007). Commensals and their associated antigens can be detected by the extending processes of lamina propria dendritic cells (DCs) or by their transport across the epithelial layer by specialized M cells in Peyer's patches and neonatal Fc receptor (FcRn)-expressing enterocytes (Figure 5.2). These interactions elicit downstream pathways, such as IgA-mediated responses, cytokine release, and the activation of immune effector cells. It has been well established that cytokines and other signaling messengers of the immune system can interact with the nervous system to bring about changes in CNS physiology and behavior (Dantzer 2001).

Additionally, the adaptive arm of the immune system has been observed to play a supportive role in learning behavior and cognition

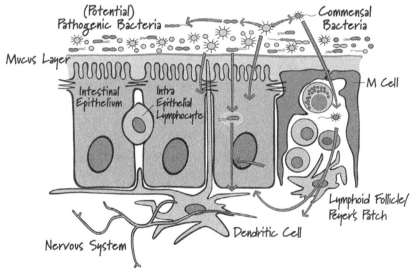

FIGURE 5.2. Host-microbiome interactions at the level of the mucosal lining in the gut that enable communication along the microbiota-gut-brain axis. Lamina propria DCs can detect commensal organisms via dendritic processes that extend into the lumen or by processing organisms and antigens that have been transcytosed across epithelial cells and specialized M cells. Such interactions activate these antigen-presenting cells to modulate downstream immune responses—such as those of T and B cells—and influence the function of the nervous system through ENS afferents. Adapted from "Immunomodulation by commensal and probiotic bacteria", by P. Forsythe and J. Bienenstock, 2010, Immunological Investigations, 39 (4-5), p. 431. Copyright 2010 by Informa Healthcare USA, Inc. Adapted with permission. Labeling of this figure has been modified from the original work to reflect the scope of this chapter.

(Brynskikh et al. 2008). In accordance with such observations, there exists evidence of immune-mediated bidirectional signaling between the CNS and the microbiome. Probiotic administration normalizes the aberrant cytokine profile and behavioral deficits elicited by maternal separation stress—induction of which resulted in changes in the composition of the microbiome (Desbonnet et al. 2008; O'Mahony et al. 2009). Similarly, irritable bowel syndrome (IBS) (O'Mahony et al. 2005) is highly comorbid with depressogenic and anxiogenic symptoms and is associated with an altered microbiome (Henningsen et al. 2003; Wu 2012). Administration of B. infantis to human subjects diagnosed with IBS engendered immunomodulatory effects by influencing the delicate balance between pro- and anti-inflammatory cytokines that is often disrupted in instances of microbial dysbiosis (O'Mahony et al. 2005). It thus appears that local interactions between the immune system and the commensal flora mediate gut-brain homeostasis on a macroscopic scale—deviations from which result in pathology and behavioral changes.

5.3.2. Vagus

With bidirectional projections that arise from the brainstem and traverse the body until they innervate the abdominal viscera, the vagus nerve serves as an important source of sensory information to the CNS regarding the internal milieu. Ascending sensory projections from visceral organs terminate in the nucleus of the solitary tract (NTS) before diffusing towards widespread areas of the brain, including the forebrain and the cerebral cortex (Travagli et al. 2003). This cranial nerve has thus emerged as the subject of numerous investigations into the communications infrastructure of the microbiota-gut-brain axis (Figure 5.3). These efforts have been further fueled by the observation in mice that intraluminal addition of *Lactobacillus rhamnosus* (JB-1) elicits rapid single- and multi-unit firing of vagus nerve afferents in the mesenteric nerve bundle (Perez-Burgos et al. 2013). Furthermore, evidence from neurophysiological experiments suggests that the enteric nervous system (ENS) may mediate interactions between gut bacteria and extrinsic afferent neurons, including those of the vagus nerve (Mao et al. 2013). In this particular study, application of psychotropic probiotic strains to the intestinal epithelium elicited sensory responses in intrinsic primary afferent neurons (IPANs) within 8 seconds, and after 15 minutes, increased their excitability. These effects were replicated upon application of polysaccharide A, the capsular exopolysaccharide of *B. fragilis*, thereby implicating the ENS in direct interactions with gut bacteria and their products.

Early evidence for the role of the vagus in gut-brain communication arises from exposure studies using c-Fos expression as a marker of neuronal cell activation. Consistent with the anatomy of the vagal pathway, *C. jejuni* infection elicited c-Fos levels in areas of the NTS—the visceral sensory nuclei—and the PVN in the forebrain (Gaykema *et al.* 2004; Goehler *et al.* 2005). Furthermore, neuronal activation appeared to occur independently of immune signaling, as suggested by a lack of elevation in proinflammatory cytokine levels at various time points following inoculation. Similar immune-independent vagal activation was also observed upon the induction of anxiogenic behavior following *C. rodentium* infection (Lyte *et al.* 2006). Such behavior occurred in the absence of any overt markers of sickness behavior, suggesting that vagal and not inflammation-induced signaling drove the observed behavioral changes.

The need for vagal integrity has also been demonstrated in probiotic signaling using animals that have undergone a subdiaphragmatic vagotomy. The anxiolytic and antidepressant effects of *Lactobacillus rhamnosus* (JB-1) were abolished in vagotomized mice (Bravo *et al.* 2011). Animals undergoing the surgery also failed to exhibit changes in various neural correlates, such as GABA receptor mRNA expression, which are thought to underlie probiotic-mediated behavioral patterns. Similarly, behavioral effects associated with *B. longum* administration were likewise inhibited by this surgery (Bercik *et al.* 2011). However, it is clear that vagotomy does not impair the microbiome-gut-brain axis in every model (Bercik *et al.* 2010), underlining the fact that these effects may not all be neuronally mediated.

That the contrasting effects of pathogen inoculation and probiotic administration can both be driven by vagus signaling suggests that the cranial nerve can distinguish between members of the commensal population—often in the absence of inflammatory markers (Lyte *et al.* 1998). Although the exact downstream signals from distinct microbial stimuli remain unclear, it appears that information about the composition and balance of the microbiome can seemingly be transmitted along ascending vagal pathways. These signals accordingly elicit differential neural activity and structural changes in the brain that are reflected by distinct behavioral phenotypes.

Vagus Nerve Stimulation (VNS) is an FDA-approved therapy that has been described to treat intractable forms of depression and epilepsy (Groves and Brown 2005). This, along with evidence of probiotic bacteria modulating despair behavior via the vagus (Bravo *et al.* 2011), highlights the importance of the vagal pathway in mood-related behaviors, especially in depression- and anxiety-like phenotypes. Furthermore, the ability of probiotic bacteria to elicit vagal firing engenders a potentially novel screening method to identify gut bacterial strains that can have a beneficial effect on physiology and behavior (Perez-Burgos *et al.* 2013).

5.3.3. Short-Chain Fatty Acids

The fermentation of dietary carbohydrates by intestinal bacteria culminates in the production of short-chain fatty acids (SCFAs) such as acetic, propionic, and butyric acid (Figure 5.3). Of these metabolites, butyrate, propionate, lactate, and pyruvate possess, to varying degrees, histone deacetylase (HDAC) inhibiting abilities (Stilling *et al.* 2014). Systemic administration of sodium butyrate has been observed to potently mimic the effects of antidepressants in animal models of depression (Schroeder *et al.* 2006; Tsankova *et al.* 2006). Epigenetic mechanisms have for long been thought to underlie the stable and chronic nature of depression (Tsankova *et al.* 2007). Chronic social defeat, an animal model of depression, engenders numerous symptoms of depression along with altered expression of *Bdnf* transcripts due to changes in methylation levels (Tsankova *et al.* 2006). Administration of sodium butyrate induced antidepressant-like effects by increasing levels of histone acetylation in the frontal cortex and the hippocampus, resulting in an elevation of *Bdnf* transcript levels (Schroeder *et al.* 2006; Tsankova *et al.* 2006). Similar epigenetic changes also appear to underlie the effects of early life maternal care on programming of the stress response (Weaver *et al.* 2005). It is not clear whether the systemic levels achieved of these fermentation metabolites are sufficient to produce behavioral changes. Nevertheless, it has been suggested that the chronic effects of SCFAs produced by the microbiota may result in subtle, stable changes in chromatin regulation and gene expression by altering levels of histone acetylation in the CNS (Stilling *et al.* 2014). Given the implications of such a form of communication on neurodevelopment and behavioral programming, as well as the paucity of evidence in the literature, this is an area of research that clearly necessitates further investigation.

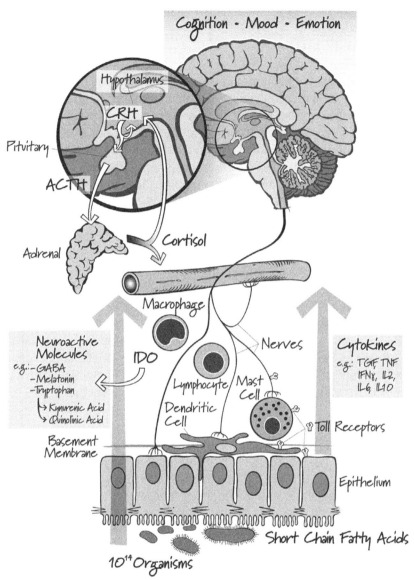

FIGURE 5.3. The microbiota-gut-brain axis along with its numerous routes of communication that enable bidirectional interplay between the host and the microbiome to maintain homeostasis and influence various aspects of behavior and physiology. The gut bacteria can synthesize a vast array of neuroactive molecules, including neurotransmitters such as GABA, and short-chain fatty acids that can alter the chemistry and functioning of the CNS. These bacteria also interact with the mucosal immune system through multiple routes to influence the production of cytokines and signaling messengers that can communicate with the nervous systems. Such complex interactions allow the microbiome to regulate the CNS and the HPAA, resulting in changes in behavior, cognition, and the stress response. Adapted from "Mood and gut feelings", by P. Forsythe, N. Sudo, T. Dinan, V. Taylor, and J. Bienenstock, 2010, Brain, Behavior and Immunity, 24 (1), p. 10. Copyright 2009 by Elsevier, Inc. Adapted with permission.

5.4. CONCLUSION

The literature is becoming replete with observations of the bidirectional interplay between the CNS and the microbiome, thus expanding the role of the gut flora beyond merely nutrition and physiological homeostasis. Rapidly emerging evidence presents a compelling case for its pivotal role in laying down the foundations of the CNS, with critical downstream consequences on the development of behavior, cognition, and neurodevelopment disorders. Moreover, the influence of gut bacteria also extends to the development and regulation of mood, anxiety, and the stress response. This relationship is hardly straightforward however, and depends on a number of variables that influence the microbiota-gut-brain axis, such as the early life environment, maternal factors, sex, and epigenetic mechanisms. Yet it is this very complexity that lends enormous potential to a field that is still in its infancy, and can help us better understand mental health and the biological underpinnings of mood and stress-related disorders—study of which has been traditionally restricted to the brain. These issues will need to be addressed in future efforts, along with a greater push towards understanding, from a molecular perspective, the complex interkingdom communication pathways that underpin the effects of the microbiome on the brain.

5.5. ACKNOWLEDGEMENTS

The authors gratefully acknowledge funding support from the Canadian Institutes of Health Research (GSM-136180) to A.B. and from the Natural Sciences and Engineering Research Council of Canada (371513-2009) to P.F. A.B. is also thankful to NSERC (USRA; 2014-2015) for an undergraduate fellowship.

5.6. REFERENCES

Adams, J.B., Johansen, L.J., Powell, L.D., Quig, D., and Rubin, R.A. 2011. Gastrointestinal flora and gastrointestinal status in children with autism—comparisons to typical children and correlation with autism severity. *BMC gastroenterology, 11*(1), 22.

Anisman, H., Zaharia, M.D., Meaney, M.J., and Merali, Z. 1998. Do early-life events permanently alter behavioral and hormonal responses to stressors? *International Journal of Developmental Neuroscience, 16*(3), 149-164.

Bercik, P., Denou, E., Collins, J., Jackson, W., Lu, J., Jury, J., et al. 2011. The intestinal microbiota affect central levels of brain-derived neurotropic factor and behavior in mice. *Gastroenterology, 141*(2), 599–609.

Bercik, P., Park, A.J., Sinclair, D., Khoshdel, A., Lu, J., Huang, X., *et al.* 2011. The anxiolytic effect of Bifidobacterium longum NCC3001 involves vagal pathways for gut–brain communication. *Neurogastroenterology and Motility, 23*(12), 1132–1139.

Bercik, P., Verdu, E.F., Foster, J.A., Macri, J., Potter, M., Huang, X., *et al.* 2010. Chronic gastrointestinal inflammation induces anxiety-like behavior and alters central nervous system biochemistry in mice. *Gastroenterology, 139*(6), 2102–2112.

Blanchard, E.B., Keefer, L., Payne, A., Turner, S.M., and Galovski, T.E. 2002. Early abuse, psychiatric diagnoses and irritable bowel syndrome. *Behaviour research and therapy, 40*(3), 289–298.

Bravo, J.A., Forsythe, P., Chew, M.V., Escaravage, E., Savignac, H.M., Dinan, T. G., *et al.* 2011. Ingestion of Lactobacillus strain regulates emotional behavior and central GABA receptor expression in a mouse via the vagus nerve. *Proceedings of the National Academy of Sciences, 108*(38), 16050–16055.

Brynskikh, A., Warren, T., Zhu, J., and Kipnis, J. 2008. Adaptive immunity affects learning behavior in mice. *Brain, behavior, and immunity, 22*(6), 861–869.

Cani, Patrice D., Cédric Dewever, and Nathalie M. Delzenne. 2004. "Inulin-type fructans modulate gastrointestinal peptides involved in appetite regulation (glucagon-like peptide-1 and ghrelin) in rats." *British Journal of Nutrition* 92.03: 521–526.

Cani, P.D., Lecourt, E., Dewulf, E.M., Sohet, F.M., Pachikian, B.D., Naslain, D., *et al.* 2009. Gut microbiota fermentation of prebiotics increases satietogenic and incretin gut peptide production with consequences for appetite sensation and glucose response after a meal. *The American journal of clinical nutrition, 90*(5), 1236–1243.

Cani, P.D., Neyrinck, A.M., Fava, F., Knauf, C., Burcelin, R.G., Tuohy, K.M., *et al.* 2007. Selective increases of bifidobacteria in gut microflora improve high-fat-diet-induced diabetes in mice through a mechanism associated with endotoxaemia. *Diabetologia, 50*(11), 2374–2383.

Clarke, G., Grenham, S., Scully, P., Fitzgerald, P., Moloney, R.D., Shanahan, F., *et al.* 2012. The microbiome-gut-brain axis during early life regulates the hippocampal serotonergic system in a sex-dependent manner. *Molecular psychiatry, 18*(6), 666–673.

Collins, S.M., Surette, M., and Bercik, P. 2012. The interplay between the intestinal microbiota and the brain. *Nature Reviews Microbiology, 10*(11), 735–742.

Crumeyrolle-Arias, M., Jaglin, M., Bruneau, A., Vancassel, S., Cardona, A., Daugé, V., *et al.* 2014. Absence of the gut microbiota enhances anxiety-like behavior and neuroendocrine response to acute stress in rats. *Psychoneuroendocrinology, 42*, 207–217.

Cryan, J.F. and Kaupmann, K. 2005. Don't worry 'B'happy!: A role for $GABA_A$ and $GABA_B$ receptors in anxiety and depression. *Trends in pharmacological sciences, 26*(1), 36–43.

Dantzer, R. 2001. Cytokine-induced sickness behavior: Where do we stand? *Brain, behavior, and immunity, 15*(1), 7–24.

de Theije, C.G., Wopereis, H., Ramadan, M., van Eijndthoven, T., Lambert, J., Knol, J., *et al.* 2013. Altered gut microbiota and activity in a murine model of autism spectrum disorders. *Brain, behavior, and immunity.*

de Theije, C.G., Wu, J., da Silva, S.L., Kamphuis, P.J., Garssen, J., Korte, S M., and Kraneveld, A.D. 2011. Pathways underlying the gut-to-brain connection in autism spectrum disorders as future targets for disease management. *European journal of pharmacology, 668*, S70–S80.

Desbonnet, L., Clarke, G., Shanahan, F., Dinan, T.G., and Cryan, J.F. 2013. Microbiota is essential for social development in the mouse. *Molecular psychiatry.*

Desbonnet, L., Garrett, L., Clarke, G., Bienenstock, J., and Dinan, T.G. 2008. The probiotic Bifidobacteria infantis: An assessment of potential antidepressant properties in the rat. *Journal of psychiatric research, 43*(2), 164–174.

Desbonnet, L., Garrett, L., Clarke, G., Kiely, B., Cryan, J.F., and Dinan, T.G. 2010. Effects of the probiotic Bifidobacterium infantis in the maternal separation model of depression. *Neuroscience, 170*(4), 1179–1188.

Dillon, R.J., Vennard, C.T., Charnley, A.K. 2000. Pheromones: Exploitation of gut bacteria in the locust. *Nature 403.*6772: 851–851.

Dillon, R.J., Vennard, C.T., and Charnley, A.K. 2002. A note: Gut bacteria produce components of a locust cohesion pheromone. *Journal of applied microbiology, 92*(4), 759–763.

Forsythe, P. and Bienenstock, J. 2010. Immunomodulation by commensal and probiotic bacteria. *Immunological investigations, 39*(4–5), 429–448.

Forsythe, P. and Kunze, W.A. 2013. Voices from within: Gut microbes and the CNS. *Cellular and Molecular Life Sciences, 70*(1), 55–69.

Forsythe, P., Sudo, N., Dinan, T., Taylor, V.H., and Bienenstock, J. 2010. Mood and gut feelings. *Brain, behavior, and immunity, 24*(1), 9–16.

Frankenhuis, W.E. and de Weerth, C. 2013. Does Early-Life Exposure to Stress Shape or Impair Cognition?. *Current Directions in Psychological Science, 22*(5), 407–412.

García-Ródenas, C.L., Bergonzelli, G.E., Nutten, S., Schumann, A., Cherbut, C., Turini, M., et al. 2006. Nutritional approach to restore impaired intestinal barrier function and growth after neonatal stress in rats. *Journal of pediatric gastroenterology and nutrition, 43*(1), 16–24.

Gareau, M.G., Wine, E., Rodrigues, D.M., Cho, J.H., Whary, M.T., Philpott, D.J., et al. 2011. Bacterial infection causes stress-induced memory dysfunction in mice. *Gut, 60*(3), 307–317.

Gaykema, R., Goehler, L.E., and Lyte, M. 2004. Brain response to cecal infection with *Campylobacter jejuni*: Analysis with Fos immunohistochemistry. *Brain, behavior, and immunity, 18*(3), 238–245.

Goehler, L.E., Gaykema, R., Opitz, N., Reddaway, R., Badr, N., and Lyte, M. 2005. Activation in vagal afferents and central autonomic pathways: Early responses to intestinal infection with *Campylobacter jejuni*. *Brain, behavior, and immunity, 19*(4), 334–344.

Goehler, L.E., Park, S.M., Opitz, N., Lyte, M., and Gaykema, R. 2008. *Campylobacter jejuni* infection increases anxiety-like behavior in the holeboard: Possible anatomical substrates for viscerosensory modulation of exploratory behavior. *Brain, behavior, and immunity, 22*(3), 354–366.

Groves, D.A. and Brown, V.J. 2005. Vagal nerve stimulation: A review of its applications and potential mechanisms that mediate its clinical effects. *Neuroscience and Biobehavioral Reviews, 29*(3), 493–500.

Hanstock, T.L., Clayton, E.H., Li, K.M., and Mallet, P.E. 2004. Anxiety and aggression associated with the fermentation of carbohydrates in the hindgut of rats. *Physiology and behavior,* 82(2), 357–368.

Hedges, S.B., Dudley, J., and Kumar, S. 2006. TimeTree: A public knowledge-base of divergence times among organisms. *Bioinformatics,* 22(23), 2971–2972.

Heijtz, R.D., Wang, S., Anuar, F., Qian, Y., Björkholm, B., Samuelsson, A., *et al.* 2011. Normal gut microbiota modulates brain development and behavior. *Proceedings of the National Academy of Sciences,* 108(7), 3047–3052.

Henningsen, P., Zimmermann, T., and Sattel, H. 2003. Medically unexplained physical symptoms, anxiety, and depression: A meta-analytic review. *Psychosomatic medicine,* 65(4), 528–533.

Hsiao, E.Y., McBride, S.W., Hsien, S., Sharon, G., Hyde, E.R., McCue, T., *et al.* 2013. Microbiota modulate behavioral and physiological abnormalities associated with neurodevelopmental disorders. *Cell,* 155(7), 1451–1463.

Hyman, S.L., Arndt, T.L., and Rodier, P.M. 2006. Environmental agents and autism: Once and future associations. *Intern Rev Res Ment Retard,* 30, 171–94.

Iwata, K., Matsuzaki, H., Takei, N., Manabe, T., and Mori, N. 2010. Animal models of autism: An epigenetic and environmental viewpoint. *Journal of central nervous system disease,* 2, 37.

Lau, J.Y. and Pine, D.S. 2008. Elucidating risk mechanisms of gene–environment interactions on pediatric anxiety: Integrating findings from neuroscience. *European archives of psychiatry and clinical neuroscience,* 258(2), 97–106.

Levine, S. 2002. Enduring effects of early experience on adult behavior. *Hormones, brain and behavior,* 4, 535–542.

Li, W., Dowd, S.E., Scurlock, B., Acosta-Martinez, V., and Lyte, M. 2009. Memory and learning behavior in mice is temporally associated with diet-induced alterations in gut bacteria. *Physiology and behavior,* 96(4), 557–567.

Lyte, M., Li, W., Opitz, N., Gaykema, R., and Goehler, L.E. 2006. Induction of anxiety-like behavior in mice during the initial stages of infection with the agent of murine colonic hyperplasia *Citrobacter rodentium*. *Physiology and behavior,* 89(3), 350–357.

Lyte, M., Varcoe, J.J., and Bailey, M.T. 1998. Anxiogenic effect of subclinical bacterial infection in mice in the absence of overt immune activation. *Physiology and behavior,* 65(1), 63–68.

Malkova, N.V., Yu, C.Z., Hsiao, E.Y., Moore, M.J., and Patterson, P.H. 2012. Maternal immune activation yields offspring displaying mouse versions of the three core symptoms of autism. *Brain, behavior, and immunity,* 26(4), 607–616.

Mao, Y.K., Kasper, D.L., Wang, B., Forsythe, P., Bienenstock, J., and Kunze, W.A. 2013. Bacteroides fragilis polysaccharide A is necessary and sufficient for acute activation of intestinal sensory neurons. *Nature communications,* 4, 1465.

Martinowich, K. and Lu, B. 2007. Interaction between BDNF and serotonin: Role in mood disorders. *Neuropsychopharmacology,* 33(1), 73–83.

Meaney, M.J., Bhatnagar, S., Diorio, J., Larocque, S., Francis, D., O'Donnell, D., *et al.* 1993. Molecular basis for the development of individual differences in the hypothalamic-pituitary-adrenal stress response. *Cellular and molecular neurobiology,* 13(4), 321–347.

Messaoudi, M., Lalonde, R., Violle, N., Javelot, H., Desor, D., Nejdi, A., et al. 2011. Assessment of psychotropic-like properties of a probiotic formulation (Lactobacillus helveticus R0052 and Bifidobacterium longum R0175) in rats and human subjects. *British Journal of Nutrition, 105*(05), 755–764.

Murphy, K. 2011. *Janeway's immunobiology*. Garland Science.

Neufeld, K.M., Kang, N., Bienenstock, J., and Foster, J.A. 2011. Reduced anxiety-like behavior and central neurochemical change in germ-free mice. *Neurogastroenterology and Motility, 23*(3), 255–e119.

Nishino, R., Mikami, K., Takahashi, H., Tomonaga, S., Furuse, M., Hiramoto, T., et al. 2013. Commensal microbiota modulate murine behaviors in a strictly contamination-free environment confirmed by culture-based methods. *Neurogastroenterology and Motility, 25*(6), 521–e371.

O'Mahony, S.M., Marchesi, J.R., Scully, P., Codling, C., Ceolho, A.M., Quigley, E.M., et al. 2009. Early life stress alters behavior, immunity, and microbiota in rats: Implications for irritable bowel syndrome and psychiatric illnesses. *Biological psychiatry, 65*(3), 263–267.

O'Mahony, L., McCarthy, J., Kelly, P., Hurley, G., Luo, F., Chen, K., et al. 2005. Lactobacillus and bifidobacterium in irritable bowel syndrome: Symptom responses and relationship to cytokine profiles. *Gastroenterology, 128*(3), 541–551.

Perez-Burgos, A., Wang, B., Mao, Y. K., Mistry, B., Neufeld, K.A.M., Bienenstock, J., and Kunze, W. 2013. Psychoactive bacteria *Lactobacillus rhamnosus* (JB-1) elicits rapid frequency facilitation in vagal afferents. *American Journal of Physiology-Gastrointestinal and Liver Physiology, 304*(2), G211–G220.

Ren-Patterson, R.F., Cochran, L.W., Holmes, A., Sherrill, S., Huang, S. J., Tolliver, T., et al. 2005. Loss of brain-derived neurotrophic factor gene allele exacerbates brain monoamine deficiencies and increases stress abnormalities of serotonin transporter knockout mice. *Journal of neuroscience research, 79*(6), 756–771.

Rolls, A., Shechter, R., London, A., Ziv, Y., Ronen, A., Levy, R., and Schwartz, M. 2007. Toll-like receptors modulate adult *hippocampal neurogenesis*. *Nature cell biology, 9*(9), 1081–1088.

Schroeder, F.A., Lin, C.L., Crusio, W.E., and Akbarian, S. 2007. Antidepressant-like effects of the histone deacetylase inhibitor, sodium butyrate, in the mouse. *Biological psychiatry, 62*(1), 55–64.

Sharon, G., Segal, D., Ringo, J.M., Hefetz, A., Zilber-Rosenberg, I., and Rosenberg, E. 2010. Commensal bacteria play a role in mating preference of *Drosophila melanogaster*. *Proceedings of the National Academy of Sciences, 107*(46), 20051–20056.

Stilling, R.M., Dinan, T.G., and Cryan, J.F. 2014. Microbial genes, brain and behaviour–epigenetic regulation of the gut–brain axis. *Genes, Brain and Behavior, 13*(1), 69–86.

Sudo, N., Chida, Y., Aiba, Y., Sonoda, J., Oyama, N., Yu, X. N., et al. 2004. Postnatal microbial colonization programs the hypothalamic–pituitary–adrenal system for stress response in mice. *The Journal of physiology, 558*(1), 263–275.

Tillisch, K., Labus, J., Kilpatrick, L., Jiang, Z., Stains, J., Ebrat, B., et al. 2013. Consumption of fermented milk product with probiotic modulates brain activity. *Gastroenterology, 144*(7), 1394–1401.

Travagli, R.A., Hermann, G.E., Browning, K.N., and Rogers, R.C. 2003. Musings on the Wanderer: What's New in our Understanding of Vago-Vagal Reflexes: III. Activity-dependent plasticity in vago-vagal reflexes controlling the stomach. American journal of physiology. *Gastrointestinal and liver physiology, 284*(2), G180.

Tsankova, N.M., Berton, O., Renthal, W., Kumar, A., Neve, R.L., and Nestler, E.J. 2006. Sustained hippocampal chromatin regulation in a mouse model of depression and antidepressant action. *Nature neuroscience, 9*(4), 519–525.

Tsankova, N., Renthal, W., Kumar, A., and Nestler, E.J. 2007. Epigenetic regulation in psychiatric disorders. *Nature Reviews Neuroscience, 8*(5), 355–367.

Umesaki, Y., Okada, Y., Matsumoto, S., Imaoka, A., and Setoyama, H. 1995. Segmented Filamentous Bacteria Are Indigenous Intestinal Bacteria That Activate Intraepithelial Lymphocytes and Induce MHC Class II Molecules and Fucosyl Asialo GM1 Glycolipids on the Small Intestinal Epithelial Cells in the Ex-Germ-Free Mouse. *Microbiology and immunology, 39*(8), 555–562.

Vetulani, J. 2013. Early maternal separation: A rodent model of depression and a prevailing human condition. *Pharmacological Reports, 65*(6), 1451–1461.

Weaver, I.C., Champagne, F.A., Brown, S.E., Dymov, S., Sharma, S., Meaney, M.J., and Szyf, M. 2005. Reversal of maternal programming of stress responses in adult offspring through methyl supplementation: Altering epigenetic marking later in life. *The Journal of Neuroscience, 25*(47), 11045–11054.

Weinberg, M.K. and Tronick, E.Z. 1998. The impact of maternal psychiatric illness on infant development. *Journal of clinical psychiatry.*

Wu, J.C. 2012. Psychological co-morbidity in functional gastrointestinal disorders: Epidemiology, mechanisms and management. *Journal of neurogastroenterology and motility, 18*(1), 13–18.

CHAPTER 6

Effects on Immunity

LEANDRO A. LOBO, ROSANA B.R. FERREIRA, and
L. CAETANO M. ANTUNES

6.1. INTRODUCTION

THE mammalian resident microbiota has been recognized for its contributions to host health and development for over a century. The study of the relationship between host and the microorganisms that reside in and on it began with Louis Pasteur, who postulated that microbial colonization was required for life (Pasteur 1885). Later, Nuttal and Thierfelder (1895) proved this postulate, to be incorrect by raising guinea pigs under germ free conditions. From the 1940s and on, advances in the methodology of raising animals under germ free conditions enabled scientists to start to understand more about the effects of the microbiota on its host, including how they peacefully coexist (Gordon *et al.* 1966; Reyniers and Trexler 1955). These germ-free animals have since then greatly increased our knowledge about the impact of commensals on the host. Essentially, these animals find life without the microbiota to be very difficult.

The now fully-appreciated benefits of the microbiota include the defense against pathogens, the maintenance of intestinal structure, and a critical contribution to the development and regulation of the mammalian immune system (Bäckhed *et al.* 2005; Chung and Kasper 2010; Hill and Artis 2010; Macpherson and Harris 2004; Spasova and Surh 2014). More importantly, some of the mechanisms involved have been elucidated such that we can identify certain immunological markers, cellular and molecular, to determine effects. In the context of overall health, the correct immunological information can be overwhelming; yet, thanks to this work a robust understanding is now possible.

6.2. GERM FREE MICE

The fundamental studies involving microbes and the immune system were performed on germ free (GF) mice. Physiologically, these mice have small intestines with elongated but thinner villi compared to conventionally-raised mice, resulting in increased susceptibility to infections as well as decreased vascularization and oxygenation of the villus (Cahenzli *et al.* 2013a; Savage 1977). These mice also present an underdeveloped gut-associated lymphoid tissue (GALT) compared to conventionally-raised mice. The presence of a normal gut microbiota leads to heavy recruitment of immune cells to the mucosa (Dethlefsen *et al.* 2007; Hooper and Macpherson 2010; Turnbaugh *et al.* 2007). Furthermore, the colonization of mice with their native microbiota leads to the induction of innate immune responses such as the expression of inducible nitric oxide synthase (iNOS) and antimicrobial peptides (AMPs), which can protect the mucosa from pathogens (Figure 6.1).

The intestinal microbiota also impacts the adaptive immune responses of its host. GF mice have fewer IgA-expressing B cells in the lamina propria, while B cell proliferation and induction of IgA expression occur in the presence of the microbiota, leading to increased size and number of germinal centers in the Peyer's patches (Benveniste *et al.* 1971; Hapfelmeier *et al.* 2010; Shroff *et al.* 1995) and maturation of cryptopatches into isolated lymphoid follicles (Tsuji *et al.* 2008). It has also been shown that the microbiota is responsible for an increase of intestinal lamina propria $CD4^+$ T cells (Macpherson and Harris 2004) as well as colonic regulatory T cells (Geuking *et al.* 2011) and intraepithelial lymphocytes expressing the $\alpha\beta$ T cell receptor (Bandeira *et al.* 1990). The impact of the gut microbiota on the immune system is also observed systemically, with increased total serum IgA, IgM, and IgG (Macpherson and Harris 2004).

The intestinal microbiota has also been involved in promoting systemic inflammation in the context of autoimmunity and infection. The balanced equilibrium between the mammalian immune system and the commensal microbiota is essential for optimal host health. Immune-related disorders such as inflammatory bowel disease (IBD) (Manichanh *et al.* 2006) and allergies (Penders *et al.* 2007) may result from an interruption of this balance.

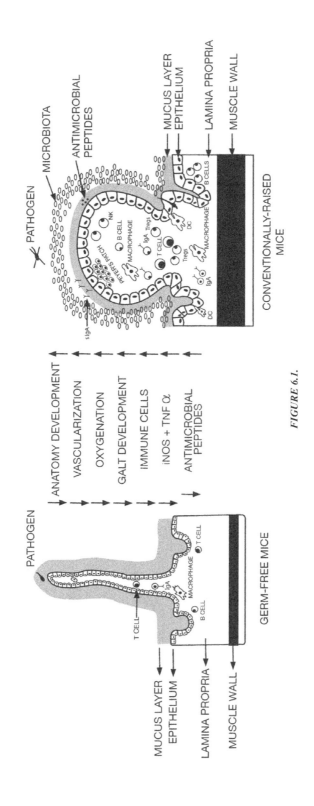

FIGURE 6.1.

The above are just some examples of the broad impact of the microbiota on the host immune system. This chapter will describe in detail the effect of the commensal microbiota on each of these mammalian immune functions. The impact of the gut microbiota on the physical development of the GALT, T and B cell differentiation and function, as well as AMP production is discussed. Small molecules produced by the microbes and their role in modulating the immune system will also be examined. Finally, the data available on the role of the microbiota from other sites of the body on the host immune response and also some diseases associated with unbalanced host immune responses to its own resident microbiota will be reviewed.

6.3. GALT FORMATION

The GALT is a major player on the immune defenses against pathogens entering the intestine (Brandtzaeg et al. 1989). It is generally divided in three compartments: the Peyer's patches, the lamina propria lymphocytes, and the intraepithelial lymphocytes, and it is populated by lymphocytes, such as B and T cells (Butcher and Picker 1996). The Peyer's patches are organized lymphoid tissues in the wall of the small intestine that contain B lymphoid follicles and interfollicular populations of $CD4^+$ and $CD8^+$ T cells. The gut lamina propria, the meshwork of connective tissue underlying the epithelium, contains a broad spectrum of myeloid and lymphoid cells, especially immunoglobulin (Ig) A plasmablasts, $CD4^+$ T cells, dendritic cells (DCs), and mast cells. The intraepithelial lymphocytes are located between intestinal epithelial cells and above the basal membrane and are populated by a variety of small, round cells, especially natural killer cells and many $CD8^+$ T cell subsets (Cebra 1999).

The functionality and development of an effective immune system at the GALT depends on stimulation by the resident microbiota, as well as temporary encounters with luminal microbes (Kamada and Núñez 2014). Studies using GF technology showed that these animals have an impaired GALT, including Peyer's patches and isolated lymphoid follicles (Bouskra et al. 2008; Gordon et al. 1966; Hamada et al. 2002). GF rats showed increased numbers of IgE-bearing B cells as well as decreased numbers of IgA-bearing B cells compared to animals raised under normal conditions (Durkin et al. 1981). Furthermore, when GF animals are colonized by commensals, immune reactions in lymphoid cell clusters and IgA development is observed (Shroff et al. 1995; Talham et al. 1999).

6.3.1. Antimicrobial Peptides (AMP)

In the heavily bacterial colonized environment of the human gut, our immune system must struggle not only to ward off potential invaders, but also to keep the number of beneficial bacteria in our microbiota in check. AMPs are a critical tool in the effort to maintain a normal and healthy homeostasis in our guts. AMPs are one of the most ancient mechanisms of protection against pathogens in our innate immune system. In fact, AMPs are produced by most living organisms as an important component of the innate immune system. Differently from classic antibiotics, such as penicillin, these molecules are encoded by specific genes and synthesized by ribosomes, and are not the product of the metabolic activity of enzymes (Ganz and Lehrer 1999). Despite their ubiquitous distribution in nature, AMPs share common features that are important for their activity and selectivity, such as the amphiphilic and cationic nature of these molecules.

In humans, AMPs are produced and secreted by several cell types, but most prominently by neutrophils and the cells lining the intestinal epithelia, in particular Paneth cells, a specialized epithelial cell lodged in the intestinal crypts in the villi of the small intestine that carry an extensive secretory activity (Ouellette 2010). The cytoplasm of these cells is marked by the presence of protein granules filled with AMPs (Figure 6.2). The granules are secreted after stimuli from the gut lumen environment. Molecules produced by bacteria in the gut, such as microbial-associated molecular patterns (MAMPs), are the main stimuli that induce the secretion of AMP granules, although Paneth cells maintain a basal level of AMP secretion in the gut (Ayabe *et al.* 2000). In the large intestine, neutrophils and epithelial cells also secrete AMPs in response to MAMPs. When AMPs and secretory IgA accumulate in the mucus layer, composed of mucin, glycopeptides, and lipids, an efficient physical-chemical barrier that protects the intestinal epithelium from pathogens is formed (Chen *et al.* 2006).

Due to the cationic nature of these peptides, AMPs are attracted to the negatively-charged microbial membranes instead of the mammalian host membranes (Figure 6.2). The architecture and composition of bacterial membranes, rich in charged phospholipids, along with the presence of Lipopolysaccharide (LPS) in Gram-negative and teichoic/teichuronic acids in Gram-positive bacteria supply the negative charges responsible for the selective toxicity of AMPs, while the outer layer of mammalian membranes are rich in neutral phospholipids and cholester-

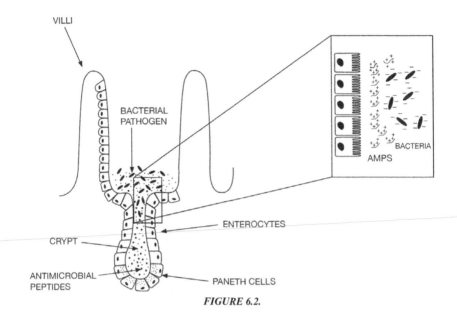

FIGURE 6.2.

ol (Yeaman and Yount 2003). The most common mechanism of action of AMPs is direct lysis of the microbial cell by disruption of the membrane integrity; this is accomplished in different ways according to the particular physical and chemical characteristics of the peptide and the target membrane. At least three mechanisms were proposed to explain the membrane disturbances that lead to cell lysis: the barrel-stave, the toroidal, and the carpet models (reviewed in Yeaman and Yount 2003). In each scenario, the insertion and accumulation of AMPs in the target membrane leads to permeabilization, with cytoplasmic leakage and interruption of essential functions such as the respiratory chain. Although membrane disruption was the first recognized mechanism of action of AMPs, recent studies indicate that vital processes that take place in the cytoplasm may be important targets, such as protein, RNA, and DNA synthesis (Guilhelmelli *et al.* 2013). Immunomodulatory actions of the AMPs also contribute to a robust activation of the adaptive immune system, which leads to microbial killing.

Defensins constitute the major class of AMPs in mammals. These molecules are small cationic peptides with up to 45 amino acids in their mature form. Based on their length and the distribution of cysteine residues and disulfide linkages in the molecule, defensins can be grouped into three different families: α-, β-, and θ-defensins. In humans, the α- and β-defensin families are present, and 17 defensins have been identified so far; six are α-defensins and 11 are β-defensins (Bevins and Salzman 2011). Four human α-defensins are produced by neutrophils, Paneth cells, and epithelial cells (DEFA1-4), whereas two (DEFA5 and 6) are constitutively produced by Paneth cells in the intestine and hence are called enteric defensins (Wehkamp *et al.* 2006). The β-defensins are produced by epithelial cells in every tissue of the human body, and are generally expressed upon stimulation. Defensins have strong chemotactic activity towards DCs, $CD4^+$ T cells, monocytes, and macrophages (Ramasundara *et al.* 2009).

The cathelicidins represent a second group of important AMPs secreted in the human gut. Human cathelicidin LL-37 exerts broad antimicrobial activity against Gram-negative and Gram-positive bacteria, exerted through the insertion of the cationic C-terminal portion of the molecule and disruption of microbial membrane integrity (Kościuczuk *et al.* 2012). Cathelicidins are produced as prepropeptides that are processed and secreted upon stimulation (Scott *et al.* 2002). Upon processing, the conserved N-terminal portion containing a cathelin-like domain (chatepsin L inhibitor) that shares similarities with cysteine protease inhibitors of the cystatin family is released from the cationic C-terminus. Antibacterial and protease inhibiting activities were demonstrated *in vitro* for this fragment (Zaiou *et al.* 2003), but no *in vivo* biological role for the cathelin domain has been demonstrated yet. Besides the bactericidal activity, cathelicidins have several other biological functions that may improve immune responses, such as angiogenesis, chemotaxis, cytokine production, and histamine release (Scott *et al.* 2002). Cathelicidin also binds the bacterial cell components LPS and lipotheicoic acid, and neutralizes proinflammatory cytokine responses that may lead to tissue damage (Bowdish *et al.* 2006).

There are several other AMPs in the repertoire of Paneth cells that play important roles to control bacterial growth in the human gut. For example, Paneth cells secrete C-type lectins and lysozyme C, both of which interact with the bacterial peptidoglycan and thus have strong activity against Gram-positive pathogens (Bevins and Salzman 2011). The main C-type lectin in humans is RegIIIα, which is also activated by trypsin digestion in Paneth cells (Iovanna and Dagorn 2005) and has high affinity for peptidoglycan repeating units. Lysozyme C is a glycan hydrolase specific for the peptidoglycan. Its activity renders the bacterial cell susceptible to osmotic lysis and also clears the infection site of peptidoglycan fragments, therefore dampening the proinflammatory responses elicited by these molecules (Ganz et al. 2003). Epithelial cells also secrete phospholipase A2, which preferentially hydrolyses two glycolipids abundant in bacterial membranes, phosphatidylglycerol and phosphatidylethanolamine (Lambeau and Gelb 2008). And finally, an unusual but nevertheless important class of AMPs are the bacteriocins. These antimicrobial molecules are not produced by humans, but rather by some species of commensal bacteria that inhabit the human intestine (Gillor et al. 2008).

An example of the importance of AMPs in intestinal homeostasis can be observed in Crohn's disease (CD). It is known that mutations in the human gene encoding an intracellular receptor for MAMPs, the nucleotide-binding oligomerization domain containing protein 2 (NOD2) receptor, result in a 40-fold increase in the risk of developing the disease. Expression of α-defensins by Paneth cells in patients with CD is lower when compared to non-CD controls, and more so in patients with defective NOD2 (Wehkamp et al. 2006). Knockout mice defective for the NOD2 gene have more severe symptoms of induced colitis and ileitis and are more susceptible to bacterial challenge with *Listeria monocytogenes* (Kobayashi et al. 2005). The same adverse effect is observed by complete ablation of Paneth cells in mice (Vaishnava et al. 2008). Eriguchi and collaborators (Eriguchi et al. 2012) corroborated the importance of α-defensins using a model of graft-versus-host-disease (GVHD) in mice. In this condition, Paneth cells are damaged and defensin production is impaired, thus reducing the diversity of the intestinal microbiota with simultaneous expansion of pathogenic bacteria. Systemic effects caused by translocation of bacterial LPS into the circulation, with a concomitant increase of proinflammatory cytokines, were also observed (Eriguchi et al. 2012). In the colon, the main AMPs are the β-defensins, and epithelial cells constitutively express β-defensin 1 (HBD1), which has broad antimicrobial spectrum against Gram-negative bacteria. Expression of HBD1 remains unaffected by the presence of inflammatory cytokines (IL-1α, TNF-α and IFN-γ) or bacterial stimuli (Cunliffe et al. 2001), whereas expression of human β-defensin 2 (HBD2) is limited in normal conditions but highly induced in inflammatory processes such as ulcerative colitis (UC).

The expression of human cathelicidin LL-37 is upregulated in patients suffering from UC, but not CD. The human cathelicidin seems to be important to combat infection by enteric pathogens as well. *Shigella* was found to suppress LL-37 expression during infection (Islam et al. 2001). In mice, the artificial induction of cathelicidin expression renders the animals less susceptible to infection by *Shigella* (Raqib *et al.* 2006). Genetically-modified mice lacking the murine cathelicidin (CRAMP) gene are more susceptible to oral challenge by the enteric pathogen *Citrobacter rodentium* (Iimura *et al.* 2005) and more susceptible in a mouse model of chemically-induced colitis by dextran sodium sulfate. The effects of the induced colitis are prevented by oral administration of a synthetic cathelicidin (Koon *et al.* 2011).

The vast number of AMPs with antimicrobial properties available in nature and their wide distribution among species is regarded as evidence of their importance in the fight against parasitic microorganisms. With the discovery of human defensins, the contribution of AMPs to gut homeostasis has become more evident in the past decade, and nowadays AMPs are considered an essential part of our innate immune system. Recently, shifts in microbiota composition have been linked to several chronic inflammatory diseases and intestinal cancer. Since AMPs are fundamental in the regulation of the human microbiota, their potential to control these diseases remains to be elucidated. AMPs also emerge as a new source of antimicrobial compounds capable of aiding in the fight against the rising threat of drug resistance, and several laboratories are already pursuing this line of investigation. These powerful molecules combine antimicrobial properties with immunomodulatory activities, which may prove useful in the development of new strategies for the control of various infectious diseases.

6.3.2. T Cells

The adaptive immune response is orchestrated by T cells, a class of lymphocytes that express a surface receptor called a T cell receptor (TCR). The T cells are activated by processed antigens bound to the major histocompatibility complex (MHC) receptor on the surface of antigen-presenting cells (APCs). The recognition of the MHC-bound antigen by the TCR takes place in secondary lymphoid tissue, such as the GALT, and induces the differentiation of the circulating naïve lymphocytes into one of the many classes of known T cells. Specialized T cells inhabit the lamina propria of the intestinal mucosa and are characterized by the cytotoxic response (CD8 T cells) or secretion of chemokines and cytokines that control the adaptive immune response (CD4 T cells) (Nijkamp and Parnham 2011).

The nature of the relationship between the adaptive immune system and the gut microbiota is delicate and unstable. In the gut, the microbiota produces a large number of antigens that may promote a strong T cell mediated response. These antigens are constantly presented to T cells by APCs, and in the gut epithelium the role of antigen presentation is taken not only by DCs, but also by enterocytes (Miron and Cristea 2012). In the face of the enormous challenge posed by the gut microbiota, our immune system must determine whether or not to elicit a response. A strong reaction is required to prevent infection by invading microorganisms, but a misdirected response towards dietary antigens or commensal bacteria may lead to inflammatory diseases. For this reason, our gut is said to be in a constant state of controlled inflammation, maintained by a balance in the population of T cells that is highly skewed towards tolerance. In this condition of equilibrium, the T cell population of the GALT is composed of T helper 1 (Th1), T helper 17 (Th17), and regulatory T cells (T_{reg}). In contrast, the immune system of GF mice is not fully mature, the GALT is not completely developed, and the numbers of $CD4^+$ T cells, epithelial T cells, DCs, and IgA producing B cells are diminished (Maynard and Weaver 2009). The Th1/Th2 balance is also atypical in these animals, with a predominant Th2 response (Smith *et al.* 2007), characteristic of individuals prone to developing allergic diseases.

The preponderance of Th1 cells in favor of Th2 does not set a standard for good health either; on the contrary, in individuals where the gut T cell population is dominated by Th1 cells, autoimmune inflammatory diseases are rampant. The maintenance of a harmonious balance between Th1 and Th2 cells is essential for an efficient and healthy gut. The maintenance of this balance is achieved through a tight control by T lymphocytes in the gut, the T_{reg} (Mazmanian *et al.* 2005). T_{reg} cells are a subset of T cells, mostly $CD4^+$, that control the immune response through direct cell-to-cell contact or the secretion of cytokines, such as TGF-β, IL-10, and IL-35.

The idea that T_{reg} $CD4^+$ cells may regulate the immune response was already considered in the 1970s (for a historical review see Sakaguchi *et al.* 2007)). The role of these cells in gut immunity was demonstrated in experiments where certain subpopulations of $CD4^+$ cells expressing CD45 were transferred into nude mice. In these experiments, a population of $CD4^+$ $CD45RB^{high}$ cells induced a wasting disease and colitis in mice, and the disease could be kept under control by a second population of $CD4^+$ $CD45RB^{low}$ cells, which were later shown to constitute the T_{reg} cells (Powrie *et al.* 1993). The best characterized T_{reg} cell subtype is the $CD25^+$ Foxp3 (forkhead family transcriptional factor 3) $CD4^+$ T cells. Expression of the transcriptional regulator Foxp3 is paramount in the development and function of T_{reg} cells. Among other functions, this T cell subtype is important in the prevention of autoimmune diseases by maintaining self tolerance, oral tolerance to dietary antigens, regulation of effector T cells, and suppression of the immune response against the gut microbiota (Corthay 2009).

The differentiation of T_{reg} cells is independent of the gut microbiota, since it also takes place in GF mice (Ostman et al. 2006). However, in these animals, fewer Foxp3$^+$ T_{reg} cells are present and their capacity to suppress inflammation is impaired. Some species of commensal bacteria that inhabit the human gut, such as *Bifidobacterium infantis*, are capable of restoring T_{reg} numbers and function in GF mice. Moreover, a single molecule purified from *Bacteroides fragilis*, a bacterium commonly found in the human gut, is able to induce Foxp3$^+$ T_{reg} cell function and restore the balance of Th1/Th2 cells. The capsular polysaccharide A (PSA) antigen is a zwitterionic molecule that is processed by antigen presenting cells (APCs) and presented via MHC class II to CD4$^+$ T lymphocytes. PSA also provides protection to induced colitis in mouse models through the suppression of Th17 cytokine responses (Mazmanian et al. 2008). The modulatory effect of T_{reg} cells over Th17 cells is largely set by the production of IL-10. In fact, spontaneous colitis is triggered by normal microbiota in IL-10 deficient mice (Sellon et al. 1998). In experimental infections, PSA purified from *B. fragilis* can induce the expression of IL-10 in T_{reg} cells and ameliorate the symptoms of colitis (Round and Mazmanian 2009).

T cells in the Peyer's patches or mesenteric lymph nodes are activated by DCs and migrate to the intestinal lamina propria. DCs extend their dendritic appendages to sample the intestinal lumen content directly or capture antigens internalized through M (microfold) cells. These cells can sense the surrounding environment with precision and signal to the cells in the adaptive immune system, so the fate of the naive T cells depends on the subset of DCs responsible for antigen presentation and the nature of the molecule presented (Kool et al. 2011). The population of DCs can be divided according to the expression of membrane markers; a subset of DCs expressing CXCR1 (CXC chemokine receptor 1) favors the differentiation of Th17 cells, whereas DCs expressing CD103 (αE integrin) secrete TGF-β and retinoic acid and induce the differentiation of T_{reg} cells (Farache et al. 2013).

The secretion of proinflammatory cytokines by Th17 cells is essential for an efficient response against pathogens and also for the management of the number of commensal bacteria in the gut. IL-17A, IL-17F and IL-22 provide several protective functions, such as AMP and mucin secretion and formation of tight junctions. In mice, IBD and experimental colitis are accompanied by an increase in IL-17 and IL-22 expression (Kamada and Núñez 2013). An increased susceptibility to *Citrobacter rodentium* is observed in mice defective for IL-22. The development of Th17 in mice is closely related to the type of microorganisms found in the microbiota of the animal. GF mice, for example, are practically devoid of Th17 cells, an effect that can be reversed by colonization with a normal microbiota. A group of as-yet uncultured bacteria known as segmented filamentous bacteria (SFB) was found to be essential for Th17 cell differentiation. DCs and macrophages seem to induce Th17 differentiation during inflammation (Gaboriau-Routhiau *et al.* 2009).

$CD8^+$ T cells constitute most of the population of the intestinal intraepithelial lymphocytes (IELs) and play an important role in both innate immunity and in the regulation of the gut adaptive immune response against infection. IELs are important in the maintenance and repair of the epithelial layer integrity, induction of oral tolerance and regulation of the immune responses (Cheroutre *et al.* 2011). Due to their proximity to the single layered intestinal epithelium, these cells also serve as the first line of defense against invading microorganisms, but when the activation of $CD8^+$ T cells is not properly regulated, this proximity poses a threat to the organism. These cells may promote tissue damage and recruit neutrophils in response to microbial and inflammatory signals, thus enhancing the damage caused in conditions like IBD and celiac disease (Hayday *et al.* 2001).

Even with the current advances in metagenomics that have led to the discovery of numerous microbial players in our gut microbiota, much remains to be discovered about the influence of these microorganisms on the host immune system and the regulation of T cell development and function. Since T cells are responsible for the government and coordination of the immune response, a deranged interaction of the microbiota with these immune cells often results in autoimmune disease and chronic inflammatory syndromes. Our understanding of gut immunology is currently evolving at a fast pace, and new T cell subsets are still being discovered. Microbial molecules with immunomodulatory activity are still largely unknown and are bound to cause major changes in the way we see the interaction of our immune system and our gut microbiota.

6.3.3. B Cells

One of the main functions of B cells is to produce antibodies to protect us from infections. These molecules have a variety of important functions, such as the opsonization of antigens, fixation of complement, and activation of cellular receptors. With such critical functions, it is not surprising that antibodies would be another important tool used by the human body to interact with the gut microbiota, in order to avoid microbial translocation and keep microbial loads in check. To perform these functions, an antibody must be able to traverse the intestinal mucosa, through active secretion. Indeed, a clear demonstration of the important functions elicited by antibodies to achieve balance at mucosal sites is the fact that almost every animal with an adaptive immune system is able to secrete these molecules across mucosal barriers (Sunyer 2013). In the recent past, much has been learned about the secretion of IgA into the intestinal lumen and its important role during interactions with the gut microbiota (Kamada and Núñez 2013; Pabst 2012).

In humans, IgA produced by plasma cells in the lamina propria is transported through the mucosal barrier in polymers, by the polymeric Ig receptor (Shimada *et al.* 1999). Once at the intestinal lumen, the receptor is cleaved, releasing polymeric IgA. The antibody can then interact with potential antigens, causing microbial agglutination and potentially neutralizing the deleterious effects of invading pathogens. These IgA-antigen complexes can also be captured by a specialized type of cells called microfold cells, or 'M' cells, for further processing (Corthésy 2007). This is an effective strategy to counteract potential pathogens; studies have shown that recognition by IgA in the intestinal lumen represents a selective disadvantage during competition with strains that are not recognized by the antibodies (Endt *et al.* 2010; Peterson *et al.* 2007).

Although B cells can exert important effects on the microbiota through the mechanisms mentioned above and many others, the opposite is also true, that is, the microbiota can also influence B cell activity. Many B cell functions are altered in GF animals, indicating that the microbiota is critical for the functioning of these cells (Smith *et al.* 2007). For instance, Cahenzli and others have shown that GF mice display abnormally-high systemic levels of IgE (Cahenzli *et al.* 2013b). This occurs early in life, when B cells undergo isotype switching to IgE, and colonization by a diverse microbiota is required to inhibit this phenomenon. The authors also showed that all other Ig isotypes (IgA,

IgG, IgM) were present at lower levels in GF mice, when compared to conventionally-raised animals. Attempts to rescue physiological levels of IgE production by colonizing GF animals with defined microbial mixtures (Schaedler flora) revealed that a microbiota with a certain degree of complexity is required for immunological health. When mice were given this set of eight different species (Dewhirst *et al.* 1999) IgE hyperproduction still occurred. To achieve physiological production of IgE, a low complexity microbiota consisting of 40 members was required (Cahenzli *et al.* 2013b).

The role of the microbiota on IgA production, uncovered by the GF studies mentioned above, is also evident when one observes the most natural model of the effects of the lack of microbiota on host physiology, that is, newborns. At birth, infants are devoid of polymeric IgA, and a certain development period after birth is required for the production of these immunoglobulins. The amount of time required is approximately 1 month, and coincides with the initial and critical phases of microbial colonization of the newborn body (Walker 2013). In an elegant study using inoculation with *Enterobacter cloacae* and other mouse commensals, Macpherson and Uhr (2004) showed that one of the mechanisms for the induction of IgA production relies on microbial sampling by DCs. Although macrophages rapidly kill microbiota cells that are engulfed, DCs can retain live cells for several days. After microbial sampling, DCs migrate to the mesenteric lymph nodes, where they can initiate antigen presentation that will culminate in antibody production. This is accomplished without activation of the systemic immune system, given that microbe-loaded DCs are confined to the mesenteric lymph nodes and do not travel to the spleen (Macpherson and Uhr 2004). By doing so, the immune system is able to interact with the resident microbiota without activating an overt immune response.

The aforementioned effects of microbial colonization on IgA production also result in an effective mechanism to control overgrowth of the intestinal microbiota. Shroff and coauthors (1995) have shown that gut colonization with the murine commensal *Morganella morganii* results in hypertrophy of germinal center reactions in Peyer's patches and the development of specific IgA responses. These immune reactions seem important for the control of microbiota translocation. In their study, Shroff and others found that *M. morganii* initially translocates to the mesenteric lymph nodes and also the spleen, but the specific IgA response generated gradually reduces the microbial translocation, indicating that the induction of IgA by the organism ultimately results in the control of its growth and translocation. Interestingly, after an initial period of induction, germinal center reactions begin to subside, indicating that some sort of tolerance was achieved. However, this does not mean that antibodies specific to the microbiota member originally sensed are no longer being produced; specific-IgA-producing cells can still be detected 314 days after the initial colonization, whereas germinal center reactions peak at 14 days post-inoculation (Shroff *et al.* 1995).

The effect of microbial gut colonization on B cell function is not specific to a certain microbe or microbial group. Besides the examples of microbes that can affect B cell physiology mentioned above, others have also been shown to modulate B cells. Perhaps one of the most potent microbial modulators of B cell function is the segmented filamentous bacteria (SFB), a group of as-yet-uncultured bacteria with a remarkable impact on the host (Ivanov and Littman 2010). Monocolonization of GF mice by SFB stimulates germinal center reactions and the production of IgA; remarkably, colonization with this single microbe can induce IgA production to levels equivalent of more than 60% of the IgA production elicited by a conventionally-raised animal (Talham *et al.* 1999). Although colonization with SFB proved to be an important activator of the host's immune responses, this activation did not compromise colonization by other members of the microbiota. In this study, colonization with SFB not only did not prevent colonization by a second member of the microbiota, *M. morganii* in this case, but also did not prevent translocation of this organism to the mesenteric lymph nodes and spleen. Additionally, secondary colonization with *M. morganii* was able to stimulate germinal center reactions and the production of specific IgA, despite the previous stimulation elicited by SFB. These observations highlight the remarkable plasticity of the mammalian adaptive immunity, which is capable of recognizing and responding to diverse antigenic stimuli in independent and specific ways, and allowing the formation and maintenance of a complex microbiota without excessive microbial growth or immune stimulation.

6.4. SMALL MOLECULES, THE MICROBIOME, AND THE IMMUNE SYSTEM

Although the microbial composition of feces among healthy subjects is diverse, metabolic capacity, as determined by metabolic gene composition, is stable, suggesting that the metabolic output is more critical to microbiota function than microbial composition (Consortium 2012).

Therefore, it is likely that the mammalian immune system would have evolved mechanisms to sense key molecules produced by genetically diverse groups of microorganisms. This would represent an advantage over the sensing of specific microbial structural molecules, which could vary widely depending on the microbiota composition of specific individuals. To this end, small metabolites may be attractive molecules from both the host and microbial points of view. From the host's perspective, these molecules could be used as general signatures of microbiota metabolic activity, and function as general immune targets. From the microbial perspective, such molecules could be used as general signals to modulate host function in order to maintain the equilibrium of diverse microbial ecosystems.

Many small metabolites have been shown to affect host immune function, and there are a few ways through which microbes can modulate host immunity using these molecules. First, members of the microbiota can directly produce small molecules that will act on the host immune system. Another possibility is the transformation of host-produced molecules, either increasing or decreasing their immune reactivity. The last possibility is through the modulation of host pathways involved in the synthesis of immunoregulatory molecules. In this case, microbes themselves do not produce the molecules. Instead, microbes are able to control host immunity through the induction or repression of metabolic pathways involved in the synthesis of immunomodulatory molecules. In some cases, the mechanisms involved in these three strategies have been elucidated and a few examples of these are discussed below.

6.4.1. Short-Chain Fatty Acids

Perhaps the most studied case of microbial-derived small molecules with effects on host immunity is that of short-chain fatty acids (SCFAs) and their immunoregulatory properties. Many complex, plant-derived, carbohydrates ingested with food are indigestible to humans. However, the gut microbiota has the metabolic capacity to digest these molecules into SCFAs, a series of biologically-active metabolites (Figure 6.3). These include acetate, propionate, and butyrate (Cummings 1983). Before discussing the role of these molecules as immune regulators, it is important to mention that SCFAs act as an important source of energy for the intestinal host cells. Colonic epithelial cells of GF mice show signs of nutrient deprivation, such as the activation of autophagy pathways. Colonization of GF animals with *Butyrivibrio fibrisolvens*, a butyrate-producing microbe can rescue these effects, showing that microbiota-derived butyrate production is a major nutrient source for host cells in the mammalian gut (Donohoe *et al.* 2011). Confirming the important role of commensal gut microbes in modulating the host's metabolic capacity, Cho and others (2012) have shown, by altering the composition of the gut microbiota through the administration of low doses of antibiotics, that the microbial composition of the gut has a significant effect on host adiposity.

As mentioned above, besides acting as nutrients, SCFAs can also act as specific signals, activating several host G-protein coupled receptors, such as GPR41, GPR43, and GPR109A (encoded by the *Ffar3*, *Ffar2*, and *Niacr1* genes, respectively) (Brown *et al.* 2003; Thangaraju *et al.* 2009). The interactions between SCFAs and these host receptors have important consequences for host immunity. Maslowski and others have used the administration of dextran sodium sulfate in drinking water to show that GF mice are more susceptible to colitis than their conventionally-raised counterparts (Maslowski *et al.* 2009). This confirmed

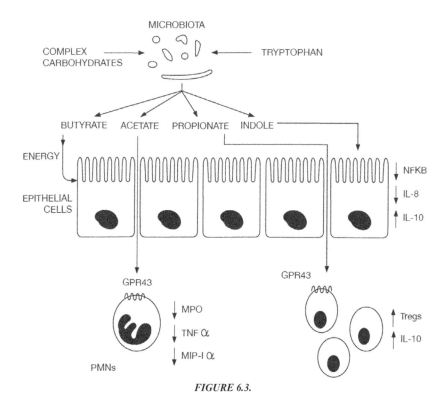

FIGURE 6.3.

that the intestinal microbiota is a critical factor in controlling immune function in the gut.

Because GF animals do not produce SCFAs, due to the lack of microbiota, the authors set out to test if the administration of these compounds in drinking water would have an effect on the observed colitis. Indeed, administration of acetate in drinking water resulted in a significant improvement of colitis parameters in the animals; colon length was increased in the treated animals, and the production of the inflammatory mediator myeloperoxidase was decreased, together with the levels of tumor necrosis factor α (TNF-α) and macrophage inflammatory protein 1α (MIP-1α). Because acetate had been previously shown to bind GPR43, the authors postulated that the observed effects could be determined by this interaction (Brown et al. 2003). In fact, gene expression analyses showed that GPR43 expression was increased in neutrophils and eosinophils, potentially linking the activity of this receptor with immune function (Maslowski et al. 2009). Furthermore, patterns of expression of GPR43 correlated with the expression of many other immune receptors such as the Toll-like receptors (TLRs) 2 and 4, as well as formyl peptide receptors (FPRs) 1 and 2, among others. In order to definitively determine the role of GPR43 in the regulation of immune activity in the gut in response to SCFAs, the authors tested the effect of acetate administration in drinking water in $Gpr43^{-/-}$ mice. As expected, acetate did not protect $Gpr43^{-/-}$ mice from the colitis elicited by dextran sodium sulfate. Although this confirmed that acetate sensing through GPR43 controlled gut inflammation, due to the other effects SCFAs can have on enterocytes and other host cells, the authors sought direct evidence that immune cells were involved in this phenomenon. Thus, bone marrow chimeras were created; bone marrow cells from $Gpr43^{-/-}$ were introduced in both wildtype and $Gpr43^{-/-}$ mice, and dextran sodium sulfate-induced colitis was evaluated. As expected, wildtype animals reconstituted with $Gpr43^{-/-}$ bone marrow cells were highly susceptible to colitis, to a similar extent as $Gpr43^{-/-}$ animals reconstituted with $Gpr43^{-/-}$ cells. Altogether, these data showed that SCFA sensing through the GPR43 receptor in immune cells is responsible for the impact of the microbiota on gut immune function.

Another example of the impact SCFAs have on immune function involves the expansion and function of T_{reg} cells. As previously mentioned, colonic Foxp3$^+$ T_{reg} cells are critical for the maintenance of immune homeostasis in the gut. Because GF animals have both lower levels of SCFAs in the gut as well as reduced numbers of colonic T_{reg} cells, Smith and others (2013) hypothesized that the decreased levels of SCFAs may contribute to the immune defects observed in GF animals, the authors provided propionate, acetate, and butyrate, or a mixture of these three SCFAs, to GF mice in drinking water and found that these SCFAs, individually or as a mixture, increased the frequency and numbers of colonic T_{reg} cells, without significantly affecting the numbers of Th1 and Th17 cells. Besides affecting T_{reg} numbers, the authors showed that SCFAs also altered T_{reg} function. Propionate stimulation of T_{reg} cells, both *in vivo* and *in vitro*, resulted in higher expression levels of interleukin 10 (IL-10), a cytokine critical for the anti-inflammatory T_{reg} function. The authors also showed the involvement of GPR43 on this phenomenon by treating $Gpr43^{+/+}$ and $Gpr43^{-/-}$ mice with propionate and monitoring the effects on colonic T_{reg} numbers. Propionate treatment resulted in increased number and frequency of colonic T_{reg} cells in $Gpr43^{+/+}$ but not in $Gpr43^{-/-}$ animals. Additionally, the upregulation of IL-10 production elicited by propionate was only seen in $Gpr43^{+/+}$ cells, confirming that the effect of SCFAs on T_{reg} function works through GPR43 signaling. Lastly, the authors used a T cell transfer model of colitis to further demonstrate the effect of SCFAs on T_{reg} cell function. In these experiments, $Rag2^{-/-}$ animals, which lack mature T and B lymphocytes (Shinkai *et al.* 1992), were injected with $CD4^+$ naïve T cells alone or together with T_{reg} cells. These animals were then treated with water, propionate or a SCFA mixture. Both propionate and the SCFA mixture attenuated the severity of colitis in animals that received a combination of naïve T cells and T_{reg} cells but failed to prevent colitis in animals that had received only naïve T cells. Once again, the phenotype elicited by SCFAs was dependent upon GPR43; experiments involving the transfer of $Gpr43^{-/-}$ T_{reg} cells to $Rag2^{-/-}$ animals resulted in the absence of an effect of propionate on intestinal inflammation (Smith *et al.* 2013).

The studies described above, together with many others, have undoubtedly established that SCFAs can have significant effects on host immunity in a laboratory setting, that is, using *in vitro* systems and laboratory animals.

In order to assess the significance of some of these findings for host-microbiota interactions in a natural setting it was critical that studies using human samples were performed. Indeed, supporting the notion that SCFAs have important immunomodulatory activities stemming from the aforementioned studies, Frank and colleagues

(2007) have shown that butyrate-producing bacteria are decreased in the microbiota of patients with IBD, is a chronic, relapsing condition characterized by extensive inflammation in the lining of the digestive tract. Although the etiology of IBD is complex and not completely understood, it is known to involve an excessive reaction of the host's immune system to antigens provided by the gut microbiota (Fakhoury *et al.* 2014). The work by Frank and colleagues (2007) hints to a potential role of SCFA production in controlling the development of this disease. However, it is important to point out that it is still unknown whether the decrease in butyrate-producing microbes in IBD patients is a cause or consequence of the intestinal inflammation. In any case, the observation that butyrate may be involved in the control of an excessive immune reaction in the gut unraveled a potential application of this SCFA in the control of the symptoms of this debilitating condition. To this end, studies have shown that treatment of patients with IBD with butyrate can in fact ameliorate disease (Di Sabatino *et al.* 2005; Hallert *et al.* 2003). Therefore, SCFAs can not only act as critical signals during interactions between the host immune system and the gut microbiota but also be mined for potential therapeutic properties.

6.4.2. Tryptophan Metabolites

Another group of microbial-derived molecules that have been implicated in the regulation of host immune function is that of metabolic byproducts of tryptophan (Figure 6.3). One such metabolite is indole. Indole has been shown to function as an interkingdom signaling molecule produced by commensal *Escherichia coli* (Lee *et al.* 2007), and can be detected in human feces (Karlin *et al.* 1985). As with the other molecules cited above, indole has been added to the growing list of microbial metabolites that have important effects on host immunity. For instance, by investigating the effect of indole exposure on gene expression of a human enterocyte cell line, Bansal and others (2010) have shown that this microbial metabolite elicits an anti-inflammatory effect on host cells. Indole exposure resulted in a decrease in TNF-α-mediated NFκB activation and expression of the pro-inflammatory chemokine IL-8. Additionally, treatment with indole caused an increase in the expression of the anti-inflammatory cytokine IL-10.

Besides indole, other tryptophan metabolites, such as indole acetic acid and tryptamine, can also have effects on host immunity. These three compounds are ligands for the aryl hydrocarbon receptor (AhR) (Heath-Pagliuso *et al.* 1998), a transcriptional factor that controls multiple facets of host immunity. For example, AhR activation has been shown to control T_{reg} cell development (Fallarino *et al.* 2006). Additionally, AhR activation provides colonization resistance against pathogens through the production of IL-22 and the resulting balance in mucosal inflammation (Zelante *et al.* 2013). Lastly, AhR signaling has been associated with increased expression of the AMPs RegIIIβ and RegIIIγ, and has been shown to inhibit chemically-induced colitis (Fukumoto *et al.* 2014). Therefore, the aforementioned indole derivatives may act as important microbiota-derived signals that modulate the host's immune system.

6.4.3. Bile Acids

Bile acids are another class of molecules with a well-established link to the gut microbiota as well as the immune system. They are produced in the liver from the metabolism of cholesterol, are stored in the gallbladder and released in the small intestine upon feeding. These molecules have detergent-like properties that aid in the digestion of fatty foods. Liver-produced bile acids are called primary bile acids, and these can be transformed by the gut microbiota when released in the intestinal lumen, generating the secondary bile acids (Dawson and Karpen 2014). Besides aiding in digestion, bile acids have important signaling functions. They can bind to and modulate the function of several nuclear receptors, such as the farnesoid X receptor (FXR), the pregnane X receptor (PXR), and the vitamin D receptor (VDR) (Makishima *et al.* 1999, 2002; Staudinger *et al.* 2001). These regulators, in turn, play important roles in the control of several physiological parameters, including bile acid synthesis and excretion, as well as lipid and sugar metabolism, among others (Chiang 2009).

Besides the aforementioned roles of bile acids, and relevant to this article, bile acid signaling controls many facets of the host's immune response. For instance, it has been shown that feeding mice an atherogenic diet containing a significant amount of a bile acid induces the expression of inflammatory genes (Liao *et al.* 1993; Miyake *et al.* 2000). Given the capacity of the gut microbiota to transform these molecules, it was reasonable to predict that gut microbiota activity would impact on the immune control elicited by bile acids. Indeed, in a study comparing IBD patients and healthy controls, Duboc and others (2013) have shown that although total bile acid concentrations were comparable, concentrations of primary bile acids were higher in IBD patients, while concentrations of secondary bile acids were lower. This suggested that the microbial transformation of bile acids was reduced in IBD patients. Indeed, the gut microbiota of patients with active IBD showed a reduced capacity for deconjugation, transformation, and desulfation of bile acids. Using a Caco-2 intestinal epithelial cell culture model, the authors showed that primary bile acids had no effect on IL-1β-induced IL-8 secretion, while secondary bile acids inhibited secretion. Therefore, the reduced capacity of the IBD gut microbiota to produce secondary bile acids may be associated with a reduced ability to control gut inflammation. Interestingly, sulfation of a secondary bile acid by the host can abolish the anti-inflammatory effect, further supporting the notion that modification (desulfation in this case) of bile acids by the gut microbiota is important for the control of gut inflammation (Duboc *et al.* 2013).

6.4.4. Steroids and Eicosanoids

So far, the activity of metabolites that are either produced directly by microbes or produced by host cells and modified by microbes has been discussed. However, as mentioned above, a third possible mechanism for the modulation of host immunity by microbes relies on the regulation of host pathways by the host microbiota. Recently, Antunes and colleagues (2011) have addressed this topic by using a high-throughput metabolomics approach to determine the impact of the gut microbiota on the chemical composition of feces. In this study, mice were treated with a high dose of streptomycin, which resulted in the elimination of approximately 95% of their gut microbiota, and small molecules from their feces were extracted with an organic solvent and analyzed through Fourier Transform Ion Cyclotron Resonance Mass Spectrometry. As a result, the authors detected over 2,000 metabolites, and determined that most of these were affected by the antibiotic treatment and, therefore, are potentially modulated by the intestinal microbiota. Of interest, two of the most overrepresented classes of molecules affected by antibiotic treatment were the steroids and eicosanoids. These are important lipid mediators, with critical functions for the host's immune system, among others (Butts and Sternberg 2008; Calder 2009).

The first class of mammalian hormones affected by the gut microbiota, the steroids, is widely known for its immunoregulatory activities (Busillo and Cidlowski 2013; Cruz-Topete and Cidlowski 2014).

For instance, glucocorticoids have been shown to attenuate the activation of TLR4 by LPS, reducing the production of TNF-α and IL-6 in response to this potent immune stimulator (Bhattacharyya et al. 2007). The authors also showed that this effect was due to the inhibition of p38 MAPK activation by LPS. Additionally, Tuckermann and others (2007) showed that glucocorticoids inhibit the production of IL-1β, monocyte chemoattractant protein 1 (MCP-1), MIP-2, and IFN-γ-inducible protein 10 (IP-10) by LPS-stimulated bone marrow derived macrophages. The second class of hormones identified by Antunes and colleagues as being affected by the microbiota, the eicosanoids, also has critical roles in the regulation of immunity (Calder 2009; Funk 2001). Buckner et al. (2013) have recently shown that 15-deoxy-$\Delta^{12,14}$-prostaglandin J_2 exerts a significant anti-inflammatory activity during *Salmonella* infection of cultured macrophages, inhibiting the production of TNF-α, MCP-1, IL-6, and IL-10 and altering the progression of the infection process. Conversely, Xue et al. (2005) have shown that prostaglandin D_2 exerts a proinflammatory activity, by inducing the production of IL-4, IL-5, and IL-13 by human Th2 cells. Not surprisingly, and due to their immunoregulatory activities, many agonists and antagonists of steroid and eicosanoid metabolism are commonly prescribed as anti-inflammatories. That antibiotic treatment affects the production of these molecules suggests that the microbiota is intricately connected with host endocrine systems, opening an avenue of investigation into novel, thus far unrecognized processes of host-microbial interactions.

6.5. THE IMPACT OF OTHER MICROBIOMES ON HOST IMMUNITY

Although a great deal of knowledge about the impact of the gut microbiota on the host immune system has been acquired in the recent past, still very little is known about the effect of microbial communities from other sites of the body, such as the skin, mouth, vagina, and respiratory tract, on the host's immune responses.

The microbiota of the skin is highly diverse, with distinct communities on each site (Costello et al. 2009; Grice et al. 2009). Recently, cutaneous inflammatory disorders such as psoriasis, atopic dermatitis, and rosacea have been associated with a shift in the skin microbiota composition (Gallo and Nakatsuji 2011; Kong et al. 2012). In atopic dermatitis, for example, patients with mutations in the gene encoding filaggrin

(FLG) display a marked susceptibility to the disease due to the lack of hydration and higher pH of the skin. Furthermore, the expression of the AMP LL-37 is significantly lower in the lesions of individuals suffering from atopic dermatitis (Mallbris *et al.* 2010), although the same is not true for healthy skin areas of the same individuals. Conversely, the expression of other AMPs (RNase 7, psoriasin, and human beta defensin-2) is up-regulated in the skin of both atopic dermatitis and psoriasis patients when compared to healthy skin from affected individuals as well as healthy controls (Harder *et al.* 2010). Such changes in the skin environment are likely to cause shifts in the microbial composition; for instance, an increase in *Staphylococcus aureus* colonization at affected skin sites may occur. Although Harder and colleagues did not find a direct correlation between AMP production and *S. aureus* colonization at affected sites, it is known that *S. aureus* colonizes the skin lesions of more than 90% of atopic dermatitis patients (Cho *et al.* 2001). An imbalance in the skin microbiota caused by alterations in AMP production may indirectly result in increased colonization or perhaps increased virulence by *S. aureus* strains thriving in skin lesions of atopic dermatitis and psoriasis patients. It is thought that the ability of S. *aureus* to produce toxins, such as δ-toxin, results in marked activation of T cells and other immune cells, supporting the existence of a link between skin microbiota composition and local immune responses (Nakamizo *et al.* 2014).

Lipotheicoic acid from *Staphylococcus epidermidis*, one of the major species to colonize the skin, has been shown to inhibit uncontrolled skin inflammation during skin injury, by modulating the immune response (Lai *et al.* 2009). Staphylococcal lipotheicoic acid acts selectively on keratinocytes through Toll-like receptor (TLR) 2 and inhibits TLR3 signaling. After skin injury, TLR3 in keratinocytes is activated by host RNA from damaged cells, resulting in the release of inflammatory cytokines and, consequently, inflammation. Therefore, these findings reveal that a defined compound from one of the members of the skin microbiota is able to inhibit both inflammatory cytokine release from keratinocytes and inflammation triggered by injury through a TLR2-dependent mechanism.

The commensals present on the skin have also been shown to induce Th17 and Th1 differentiation to protect the host from pathogens (Naik *et al.* 2012). Similar to what happens in the intestine of GF compared to conventionally-raised mice, the skin of GF animals displays a significant reduction in IFN-γ and IL-17A production (Naik *et al.* 2012), indicating that the production of these cytokines is dependent on the presence of the skin microbiota. The higher prevalence of cutaneous Foxp3$^+$ T_{reg} cells in GF compared to conventionally-raised mice also indicates the role of the skin microbiota on the development of the immune responses (Naik *et al.* 2012).

The presence of a stable commensal microbiota in the airways was recently described (Hilty *et al.* 2010; Twigg *et al.* 2013). Gollwitzer and colleagues (2014) showed that colonization of the airways with the resident microbiota increases and changes soon after birth and continues with age. The authors also showed that the formation of the microbiota is essential for the maturation process of the immune system of the lungs during the first 2 weeks of life. The development of the airway microbiota was linked to a decreased responsiveness to house dust mite allergens and induction of regulatory cells early in life, which could prevent allergic airway inflammation in adulthood.

6.6. THE IMPACT OF THE MICROBIOTA ON IMMUNE DISORDERS

In the Western world, the prevalence of autoimmune and allergic immune diseases, such as IBD, diabetes and obesity, is rapidly increasing (Cahenzli *et al.* 2013a). Host genetics may play a role on the development of such disorders but it cannot be used to explain the increase in incidence on its own. The role of the microbiota on the development of these diseases and their rising incidence over the past decades has just started to be studied. As such, a growing number of questions still need to be addressed. The hygiene hypothesis argues that increased sanitation in industrialized countries led to decreased infections with common pathogens and a concomitant rise in allergic disorders (Strachan 1989). Actions to increase basic health in Western countries, such as food pasteurization and sterilization, vaccination, as well as the widespread use of antibiotics, had the unwilling side effect of shifting our microbiota composition (Walter and Ley 2011). These changes might have caused an imbalance between the immune system and commensal microbes, leading to an increase on the incidence of immune-related disorders. Here, we describe some examples of the data available linking the microbiota to the development of some of the most prevalent immune disorders.

6.6.1. Type 1 Diabetes

Type 1 diabetes is a disorder where the pancreatic insulin-producing beta cells are destroyed by the immune system of genetically-predisposed individuals, resulting in an inability to regulate blood sugar levels (Atkinson *et al.* 2014). Different factors may contribute to the disease; however, the specific environmental factors responsible for the autoimmune destruction of insulin-producing beta cells are unknown (Dunne *et al.* 2014). Different animal models for type 1 diabetes have been used to address the role of the microbiota on disease status. Interestingly, the incidence of diabetes on the nonobese diabetic mouse and the biobreeding diabetes-prone rat, which develop a disease similar to type 1 diabetes, correlates with the microbial exposure status of the animal. Antibiotic treatment of these rats decreased the risk of developing diabetes (Brugman *et al.* 2006; Schwartz *et al.* 2007). Further studies revealed that animals raised under conventional conditions present diabetes at a lower rate than animals hosted under specific pathogen-free conditions (Bach 2002; Cahenzli *et al.* 2013a). In another model, nonobese diabetic mice deficient for the adaptor molecule for innate immune receptors, MyD88 (myeloid differentiation primary response gene 88), did not develop diabetes, whereas their GF counterparts showed a high incidence rate for the disease. The same mice raised under specific-pathogen free conditions did not develop the disease, suggesting that the microbiota protected those animals from diabetes independently of their MyD88 status (Wen *et al.* 2008). *Lactobacillus johnsonii* colonization was associated with resistance to diabetes in biobreeding diabetes-prone rats, whereas *Lactobacillus reuteri* did not affect disease development (Valladares *et al.* 2010). This resistance was correlated to a sustained Th17 cell differentiation (Lau *et al.* 2011).

Clinical studies have revealed a trend where colonization with *Bacteroides* is correlated with type 1 diabetes development whereas a higher rate of Firmicutes is correlated with resistance to the disease (Giongo *et al.* 2011). The gut microbiota was also much less diverse in individuals that would develop type 1 diabetes compared to controls. Furthermore, commensal bacteria that produce butyrate as a fermentation product have been shown to protect against the disease, whereas other SCFAs, such as propionate, acetate, and succinate, have the opposite effect (Brown *et al.* 2011).

6.6.2. Asthma and Allergies

The microbiota has also been implicated in the development of atopic diseases, such as asthma and allergies. For example, exposure of children growing up on farms to a vast range of microbes was correlated with a better protection against childhood asthma and other atopic diseases (Ege *et al.* 2011). The presence of the microbiota was also linked to the reduction of symptoms in a model of ovalbumin-induced asthma. GF mice showed a higher number of infiltrating lymphocytes and eosinophils, with more pronounced secretion of Th2 cytokines than mice raised under specific-pathogen free conditions, demonstrating that the commensals are critical for controlling allergic airway inflammation (Herbst *et al.* 2011). More recently, another study using an ovalbumin-induced asthma model showed that perinatal antibiotic treatment, which disrupts the intestinal microbiota, causes an increase in serum IgE levels and a decrease in the number of T_{reg} cells in the colon, indicating that shifts in the microbiota composition early in life might aggravate asthma-related immune responses (Russell *et al.* 2012). A lower diversity of the gut microbiota in early life was also linked to increased mast cell surface-bound IgE and augmented systemic anaphylaxis (Cahenzli *et al.* 2013b), indicating that a significant microbiota exposure after birth is essential to control IgE induction at mucosal sites. In another study, using a peanut allergy model, allergen-specific IgE, elevated plasma histamine levels, and anaphylactic symptoms were induced in three different strains of mice lacking a functional receptor for bacterial LPS (TLR4) (Bashir *et al.* 2004). The same symptoms were observed when wildtype mice were treated with antibiotics to reduce their microbiota. Allergen-specific responses were reduced when the commensals were allowed to repopulate, suggesting that the intestinal microbiota may act through TLR4 to inhibit the development of allergic responses to food allergens (Bashir *et al.* 2004).

6.6.3. Inflammatory Bowel Disease

IBD, such as CD and UC, are chronic disorders caused by genetic, immunological, and environmental factors. Hundreds of alleles have been associated with IBD, such as MUC19 (intestine barrier function), NOD2 (bacterial sensing), CARD9 (mucosal defense), IL23R (mucosal T cell responses), CCL8, and IL8R (innate cell recruitment) (Arrieta *et al.* 2014). In addition to these factors, interactions with the gut microbiota have been implicated as contributors for the development of the disease (Leone *et al.* 2013). IBD is believed to result from a mucosal inappropriate response to the intestinal microbiota in genetically-susceptible individuals (Kaser et al. 2010).

Studies with animal models have supported the hypothesis that the microbiota is key to the development of IBD. In a model of spontaneous colitis, IL-10-deficient mice only develop colitis if they are colonized by commensal bacteria (GF animals do not), indicating that bacterial presence and immune activation are essential for the induction of colitis (Kim et al. 2005; Sellon et al. 1998). Recently, an adherent-invasive strain of *Escherichia coli* (LF82), which colonizes the small intestine of CD patients, was shown to increase the production of proinflammatory cytokines by itself in a small bowel CD model. The inflammatory response was reverted by lactoferrin, without affecting adherence of the bacteria (Bertuccini et al. 2014).

In patients suffering from IBD, the normal immune homeostasis is disturbed, leading to intestinal inflammation (Strober 2013). Both UC and CD patients show lower microbial diversity in their intestines compared to controls, with more adherent bacteria showing up in their biopsy samples. CD patients repeatedly show lower diversity, especially on the Firmicutes phylum (Frank et al. 2007; Manichanh et al. 2006). One of the species within this phylum to show lower numbers in IBD patients was *Faecalibacterium prausnitzii* (Cao et al. 2014). This species was shown to have immunomodulatory effects both *in vivo* as well as in a murine colitis model, ameliorating disease in those mice (Sokol et al. 2008). Recently, the ability of *F. prausnitzii* to induce human T_{reg} cells and anti-inflammatory cytokines *in vitro* (Qiu et al. 2013) and to improve intestinal barrier function in a colitis model (Carlsson et al. 2013) was demonstrated.

The colons of UC patients display a thinner and less sulfated mucus layer, which may account for an overgrowth of bacteria in the colon mucosa, enhancing the presentation of bacterial antigens to the immune system (Corfield et al. 1996; Pullan et al. 1994). In UC patients, the frequent colonization by *Clostridium histolyticum*, *Clostridium lituseburense*, *Escherichia*, and *Klebsiella* have been implicated in the development of the disease, supposedly due to their ability to adhere to enterocytes and penetrate the mucus layer (Kleessen et al. 2002; Kotlowski et al. 2007).

Data from clinical studies further suggests a role of the gut microbiota in IBD. Prevention and treatment of CD and pouchitis have been performed by fecal stream diversion as well as the use of antibiotics and probiotics, which have been shown to improve IBD progression (Sartor 2008).

6.7. CONCLUSION

The interactions between the mammalian immune system and the resident microbiota, are binary; the mammalian immune system controls the composition and growth of the microbiota, and the microbiota, in turn, controls many aspects of the development of the host's immune system. Although we have presented many examples of interactions between the microbiota and the host, it is important to mention that this chapter is not comprehensive. It does not present all cases where immune functions have been shown to be affected by the microbiota. Additionally, although our understanding of the mammalian immune system has increased dramatically in the last few decades, it is safe to assume that it is still far from complete. As such, it is likely that many other aspects of the host immune system that are still to be discovered will also be affected by the resident microbiota. For instance, new immune cell types continue to be discovered (Bandala-Sanchez *et al.* 2013; Gardner *et al.* 2008; Neill *et al.* 2010; Rauch *et al.* 2012), and it will be interesting to investigate if these are also affected by the commensal microbes they interact with. Our knowledge of the microbiota and its importance for human health has increased exponentially in the last few decades, and there has been an explosion in the number of articles dealing with the microbiota in the last 15 years or so. To illustrate the increasing interest in the interactions between the microbiota and the immune system, we have calculated the percentage of articles published in the last 20 years that deal with

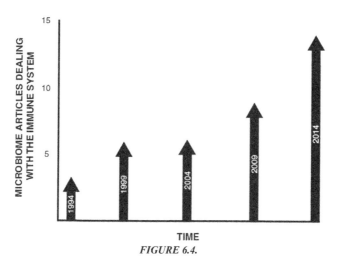

FIGURE 6.4.

the microbiota and also mention the immune system (Figure 6.4). Although some of these may not deal with host immunity directly, this illustrates the intrinsic connection between discoveries related to the microbiota and the immune system. As such, we should look forward to many more exciting discoveries in this field of research for years to come.

6.8. REFERENCES

Antunes, L.C., Han, J., Ferreira, R.B., Lolić, P., Borchers, C.H., and Finlay, B.B. 2011. Effect of antibiotic treatment on the intestinal metabolome. *Antimicrob Agents Chemother 55*, 1494–1503.

Arrieta, M.C., Stiemsma, L.T., Amenyogbe, N., Brown, E.M., and Finlay, B. 2014. The intestinal microbiome in early life: Health and disease. *Front Immunol 5*, 427.

Atkinson, M.A., Eisenbarth, G.S., and Michels, A.W. 2014. Type 1 diabetes. *Lancet 383*, 69–82.

Ayabe, T., Satchell, D.P., Wilson, C.L., Parks, W.C., Selsted, M.E., and Ouellette, A.J. 2000. Secretion of microbicidal alpha-defensins by intestinal Paneth cells in response to bacteria. *Nat Immunol 1*, 113–118.

Bach, J.F. 2002. The effect of infections on susceptibility to autoimmune and allergic diseases. *N Engl J Med 347*, 911–920.

Bandala-Sanchez, E., Zhang, Y., Reinwald, S., Dromey, J.A., Lee, B.H., Qian, J., Böhmer, R.M., and Harrison, L.C. 2013. T cell regulation mediated by interaction of soluble CD52 with the inhibitory receptor Siglec-10. *Nat Immunol 14*, 741–748.

Bandeira, A., Mota-Santos, T., Itohara, S., Degermann, S., Heusser, C., Tonegawa, S., and Coutinho, A. 1990. Localization of gamma/delta T cells to the intestinal epithelium is independent of normal microbial colonization. *J Exp Med 172*, 239–244.

Bansal, T., Alaniz, R.C., Wood, T.K., and Jayaraman, A. 2010. The bacterial signal indole increases epithelial-cell tight-junction resistance and attenuates indicators of inflammation. *Proc Natl Acad Sci USA 107*, 228–233.

Bashir, M.E., Louie, S., Shi, H.N., and Nagler-Anderson, C. 2004. Toll-like receptor 4 signaling by intestinal microbes influences susceptibility to food allergy. *J Immunol 172*, 6978–6987.

Benveniste, J., Lespinats, G., Adam, C., and Salomon, J.C. 1971. Immunoglobulins in intact, immunized, and contaminated axenic mice: Study of serum IgA. *J Immunol 107*, 1647–1655.

Bertuccini, L., Costanzo, M., Iosi, F., Tinari, A., Terruzzi, F., Stronati, L., *et al.* 2014. Lactoferrin prevents invasion and inflammatory response following *E. coli* strain LF82 infection in experimental model of Crohn's disease. *Dig Liver Dis 46*, 496–504.

Bevins, C.L. and Salzman, N.H. 2011. Paneth cells, antimicrobial peptides and maintenance of intestinal homeostasis. *Nat Rev Microbiol 9*, 356–368.

Bhattacharyya, S., Brown, D.E., Brewer, J.A., Vogt, S.K., and Muglia, L.J. 2007. Macrophage glucocorticoid receptors regulate Toll-like receptor 4-mediated inflammatory responses by selective inhibition of p38 MAP kinase. *Blood 109*, 4313–4319.

Bouskra, D., Brézillon, C., Bérard, M., Werts, C., Varona, R., Boneca, I.G., and Eberl, G. 2008. Lymphoid tissue genesis induced by commensals through NOD1 regulates intestinal homeostasis. *Nature 456*, 507–510.

Bowdish, D.M., Davidson, D.J., and Hancock, R.E. 2006. Immunomodulatory properties of defensins and cathelicidins. *Curr Top Microbiol Immunol 306*, 27–66.

Brandtzaeg, P., Halstensen, T.S., Kett, K., Krajci, P., Kvale, D., Rognum, T.O., Scott, H., and Sollid, L.M. 1989. Immunobiology and immunopathology of human gut mucosa: Humoral immunity and intraepithelial lymphocytes. *Gastroenterology 97*, 1562–1584.

Brown, A.J., Goldsworthy, S.M., Barnes, A.A., et al. 2003. The Orphan G protein-coupled receptors GPR41 and GPR43 are activated by propionate and other short chain carboxylic acids. *J Biol Chem 278*, 11312–11319.

Brown, C.T., Davis-Richardson, A.G., Giongo, A., et al. 2011. Gut microbiome metagenomics analysis suggests a functional model for the development of autoimmunity for type 1 diabetes. *PLoS One 6*, e25792.

Brugman, S., Klatter, F.A., Visser, J.T., Wildeboer-Veloo, A.C., Harmsen, H.J., Rozing, J., and Bos, N.A. 2006. Antibiotic treatment partially protects against type 1 diabetes in the Bio-Breeding diabetes-prone rat. Is the gut flora involved in the development of type 1 diabetes? *Diabetologia 49*, 2105–2108.

Buckner, M.M., Antunes, L.C., Gill, N., Russell, S.L., Shames, S.R., and Finlay, B.B. 2013. 15-Deoxy-Δ12,14-prostaglandin J2 inhibits macrophage colonization by *Salmonella* enterica serovar Typhimurium. *PLoS One 8*, e69759.

Busillo, J.M. and Cidlowski, J.A. 2013. The five Rs of glucocorticoid action during inflammation: Ready, reinforce, repress, resolve, and restore. *Trends Endocrinol Metab 24*, 109–119.

Butcher, E.C. and Picker, L.J. 1996. Lymphocyte homing and homeostasis. *Science 272*, 60–66.

Butts, C.L. and Sternberg, E.M. 2008. Neuroendocrine factors alter host defense by modulating immune function. *Cell Immunol 252*, 7–15.

Bäckhed, F., Ley, R.E., Sonnenburg, J.L., Peterson, D.A., and Gordon, J.I. 2005. Host-bacterial mutualism in the human intestine. *Science 307*, 1915–1920.

Cahenzli, J., Balmer, M.L., and McCoy, K.D. 2013a. Microbial-immune cross-talk and regulation of the immune system. *Immunology 138*, 12–22.

Cahenzli, J., Köller, Y., Wyss, M., Geuking, M.B., and McCoy, K.D. 2013b. Intestinal microbial diversity during early-life colonization shapes long-term IgE levels. *Cell Host Microbe 14*, 559–570.

Calder, P.C. 2009. Polyunsaturated fatty acids and inflammatory processes: New twists in an old tale. *Biochimie 91*, 791–795.

Cao, Y., Shen, J., and Ran, Z.H. 2014. Association between Faecalibacterium prausnitzii Reduction and Inflammatory Bowel Disease: A Meta-Analysis and Systematic Review of the Literature. *Gastroenterol Res Pract 2014*, 872725.

Carlsson, A.H., Yakymenko, O., Olivier, I., Håkansson, F., Postma, E., Keita, A.V., and

Söderholm, J.D. 2013. Faecalibacterium prausnitzii supernatant improves intestinal barrier function in mice DSS colitis. *Scand J Gastroenterol 48*, 1136–1144.

Cebra, J.J. 1999. Influences of microbiota on intestinal immune system development. *Am J Clin Nutr 69*, 1046S–1051S.

Chen, H., Xu, Z., Peng, L., Fang, X., Yin, X., Xu, N., and Cen, P. 2006. Recent advances in the research and development of human defensins. *Peptides 27*, 931–940.

Cheroutre, H., Lambolez, F., and Mucida, D. 2011. The light and dark sides of intestinal intraepithelial lymphocytes. *Nat Rev Immunol 11*, 445–456.

Chiang, J.Y. 2009. Bile acids: Regulation of synthesis. *J Lipid Res 50*, 1955–1966.

Cho, I., Yamanishi, S., Cox, L., et al. 2012. Antibiotics in early life alter the murine colonic microbiome and adiposity. *Nature 488*, 621–626.

Cho, S. H., Strickland, I., Boguniewicz, M., and Leung, D.Y. 2001. Fibronectin and fibrinogen contribute to the enhanced binding of *Staphylococcus aureus* to atopic skin. *J Allergy Clin Immunol 108*, 269–274.

Chung, H. and Kasper, D.L. 2010. Microbiota-stimulated immune mechanisms to maintain gut homeostasis. *Curr Opin Immunol 22*, 455–460.

Consortium, H.M.P. 2012. Structure, function and diversity of the healthy human microbiome. *Nature 486*, 207–214.

Corfield, A.P., Myerscough, N., Bradfield, N., et al. 1996. Colonic mucins in ulcerative colitis: Evidence for loss of sulfation. *Glycoconj J 13*, 809–822.

Corthay, A. 2009. How do regulatory T cells work? *Scand J Immunol 70*, 326–336.

Corthésy, B. 2007. Roundtrip ticket for secretory IgA: Role in mucosal homeostasis? *J Immunol 178*, 27–32.

Costello, E.K., Lauber, C.L., Hamady, M., Fierer, N., Gordon, J.I., and Knight, R. 2009. Bacterial community variation in human body habitats across space and time. *Science 326*, 1694–1697.

Cruz-Topete, D. and Cidlowski, J.A. 2014. One Hormone, Two Actions: Anti- and Pro-Inflammatory Effects of Glucocorticoids. *Neuroimmunomodulation 22*, 20–32.

Cummings, J.H. 1983. Fermentation in the human large intestine: Evidence and implications for health. *Lancet 1*, 1206–1209.

Cunliffe, R.N., Rose, F.R., Keyte, J., Abberley, L., Chan, W.C., and Mahida, Y.R. 2001. Human defensin 5 is stored in precursor form in normal Paneth cells and is expressed by some villous epithelial cells and by metaplastic Paneth cells in the colon in inflammatory bowel disease. *Gut 48*, 176–185.

Dawson, P.A. and Karpen, S.J. 2014. Intestinal Transport and Metabolism of Bile Acids. *J Lipid Res*.

Dethlefsen, L., McFall-Ngai, M., and Relman, D.A. 2007. An ecological and evolutionary perspective on human-microbe mutualism and disease. *Nature 449*, 811–818.

Dewhirst, F.E., Chien, C.C., Paster, B.J., Ericson, R.L., Orcutt, R.P., Schauer, D.B., and Fox, J.G. 1999. Phylogeny of the defined murine microbiota: Altered Schaedler flora. *Appl Environ Microbiol 65*, 3287–3292.

Di Sabatino, A., Morera, R., Ciccocioppo, R., Cazzola, P., Gotti, S., Tinozzi, F.P., Tinozzi, S., and Corazza, G.R. 2005. Oral butyrate for mildly to moderately active Crohn's disease. *Aliment Pharmacol Ther 22*, 789–794.

Donohoe, D.R., Garge, N., Zhang, X., Sun, W., O'Connell, T.M., Bunger, M.K., and Bultman, S.J. 2011. The microbiome and butyrate regulate energy metabolism and autophagy in the mammalian colon. *Cell Metab 13*, 517–526.

Duboc, H., Rajca, S., Rainteau, D., *et al.* 2013. Connecting dysbiosis, bile-acid dysmetabolism and gut inflammation in inflammatory bowel diseases. *Gut 62*, 531–539.

Dunne, J.L., Triplett, E.W., Gevers, D., Xavier, R., Insel, R., Danska, J., and Atkinson, M.A. 2014. The intestinal microbiome in type 1 diabetes. *Clin Exp Immunol 177*, 30–37.

Durkin, H.G., Bazin, H., and Waksman, B.H. 1981. Origin and fate of IgE-bearing lymphocytes. I. Peyer's patches as differentiation site of cells. Simultaneously bearing IgA and IgE. *J Exp Med 154*, 640–648.

Ege, M.J., Strachan, D.P., Cookson, W.O., *et al.* 2011. Gene-environment interaction for childhood asthma and exposure to farming in Central Europe. *J Allergy Clin Immunol 127*, 138–144, 144.e131–134.

Endt, K., Stecher, B., Chaffron, S., *et al.* 2010. The microbiota mediates pathogen clearance from the gut lumen after non-typhoidal *Salmonella* diarrhea. *PLoS Pathog 6*, e1001097.

Eriguchi, Y., Takashima, S., Oka, H., *et al.* 2012. Graft-versus-host disease disrupts intestinal microbial ecology by inhibiting Paneth cell production of α-defensins. *Blood 120*, 223–231.

Fakhoury, M., Negrulj, R., Mooranian, A., and Al-Salami, H. 2014. Inflammatory bowel disease: Clinical aspects and treatments. *J Inflamm Res 7*, 113–120.

Fallarino, F., Grohmann, U., You, S., *et al.* 2006. The combined effects of tryptophan starvation and tryptophan catabolites down-regulate T cell receptor zeta-chain and induce a regulatory phenotype in naive T cells. *J Immunol 176*, 6752–6761.

Farache, J., Koren, I., Milo, I., Gurevich, I., Kim, K.W., Zigmond, E., *et al.* 2013. Luminal bacteria recruit CD103+ dendritic cells into the intestinal epithelium to sample bacterial antigens for presentation. *Immunity 38*, 581–595.

Frank, D.N., St Amand, A.L., Feldman, R.A., Boedeker, E.C., Harpaz, N., and Pace, N.R. 2007. Molecular-phylogenetic characterization of microbial community imbalances in human inflammatory bowel diseases. *Proc Natl Acad Sci USA 104*, 13780–13785.

Fukumoto, S., Toshimitsu, T., Matsuoka, S., *et al.* 2014. Identification of a probiotic bacteria-derived activator of the aryl hydrocarbon receptor that *inhibits colitis*. *Immunol Cell Biol 92*, 460–465.

Funk, C.D. 2001. Prostaglandins and leukotrienes:Advances in eicosanoid biology. *Science 294*, 1871–1875.

Gaboriau-Routhiau, V., Rakotobe, S., Lécuyer, E., *et al.* 2009. The key role of segmented filamentous bacteria in the coordinated maturation of gut helper T cell responses. *Immunity 31*, 677–689.

Gallo, R.L. and Nakatsuji, T. 2011. Microbial symbiosis with the innate immune defense system of the skin. *J Invest Dermatol 131*, 1974–1980.

Ganz, T. and Lehrer, R.I. 1999. Antibiotic peptides from higher eukaryotes: Biology and applications. *Mol Med Today 5*, 292–297.

Ganz, T., Gabayan, V., Liao, H.I., Liu, L., Oren, A., Graf, T., and Cole, A.M. 2003. Increased inflammation in lysozyme M-deficient mice in response to Micrococcus luteus and its peptidoglycan. *Blood 101*, 2388–2392.

Gardner, J.M., Devoss, J.J., Friedman, R.S., *et al.* 2008. Deletional tolerance mediated by extrathymic Aire-expressing cells. *Science 321*, 843–847.

Geuking, M.B., Cahenzli, J., Lawson, M.A., Ng, D.C., Slack, E., Hapfelmeier, S., McCoy, K.D., and Macpherson, A.J. 2011. Intestinal bacterial colonization induces mutualistic regulatory T cell responses. *Immunity 34*, 794–806.

Gillor, O., Etzion, A., and Riley, M.A. 2008. The dual role of bacteriocins as anti- and probiotics. *Appl Microbiol Biotechnol 81*, 591–606.

Giongo, A., Gano, K.A., Crabb, D.B., *et al.* 2011. Toward defining the autoimmune microbiome for type 1 diabetes. *ISME J 5*, 82–91.

Gollwitzer, E.S., Saglani, S., Trompette, A., Yadava, K., Sherburn, R., McCoy, K.D., Nicod, L.P., Lloyd, C.M., and Marsland, B.J. 2014. Lung microbiota promotes tolerance to allergens in neonates via PD-L1. *Nat Med 20*, 642–647.

Gordon, H.A., Bruckner-Kardoss, E., and Wostmann, B.S. 1966. Aging in germ-free mice: Life tables and lesions observed at natural death. *J Gerontol 21*, 380–387.

Grice, E.A., Kong, H.H., Conlan, S., *et al.* 2009. Topographical and temporal diversity of the human skin microbiome. *Science 324*, 1190–1192.

Guilhelmelli, F., Vilela, N., Albuquerque, P., Derengowski, L.A.S., Silva-Pereira, I., and Kyaw, C.M. 2013. Antibiotic development challenges: The various mechanisms of action of antimicrobial peptides and of bacterial resistance. *Front Microbiol 4*, 353.

Hallert, C., Björck, I., Nyman, M., Pousette, A., Grännö, C., and Svensson, H. 2003. Increasing fecal butyrate in ulcerative colitis patients by diet: Controlled pilot study. *Inflamm Bowel Dis 9*, 116–121.

Hamada, H., Hiroi, T., Nishiyama, Y., *et al.* 2002. Identification of multiple isolated lymphoid follicles on the antimesenteric wall of the mouse small intestine. *J Immunol 168*, 57–64.

Hapfelmeier, S., Lawson, M.A., Slack, E., *et al.* 2010. Reversible microbial colonization of germ-free mice reveals the dynamics of IgA immune responses. *Science 328*, 1705–1709.

Harder, J., Dressel, S., Wittersheim, M., *et al.* 2010. Enhanced expression and secretion of antimicrobial peptides in atopic dermatitis and after superficial skin injury. *J Invest Dermatol 130*, 1355–1364.

Hayday, A., Theodoridis, E., Ramsburg, E., and Shires, J. 2001. Intraepithelial lymphocytes: Exploring the Third Way in immunology. *Nat Immunol 2*, 997–1003.

Heath-Pagliuso, S., Rogers, W.J., Tullis, K., Seidel, S.D., Cenijn, P.H., Brouwer, A., and Denison, M.S. 1998. Activation of the Ah receptor by tryptophan and tryptophan metabolites. *Biochemistry 37*, 11508–11515.

Herbst, T., Sichelstiel, A., Schär, C., Yadava, K., Bürki, K., Cahenzli, J., McCoy, K., Marsland, B.J., and Harris, N.L. 2011. Dysregulation of allergic airway inflammation in the absence of microbial colonization. *Am J Respir Crit Care Med 184*, 198–205.

Hill, D.A. and Artis, D. 2010. Intestinal bacteria and the regulation of immune cell homeostasis. *Annu Rev Immunol 28*, 623–667.

Hilty, M., Burke, C., Pedro, H., *et al.* 2010. Disordered microbial communities in asthmatic airways. *PLoS One 5*, e8578.

Hooper, L.V. and Macpherson, A.J. 2010. Immune adaptations that maintain homeostasis with the intestinal microbiota. *Nat Rev Immunol 10*, 159–169.

Iimura, M., Gallo, R.L., Hase, K., Miyamoto, Y., Eckmann, L., and Kagnoff, M.F. 2005. Cathelicidin mediates innate intestinal defense against colonization with epithelial adherent bacterial pathogens. *J Immunol 174*, 4901–4907.

Iovanna, J.L. and Dagorn, J.C. 2005. The multifunctional family of secreted proteins containing a C-type lectin-like domain linked to a short N-terminal peptide. *Biochim Biophys Acta 1723*, 8–18.

Islam, D., Bandholtz, L., Nilsson, J., Wigzell, H., Christensson, B., Agerberth, B., and Gudmundsson, G. 2001. Downregulation of bactericidal peptides in enteric infections: A novel immune escape mechanism with bacterial DNA as a potential regulator. *Nat Med 7*, 180–185.

Ivanov, I.I. and Littman, D.R. 2010. Segmented filamentous bacteria take the stage. *Mucosal Immunol 3*, 209–212.

Kamada, N. and Núñez, G. 2013. Role of the gut microbiota in the development and function of lymphoid cells. *J Immunol 190*, 1389–1395.

Kamada, N. and Núñez, G. 2014. Regulation of the immune system by the resident intestinal bacteria. *Gastroenterology 146*, 1477–1488.

Karlin, D.A., Mastromarino, A.J., Jones, R.D., Stroehlein, J.R., and Lorentz, O. 1985. Fecal skatole and indole and breath methane and hydrogen in patients with large bowel polyps or cancer. *J Cancer Res Clin Oncol 109*, 135–141.

Kaser, A., Zeissig, S., and Blumberg, R.S. 2010. Inflammatory bowel disease. *Annu Rev Immunol 28*, 573–621.

Kim, S.C., Tonkonogy, S.L., Albright, C.A., Tsang, J., Balish, E.J., Braun, J., Huycke, M.M., and Sartor, R.B. 2005. Variable phenotypes of enterocolitis in interleukin 10-deficient mice monoassociated with two different commensal bacteria. *Gastroenterology 128*, 891–906.

Kleessen, B., Kroesen, A.J., Buhr, H.J., and Blaut, M. 2002. Mucosal and invading bacteria in patients with inflammatory bowel disease compared with controls. *Scand J Gastroenterol 37*, 1034–1041.

Kobayashi, K.S., Chamaillard, M., Ogura, Y., Henegariu, O., Inohara, N., Nuñez, G., and Flavell, R.A. 2005. Nod2-dependent regulation of innate and adaptive immunity in the intestinal tract. *Science 307*, 731–734.

Kong, H.H., Oh, J., Deming, C., *et al.* 2012. Temporal shifts in the skin microbiome associated with disease flares and treatment in children with atopic dermatitis. *Genome Res 22*, 850–859.

Kool, M., Willart, M.A., van Nimwegen, M., *et al.* 2011. An unexpected role for uric acid as an inducer of T helper 2 cell immunity to inhaled antigens and inflammatory mediator of allergic asthma. *Immunity 34*, 527–540.

Koon, H.W., Shih, D.Q., Chen, J., *et al.* 2011. Cathelicidin signaling via the Toll-like receptor protects against colitis in mice. *Gastroenterology 141*, 1852–1863.e1851–1853.

Kotlowski, R., Bernstein, C.N., Sepehri, S., and Krause, D.O. 2007. High prevalence of Escherichia coli belonging to the B2+D phylogenetic group in inflammatory bowel disease. *Gut 56*, 669–675.

Kościuczuk, E.M., Lisowski, P., Jarczak, J., Strzałkowska, N., Jóźwik, A., Horbańczuk, J., Krzyżewski, J., Zwierzchowski, L., and Bagnicka, E. 2012. Cathelicidins: Family of antimicrobial peptides. A review. *Mol Biol Rep 39*, 10957–10970.

Lai, Y., Di Nardo, A., Nakatsuji, T., *et al.* 2009. Commensal bacteria regulate Toll-like receptor 3-dependent inflammation after skin injury. *Nat Med 15*, 1377–1382.

Lambeau, G. and Gelb, M.H. 2008. Biochemistry and physiology of mammalian secreted phospholipases A2. *Annu Rev Biochem 77*, 495–520.

Lau, K., Benitez, P., Ardissone, A., *et al.* 2011. Inhibition of type 1 diabetes correlated to a Lactobacillus johnsonii N6.2-mediated Th17 bias. *J Immunol 186*, 3538–3546.

Lee, J., Jayaraman, A., and Wood, T.K. 2007. Indole is an inter-species biofilm signal mediated by SdiA. *BMC Microbiol 7*, 42.

Leone, V., Chang, E.B., and Devkota, S. 2013. Diet, microbes, and host genetics: The perfect storm in inflammatory bowel diseases. *J Gastroenterol 48*, 315–321.

Liao, F., Andalibi, A., deBeer, F.C., Fogelman, A.M., and Lusis, A.J. 1993. Genetic control of inflammatory gene induction and NF-κ B-like transcription factor activation in response to an atherogenic diet in mice. *J Clin Invest 91*, 2572–2579.

Macpherson, A.J. and Harris, N.L. 2004. Interactions between commensal intestinal bacteria and the immune system. *Nat Rev Immunol 4*, 478–485.

Macpherson, A.J. and Uhr, T. 2004. Induction of protective IgA by intestinal dendritic cells carrying commensal bacteria. *Science 303*, 1662–1665.

Makishima, M., Okamoto, A.Y., Repa, J.J., *et al.* 1999. Identification of a nuclear receptor for bile acids. *Science 284*, 1362–1365.

Makishima, M., Lu, T.T., Xie, W., Whitfield, G.K., Domoto, H., Evans, R.M., Haussler, M.R., and Mangelsdorf, D.J. 2002. Vitamin D receptor as an intestinal bile acid sensor. *Science 296*, 1313–1316.

Mallbris, L., Carlén, L., Wei, T., Heilborn, J., Nilsson, M.F., Granath, F., and Ståhle, M. 2010. Injury downregulates the expression of the human cathelicidin protein hCAP18/LL-37 in atopic dermatitis. *Exp Dermatol 19*, 442–449.

Manichanh, C., Rigottier-Gois, L., Bonnaud, E., *et al.* 2006. Reduced diversity of faecal microbiota in Crohn's disease revealed by a metagenomic approach. *Gut 55*, 205–211.

Maslowski, K.M., Vieira, A.T., Ng, A., *et al.* 2009. Regulation of inflammatory responses by gut microbiota and chemoattractant receptor GPR43. *Nature 461*, 1282–1286.

Maynard, C.L. and Weaver, C.T. 2009. Intestinal effector T cells in health and disease. *Immunity 31*, 389–400.

Mazmanian, S.K., Liu, C.H., Tzianabos, A.O., and Kasper, D.L. 2005. An immunomodulatory molecule of symbiotic bacteria directs maturation of the host immune system. *Cell 122*, 107–118.

Mazmanian, S.K., Round, J.L., and Kasper, D.L. 2008. A microbial symbiosis factor prevents intestinal inflammatory disease. *Nature 453*, 620–625.

Miron, N. and Cristea, V. 2012. Enterocytes:Active cells in tolerance to food and microbial antigens in the gut. *Clin Exp Immunol 167*, 405–412.

Miyake, J.H., Wang, S.L., and Davis, R.A. 2000. Bile acid induction of cytokine expression by macrophages correlates with repression of hepatic cholesterol 7α-hydroxylase. *J Biol Chem 275*, 21805–21808.

Naik, S., Bouladoux, N., Wilhelm, C., et al. 2012. Compartmentalized control of skin immunity by resident commensals. *Science 337*, 1115–1119.

Nakamizo, S., Egawa, G., Honda, T., Nakajima, S., Belkaid, Y., and Kabashima, K. 2014. Commensal bacteria and cutaneous immunity. *Semin Immunopathol.*

Neill, D.R., Wong, S.H., Bellosi, A., et al. 2010. Nuocytes represent a new innate effector leukocyte that mediates type-2 immunity. *Nature 464*, 1367–1370.

Nijkamp, F.P. and Parnham, M.J. 2011. *Principles of immunopharmacology*, 3rd edn. Basel: Springer.

Nuttal, G.H.F. and Thierfelder, H. 1895. Tierisches Leben ohne Bakterien im Verdauungskanal. *Hoppe Seyler's Zeitschrift Physiol Chem 21*, 109–112.

Ostman, S., Rask, C., Wold, A.E., Hultkrantz, S., and Telemo, E. 2006. Impaired regulatory T cell function in germ-free mice. *Eur J Immunol 36*, 2336–2346.

Ouellette, A.J. 2010. Paneth cells and innate mucosal immunity. *Curr Opin Gastroenterol 26*, 547–553.

Pabst, O. 2012. New concepts in the generation and functions of IgA. *Nat Rev Immunol 12*, 821–832.

Pasteur, L. 1885. Observations relatives à la note de M. Duclaux. *C R Acad Sci 100*, 68–69.

Penders, J., Stobberingh, E.E., van den Brandt, P.A., and Thijs, C. 2007. The role of the intestinal microbiota in the development of atopic disorders. *Allergy 62*, 1223–1236.

Peterson, D.A., McNulty, N.P., Guruge, J.L., and Gordon, J.I. 2007. IgA response to symbiotic bacteria as a mediator of gut homeostasis. *Cell Host Microbe 2*, 328–339.

Powrie, F., Leach, M.W., Mauze, S., Caddle, L.B.,and Coffman, R.L. 1993. Phenotypically distinct subsets of CD4+ T cells induce or protect from chronic intestinal inflammation in C. B-17 scid mice. *Int Immunol 5*, 1461–1471.

Pullan, R.D., Thomas, G.A., Rhodes, M., Newcombe, R.G., Williams, G.T., Allen, A., and Rhodes, J. 1994. Thickness of adherent mucus gel on colonic mucosa in humans and its relevance to colitis. *Gut 35*, 353–359.

Qiu, X., Zhang, M., Yang, X., Hong, N., and Yu, C. 2013. Faecalibacterium prausnitzii upregulates regulatory T cells and anti-inflammatory cytokines in treating TNBS-induced colitis. *J Crohns Colitis 7*, e558–568.

Ramasundara, M., Leach, S.T., Lemberg, D.A., and Day, A.S. 2009. Defensins and inflammation: The role of defensins in inflammatory bowel disease. *J Gastroenterol Hepatol 24*, 202–208.

Raqib, R., Sarker, P., Bergman, P., et al. 2006. Improved outcome in shigellosis associated with butyrate induction of an endogenous peptide antibiotic. *Proc Natl Acad Sci USA 103*, 9178–9183.

Rauch, P.J., Chudnovskiy, A., Robbins, C.S., et al. 2012. Innate response activator B cells protect against microbial sepsis. *Science 335*, 597–601.

Reyniers, J.A. and Trexler, P.C. 1955. Germfree research: A basic study in host-contaminant relationship. I. General and theoretical aspects of the problem. *Bull N Y Acad Med 31*, 231–235.

Round, J.L. and Mazmanian, S.K. 2009. The gut microbiota shapes intestinal immune responses during health and disease. *Nat Rev Immunol 9*, 313–323.

Russell, S.L., Gold, M.J., Hartmann, M., *et al.* 2012. Early life antibiotic-driven changes in microbiota enhance susceptibility to allergic asthma. *EMBO Rep 13*, 440–447.

Sakaguchi, S., Wing, K., and Miyara, M. 2007. Regulatory T cells—A brief history and perspective. *Eur J Immunol 37 Suppl 1*, S116–123.

Sartor, R.B. 2008. Therapeutic correction of bacterial dysbiosis discovered by molecular techniques. *Proc Natl Acad Sci USA 105*, 16413–16414.

Savage, D.C. 1977. Microbial ecology of the gastrointestinal tract. *Annu Rev Microbiol 31*, 107–133.

Schwartz, R.F., Neu, J., Schatz, D., Atkinson, M.A. and Wasserfall, C. 2007. Comment on: Brugman, S. *et al.* (2006) Antibiotic treatment partially protects against type 1 diabetes in the Bio-Breeding diabetes-prone rat. Is the gut flora involved in the development of type 1 diabetes? *Diabetologia* 49, 2105–2108. *Diabetologia 50*, 220–221.

Scott, M.G., Davidson, D.J., Gold, M.R., Bowdish, D., and Hancock, R.E. 2002. The human antimicrobial peptide LL-37 is a multifunctional modulator of innate immune responses. J Immunol 169, 3883–3891.

Sellon, R.K., Tonkonogy, S., Schultz, M., Dieleman, L.A., Grenther, W., Balish, E., Rennick, D.M., and Sartor, R.B. 1998. Resident enteric bacteria are necessary for development of spontaneous colitis and immune system activation in interleukin-10-deficient mice. *Infect Immun 66*, 5224–5231.

Shimada, S., Kawaguchi-Miyashita, M., Kushiro, A., *et al.* 1999. Generation of polymeric immunoglobulin receptor-deficient mouse with marked reduction of secretory IgA. *J Immunol 163*, 5367–5373.

Shinkai, Y., Rathbun, G., Lam, K.P., *et al.* 1992. RAG-2-deficient mice lack mature lymphocytes owing to inability to initiate V(D)J rearrangement. *Cell 68*, 855–867.

Shroff, K.E., Meslin, K., and Cebra, J.J. 1995. Commensal enteric bacteria engender a self-limiting humoral mucosal immune response while permanently colonizing the gut. *Infect Immun 63*, 3904–3913.

Smith, K., McCoy, K.D., and Macpherson, A.J. 2007. Use of axenic animals in studying the adaptation of mammals to their commensal intestinal microbiota. *Semin Immunol 19*, 59–69.

Smith, P.M., Howitt, M.R., Panikov, N., Michaud, M., Gallini, C.A., Bohlooly, Y,M., Glickman, J. N., and Garrett, W.S. 2013. The microbial metabolites, short-chain fatty acids, regulate colonic T_{reg} cell homeostasis. *Science 341*, 569–573.

Sokol, H., Pigneur, B., Watterlot, L., *et al.* 2008. Faecalibacterium prausnitzii is an anti-inflammatory commensal bacterium identified by gut microbiota analysis of Crohn disease patients. *Proc Natl Acad Sci USA 105*, 16731–16736.

Spasova, D.S. and Surh, C.D. 2014. Blowing on embers: Commensal microbiota and our immune system. *Front Immunol 5*, 318.

Staudinger, J.L., Goodwin, B., Jones, S.A., *et al.* 2001. The nuclear receptor PXR is a lithocholic acid sensor that protects against liver toxicity. *Proc Natl Acad Sci USA 98*, 3369–3374.

Strachan, D.P. 1989. Hay fever, hygiene, and household size. *BMJ 299*, 1259–1260.

Strober, W. 2013. Impact of the gut microbiome on mucosal inflammation. *Trends Immunol 34*, 423–430.

Sunyer, J.O. 2013. Fishing for mammalian paradigms in the teleost immune system. *Nat Immunol 14*, 320–326.

Talham, G.L., Jiang, H.Q., Bos, N.A., and Cebra, J.J. 1999. Segmented filamentous bacteria are potent stimuli of a physiologically normal state of the murine gut mucosal immune system. *Infect Immun 67*, 1992–2000.

Thangaraju, M., Cresci, G.A., Liu, K., et al. 2009. GPR109A is a G-protein-coupled receptor for the bacterial fermentation product butyrate and functions as a tumor suppressor in colon. *Cancer Res 69*, 2826–2832.

Tsuji, M., Suzuki, K., Kitamura, H., Maruya, M., Kinoshita, K., Ivanov, I.I., Itoh, K., Littman, D.R., and Fagarasan, S. 2008. Requirement for lymphoid tissue-inducer cells in isolated follicle formation and T cell-independent immunoglobulin A generation in the gut. *Immunity 29*, 261–271.

Tuckermann, J.P., Kleiman, A., Moriggl, R., et al. 2007. Macrophages and neutrophils are the targets for immune suppression by glucocorticoids in contact allergy. *J Clin Invest 117*, 1381–1390.

Turnbaugh, P.J., Ley, R.E., Hamady, M., Fraser-Liggett, C.M., Knight, R., and Gordon, J.I. 2007. The human microbiome project. *Nature 449*, 804–810.

Twigg, H.L., Morris, A., Ghedin, E., et al. 2013. Use of bronchoalveolar lavage to assess the respiratory microbiome: Signal in the noise. *Lancet Respir Med 1*, 354–356.

Vaishnava, S., Behrendt, C.L., Ismail, A.S., Eckmann, L., and Hooper, L.V. 2008. Paneth cells directly sense gut commensals and maintain homeostasis at the intestinal host-microbial interface. *Proc Natl Acad Sci USA 105*, 20858–20863.

Valladares, R., Sankar, D., Li, N., et al. 2010. Lactobacillus johnsonii N6.2 mitigates the development of type 1 diabetes in BB-DP rats. *PLoS One 5*, e10507.

Walker, W.A. 2013. Initial intestinal colonization in the human infant and immune homeostasis. *Ann Nutr Metab 63 Suppl 2*, 8–15.

Walter, J. and Ley, R. 2011. The human gut microbiome: Ecology and recent evolutionary changes. *Annu Rev Microbiol 65*, 411–429.

Wehkamp, J., Chu, H., Shen, B., Feathers, R.W., Kays, R.J., Lee, S.K. and Bevins, C.L. 2006. Paneth cell antimicrobial peptides: Topographical distribution and quantification in human gastrointestinal tissues. *FEBS Lett 580*, 5344–5350.

Wen, L., Ley, R.E., Volchkov, P.Y., et al. 2008. Innate immunity and intestinal microbiota in the development of Type 1 diabetes. *Nature 455*, 1109–1113.

Xue, L., Gyles, S.L., Wettey, F.R., Gazi, L., Townsend, E., Hunter, M.G., and Pettipher, R. 2005. Prostaglandin D2 causes preferential induction of proinflammatory Th2 cytokine production through an action on chemoattractant receptor-like molecule expressed on Th2 cells. *J Immunol 175*, 6531–6536.

Yeaman, M.R. and Yount, N.Y. 2003. Mechanisms of antimicrobial peptide action and resistance. *Pharmacol Rev 55*, 27–55.

Zaiou, M., Nizet, V., and Gallo, R.L. 2003. Antimicrobial and protease inhibitory functions of the human cathelicidin (hCAP18/LL-37) prosequence. *J Invest Dermatol* *120*, 810–816.

Zelante, T., Iannitti, R.G., Cunha, C., *et al.* 2013. Tryptophan catabolites from microbiota engage aryl hydrocarbon receptor and balance mucosal reactivity via interleukin-22. *Immunity 39*, 372–385.

CHAPTER 7

Microbiota-Related Modulation of Metabolic Processes in the Body

TINGTING JU, JIAYING LI and BENJAMIN P. WILLING

7.1. INTRODUCTION

IN considering how microbes affect our metabolic processes, it is important to consider how we define metabolism. Metabolism encompasses the biochemical transformation of molecules within and between cells that are required for life. These processes are made possible by enzymes that catalyze reactions and are coordinated through diverse signalling molecules and hormones. Microbes both contribute to these biochemical processes by providing and modifying nutrients, as well as by modifying how the host processes and allocates nutrients.

Mammals utilize a complex set of regulatory mechanisms to maintain equilibrium, and have evolved to include microbial contributions to this process. For example, bile acid metabolism is highly influenced by microbial bile acid deconjugation. We also have receptors for molecules that are generated through microbial fermentation (e.g., butyrate). Therefore, alterations in the gut microbiota, and how these contribute to metabolic regulation can disrupt metabolic homeostasis, resulting in the development of metabolic disease. The processes in the body impacted by the microbiota include many aspects key to metabolic outcomes including regulation of food intake, deposition, and mobilization of fat, insulin secretion, and glucose uptake.

It has long been recognized that the intestinal microbiota increases the host's ability to derive nutrients from the diet. The most obvious example is through the digestion of dietary fiber, as mammals lack the enzymes required to digest these nonstarch polysaccharides. This increased provision of energy can both contribute to obesity development and promote health. As will be discussed later in this chapter, an expansive effort continues to explore this relationship. In 2006, the research team led by Dr. Jeffrey Gordon was the first to demonstrate that not only presence, but also the composition of intestinal microbes govern the development of obesity, suggesting that there was an opportunity to treat obesity by modifying our microbes. They showed that by transplanting a microbiota from an obese mouse into a lean mouse, an obese phenotype could be imparted (Turnbaugh et al. 2006). This discovery created great excitement in the research community and spurred efforts to understand the mechanisms through which microbes could modulate host metabolic processes to result in beneficial and adverse outcomes. This chapter discusses how microbes contribute to the modulation of metabolic processes in the body, the workings of this normal process, and how disruption of these pathways can contribute to adverse outcomes such as cardiovascular disease (CVD), obesity, and diabetes.

7.2. SHORT-CHAIN FATTY ACIDS

Humans and other vertebrates have a very limited amount of glycoside hydrolases to degrade the bulk of complex dietary fiber. In the distal gut, the partially and nondigestible polysaccharides that have not been absorbed by the host in the upper gastrointestinal tract are fermented to short-chain fatty acids (SCFA) (Flint et al. 2008). The microbial community produces SCFA as necessary waste products to maintain redox equivalent production and balance pH in the intestinal lumen (van Hoek and Merks 2012). SCFAs are a subset of fatty acids with less than six carbon atoms of which acetate (C2), propionate (C3), and butyrate (C4) are the most abundant (Cook and Sellin 1998). Even though SCFAs provide a relatively small amount of energy (approximately 5–10%) for healthy people, they make a significant contribution to the body's daily energy requirements and also have profound effects on host processes, including energy utilization, host-microbe signalling, and control of pH in the colon, which affects microbial composition, gut motility, and epithelial proliferation (Musso et al. 2011). In humans, the production rate of acetate, propionate, and butyrate in the colon is in a molar ratio of around 60:25:15 (Tazoe et al. 2008). In healthy people, 95% of SCFA are absorbed rapidly while the remaining 5% are excreted in feces. Butyrate is mostly utilized by colonic epithelial cells, where it is converted to ketone bodies or oxidized to carbon dioxide (Louis et al.

2007), whereas propionate and acetate reach the liver through the portal vein (Figure 7.1). Propionate is utilized largely for gluconeogenesis in the liver, while acetate either remains in the liver or is released to the peripheral tissues as a substrate for lipogenesis and cholesterol synthesis (Vernay 1987; Wolever *et al.* 1991; Samuel *et al.* 2008). In the gut, SCFAs are transported across the apical and basolateral membranes of colonocytes through passive diffusion of undissociated SCFA and active transport of dissociated SCFA anions (Sellin 1999).

7.2.1. SCFA-Energy Source

It has been shown that either supplementation of butyrate in the diet or oral administration of acetate to rodents can promote obesity in both genetic or diet induced obesity, independent of food intake suppression (Lin *et al.* 2012). This is associated with an overall increased energy provision to the host. However, as an energy source that can contribute to weight gain, how increased microbial SCFAs, through the provision of fiber, can paradoxically reduce obesity continues to be a major area

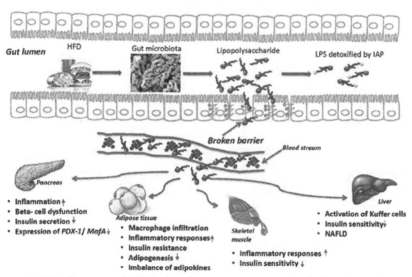

FIGURE 7.1. Summary of SCFAs function as energy source. In the large intestine, gut microbiota ferment nondigestible polysaccharides to SCFAs, including acetate, butyrate and propionate, which provide 5–10% of total energy for healthy humans. Once absorbed by the intestine epithelium, butyrate is rapidly used by colonocytes, while propionate is used for hepatic gluconeogenesis, and acetate is used for lipogenesis and cholesterol synthesis in liver and other peripheral organs.

of investigation. The hypothesis is that the effects of SCFA on obesity are dependent on the balance between their role as an energy source and their role in regulating gene expression and the release of gut hormones to inhibit satiety and regulate metabolic processes.

In germ-free (GF) mice, Bäckhed *et al.* (2004) performed a series of experiments to show the influence of microbiota on adiposity. First, they noticed 42% less fat in GF mice, which have very low SCFA production compared with conventionally raised mice even with higher food intake. Conventionalization of GF mice yielded a 57% increase in total body fat and increase in insulin resistance while these mice showed lower chow consumption compared with GF mice, suggesting better energy harvest and storage ability (Backhed *et al.* 2004). Previous studies showed that GF rodents excreted twice as many urinary and fecal calories as conventional rats fed the same diet due to the inability of GF mice to ferment dietary polysaccharides into SCFA, which account for the low body weight in GF mice (Wostmann *et al.* 1983; Hoverstad and Midtvedt 1986). In another study, leptin-deficient *ob/ob* mice showed higher concentration of cecal SCFA than the wild-type mice, more caloric extraction from the diet, and reduced energy content in their feces, suggesting that gut microbiota enhances adiposity through increasing energy extraction from diet and by modulating fat storage. The obesity phenotype was transmissible through fecal transplants from obese to GF mice (Turnbaugh *et al.* 2006). This has also been extended to human microbiota populations by introducing fecal microbes from discordant obese twins to GF mice. GF mice colonized with obese feces gained more weight and adiposity, however inconsistent with other studies showed lower concentration of cecal propionate and butyrate. When obese mice received a healthy diet (low-fat and high fiber) and were cohoused with mice containing lean microbiota, the lean microbiota finally dominated in the obese mice and prevented increased adiposity (Ridaura *et al.* 2013). They also showed that obese microbiota was not able to colonize lean mice when fed with low-fat diet. The fact that these mice remained lean suggests that diet is the prominent factor deciding which phenotype develops.

Several studies have shown that obese people have higher levels of fecal SCFA than lean, but until recently it was unclear whether this was due to increased SCFA production or reduced absorption, and whether the *Firmicutes* and *Bacteroidetes* ratio is related to fecal SCFA concentration remains inconsistent (Schwiertz *et al.* 2010; Patil *et al.* 2012, Teixeira *et al.* 2013; Fernandes *et al.* 2014). Rahat-Rozenbloom *et al.*

(2014) confirmed the hypothesis that higher fecal SCFA in overweight and obese people is not due to the differences in SCFA absorption or diet but to SCFA production. In this study, dialysis bags with a SCFA solution were used to measure the SCFA absorption; calculated by the SCFA disappearance from the dialysis bag divided by the SCFA present at the baseline. The absorption rate was the same in obese and lean individuals, and with a similar dietary intake, they further concluded that it is due to the higher production of SCFA by colonic microbiota in obese people than in lean individuals (Rahat-Rozenbloom *et al.* 2014). Further studies are needed to reveal the underlying mechanisms of how the higher SCFA production contributes to adiposity and obesity.

7.2.2. SCFA Sensing and Signal Transduction

In addition to providing energy, SCFAs enter the bloodstream and act as signalling molecules (Functions depicted in Figure 7.2). SCFAs modulate biological responses of the host largely through two major mechanisms. The first involves epigenetic modification of DNA and histones, which directly regulate gene expression. The second is via G protein-coupled receptor (GCPR) activation sending signals to cells leading to a cascade of metabolic changes.

7.2.2.1. Epigenetic

SCFA have a number of effects on cells, many of which are mediated through inhibition of histone deacetylases (HDAC). HDAC regulate gene transcription associated with pathologic processes by compacting chromatin and making it less accessible to transcriptional activators. Butyrate is well known as an inhibitor of HDAC, while propionate is less effective and acetate is completely inactive in colon cancer cell lines (Waldecker *et al.* 2008). However, the impact of SCFA is also dependent on tissue type. Oral administration of acetate has been shown to inhibit both HDAC2 activity and protein expression in the rodent brain (Soliman and Rosenberger 2011). Inhibition of HDAC activity is the main mechanism through which butyrate affects the expression of proinflammatory cytokines in humans (Zeng *et al.* 2014). HDACs are involved in the pathogenesis of diabetes and are currently of interest as targets for the treatment of the disease (Christensen *et al.* 2011). In addition to epigenetically modifying histone acetylation profiles, Remely *et al.* (2014) recently demonstrated that SCFA could regulate methyla-

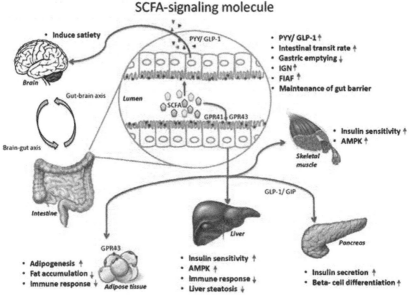

FIGURE 7.2. Schematic overview of the effects of SCFAs on host metabolism as signaling molecules. SCFAs improve metabolic related disorders by influencing gene expression as histone deacetylases (HDACs) inhibitors and via G-protein coupled receptors (GPCRs). The three main effects of SCFAs are inducing insulin secretion in pancreas, stimulating satiety in the brain, and suppressing immune responses in adipose and other tissues. In the intestine, SCFAs improve gut barrier integrity and induce production of gut hormones (PYY, GLP-1 etc.), which induce satiety in the brain and promote insulin secretion in the pancreas. SCFAs also stimulate fasting-induced adipose factor (FIAF) expression in the epithelium, which regulates downstream lipid metabolism. In the liver and skeletal muscle, SCFAs increase insulin sensitivity and increase energy expenditure by activating AMPK. In adipose tissue, FIAF suppresses lipolysis and thus inhibit fat accumulation. SCFAs also reduce the release of proinflammatory cytokines and reduce immune cells infiltration in adipose tissue. Overall, SCFAs regulate food intake, energy expenditure and reduce inflammation to promote host homeostasis.

tion of *GPR41* in obese and Type 2 diabetic patients, which might influence the satiety and hunger circle. They evaluated the methylation status of *GPR41* in obese, diabetic, and lean individuals and revealed a negative correlation between the body mass and *GPR41* methylation.

7.2.2.2. G Protein-Coupled Receptors

Recently, propionate and acetate were reported to be the ligands for two GPRs, GPR41 (free fatty acid receptor 3, FFAR3) and GPR43 (free fatty acid receptor 3, FFAR2), mainly expressed in gut epithelial cells (Brown *et al.* 2003; Le Poul *et al.* 2003). Expression of GPCRs in other cell types like adipocytes, immune cells, and sympathetic ganglion has also been reported (Hong *et al.* 2005; Maslowski *et al.* 2009; Kimura *et al.* 2011). Even though GPR43 can also be activated by other SCFA, propionate and acetate have been shown to be the most efficient for activating GPR43 (Le Poul *et al.* 2003). The fact that *Gpr43* is expressed in adipose tissue and intestines suggests that GPR43 may be involved in energy homeostasis. Hong *et al.* (2005) performed a series of studies to elucidate the functions of GPR43 in adipose tissues. They demonstrated that *Gpr43* expression was significantly greater in the white adipose tissue of mice on a high-fat diet compared with normal-fat diet fed mice, and suppression of *Gpr43* mRNA by RNA interference inhibited adipogenesis, suggesting that SCFA may promote adipogenesis via GPR43 (Hong *et al.* 2005). Li *et al.* (2014) recently showed that the effect of SCFA in enhancing adipocyte differentiation was not via GPR41 or GPR43 in the stromal vascular fraction of porcine subcutaneous fat (Li *et al.* 2014). Thus, further study is needed to identify molecular pathways of SCFA-stimulated adipogenesis.

In brown adipose tissues, which help regulate energy expenditure by thermogenesis, it was reported that *Gpr43* knockout mice fed a high-fat diet showed improved insulin sensitivity (Bjursell *et al.* 2011). However, others could not detect *Gpr43* expression in brown adipose tissues (Kimura *et al.* 2013). Recently, a series of *in vitro* and *in vivo* studies showed that *Gpr43*-deficient mice were obese even when consuming a normal diet, whereas mice overexpressing this receptor specifically in white adipose tissues remained lean, independent of calorie consumption. Importantly, GF environment and antibiotic treatment abrogated this effect leading to the hypothesis that the production of SCFA of bacterial fermentation can mediate activation of GPR43, which results in suppression of insulin signalling in the adipose tissue (Kimura *et al.* 2013).

7.2.3. SCFA, Host Metabolism, and Metabolic Diseases

SCFAs, among which butyrate is the most well studied, modulate different processes, including cell proliferation and differentiation (Zaibi *et al.* 2010), gut hormone secretion, and immune responses (Atarashi *et al.* 2008; Maslowski *et al.* 2009), and are thus involved in the regulation of host homeostasis. Although metabolic disorders are caused by multiple factors, gut microbiota dysbiosis and improper production of SCFA may play a vital part in the pathogenesis of metabolic disorders. Overall, treatments that enhance microbial production of SCFA or direct supplementation show improvement in metabolic outcomes (Yamashita *et al.* 2007; Gao *et al.* 2009; Lin *et al.* 2012).

7.2.3.1. Short-Chain Fatty Acid and Satiety

High intake of dietary fiber has been indicated to prevent the development of obesity and help reduce body weight (Liu *et al.* 2003; Grube *et al.* 2013). One of the mechanisms by which dietary fiber helps reduce obesity risk is via SCFA-mediated modulation of satiety hormones including glucagon-like peptide-1 (GLP-1) and peptide YY (PYY) that regulate food intake and energy harvest (Tolhurst *et al.* 2012). As an incretin produced by intestinal L cells, GLP-1 increases insulin secretion and decreases food intake by inducing satiety in the brain (Barrera *et al.* 2011). PYY is mainly produced by L cells in the distal gut and is known for its role in regulating pancreatic and gastric secretion, gut motility, insulin secretion, and control of appetite in the central nervous system (Manning and Batterham 2014). Lin *et al.* (2012) observed that oral administration of butyrate and propionate in mice significantly increased the plasma GLP-1 and gastric inhibitory polypeptide (GIP), accompanied by a modest increase in PYY, leading to improved insulin sensitivity. In addition, butyrate has shown the ability to improve insulin sensitivity in high-fat diet-induced obese mice by promoting energy expenditure and mitochondria function (Gao *et al.* 2009). Microbial transplantation from lean donors increases insulin sensitivity in individuals with metabolic syndrome along with the elevated levels of butyrate-producing microbes (Vrieze *et al.* 2012). The underlying mechanisms have yet to be fully established and the possible mechanisms are discussed below.

In the intestine, it is indicated that the effects of SCFA on GLP-1 and PYY production are via GPR43 (Karaki *et al.* 2006; Nohr *et al.* 2013). *Gpr43* and *Gpr41* were abundantly expressed in GLP-1 and PYY secreting L-cells. Whether GPR41 and GPR43 reside on the apical or basolateral membrane of L-cells or whether they primarily detect luminal or plasma SCFA are to be tested (Tolhurst *et al.* 2012). *Gpr43*- and *Gpr41*-knockout mice exhibited reduced SCFA-mediated GLP-1 secretion accompanied with reduced insulin secretion and impaired glucose tolerance consistent with their observation of lower GLP-1 secretion *in vitro* (Tolhurst *et al.* 2012). *Gpr41* or *Gpr43* are expressed in many types of cells including adipocytes, immune cells, and pancreatic β and α cells, therefore global knockout of these receptors may affect glucose metabolism, inflammation, or pancreatic β-cell functions, which would themselves impact on glucose tolerance. In *GPR41*-deficient mice, expression of PYY is reduced, resulting in increased intestinal transit time and reduced energy extraction from SCFA. Conventionalization of GF mice with two predominant bacteria derived from the human gut significantly increased PYY. This effect was suppressed in GF *Gpr41* KO mice when colonized with the two bacteria even though the diet consumption was similar in both groups, indicating that SCFA activation of GPR41 is required for inducing PYY secretion. Furthermore, the body weight and fat pad weight of *Gpr41* KO mice were significantly reduced compared with wild-type littermates, and this effect was abolished in a GF environment; consistent with the *Gpr43* KO mice raised in germ-free conditions, indicating that SCFA derived from metabolism of gut microbiota exert receptor-mediated effects on host adiposity (Samuel *et al.* 2008). To the contrary, another group found that the absence of GPR41 increases body fat content in male mice when fed a high-fat diet without a difference in food intake. Gut-derived SCFA actually raised energy expenditure and helped to protect against obesity by activating GPR41 (Bellahcene *et al.* 2013). In adipose tissue, it was shown that propionate-stimulated activation of GPR41 increases the release of leptin, a hormone controlling the sensation of hunger as well as energy expenditure. In addition, the secretion of leptin increased through overexpression of exogenous Gpr41 and decreased by siRNA-mediated knockdown of *Gpr41* (Xiong *et al.* 2004). However, two other groups could not detect the *Gpr41* expression in adipose tissue (Hong *et al.* 2005, Kimura, Ozawa *et al.* 2013), suggesting that the interaction between GPR41 and leptin may be indirect.

7.2.3.2. SCFA and Inflammation

Obesity and diabetes are characterized by low-grade inflammation with increased levels of cytokines, such as interleukin (IL)-1, IL-6, or tumor necrosis factor-alpha (TNF-α). These inflammatory molecules are upregulated in insulin-target tissues, including liver, adipose tissue, and muscles, thus contributing to insulin resistance (Puddu *et al.* 2014). SCFA, particularly butyrate, were shown to have anti-inflammatory effects (Kendrick *et al.* 2010 Canani *et al.* 2011) through GPCRs. The expression of Gpr43 has been observed in neutrophils, eosinophils, and activated macrophages. *Gpr43* deficient mice had a profoundly altered inflammatory response, which included exacerbated inflammation in a model of allergic airway inflammation as well as in colitis models (Maslowski *et al.* 2009). In addition, Sina *et al.* (2009) showed the importance of GPR43 on leukocyte migration and cytokine secretion in an inflammatory model. SCFA also direct the development of extrathymic anti-inflammatory regulatory T cells (Arpaia *et al.* 2013). As the only known ligands for GPR43, SCFA also control the generation of colonic regulatory T cells and protect against colitis in a Gpr43-dependent manner (Smith *et al.* 2013). SCFA are also beneficial in other ways. SCFA, particularly butyrate, inhibit activation of the transcription factor nuclear factor kappa-light-chain-enhancer of activated B cells (NF-κB) as HDAC inhibitors. GF mice with very low concentration of SCFA showed exacerbated or poorly resolving responses in many inflammatory models (Maslowski *et al.* 2009; Chervonsky 2010), similar to the responses of the *Gpr43* KO mice. SCFA were also shown to have anti-inflammatory effects by reducing chemotaxis and cell adhesion and thus SCFA at least partially prevent infiltration of immune cells into adipose tissue (Meijer *et al.* 2010; Kim *et al.* 2014). Treatment with propionic acid in overweight subjects reduced proinflammatory cytokine and chemokine secretion from human adipose tissue as well as from macrophages. However, the effects of propionate were dependent on GPR41 but not GPR43 (Al-Lahham *et al.* 2012).

7.2.3.3. SCFA Regulate Glucose Metabolism

Plasma glucose levels are determined by food intake, gluconeogenesis, and uptake by multiple organs. As discussed above, in addition to the fact that propionate can be used for gluconeogenesis in liver, the major effects of SCFA on glucose metabolism are through the GPCR by influencing the gut hormones PYY and GLP-1. Intestinal gluconeogenesis (IGN) has also been identified to have beneficial effects on glucose and energy homeostasis including improved glucose tolerance (De Vadder *et al.* 2014). SCFA, especially propionate and butyrate, may exert some of their effects by directly modulating IGN. It was reported that butyrate activated IGN gene expression through a cAMP-dependent mechanism, while propionate, itself a substrate of IGN, activates IGN gene expression via a gut-brain neural circuit involving GPR41. The benefits on glucose and energy control induced by SCFA or dietary fiber in normal mice were absent in mice deficient for IGN (intestinal specific deletion of G6Pase catalytic subunit), despite similar modifications in gut microbiota composition. Therefore, regulation of IGN seems to be key to beneficial metabolic effects associated with SCFA and soluble fiber (De Vadder *et al.* 2014).

7.2.3.4. SCFA Regulate Fatty Acid Metabolism

Adipocytes are key target cells concerning the prevention and treatment of Type 2 diabetes. Impaired regulation of lipolysis and accumulation of lipid intermediates may contribute to obesity-related insulin resistance and Type 2 diabetes. In adipose tissue, the hydrolysis of triacylglycerol into free fatty acids and glycerol, known as lipolysis, often leads to excessive release of fatty acids to the plasma and decreased insulin sensitivity in overweight or obese individuals (Nielsen *et al.* 2004; Jocken *et al.* 2013). Heimann *et al.* (2014) recently showed that SCFA inhibit lipolysis and the basal and insulin-stimulated lipogenesis as well as enhancing the insulin-stimulated glucose uptake in primary adipocytes that might be beneficial for prevention of dysfunctional adipocytes associated with insulin resistance as seen in obese and diabetic individuals.

The series of experiments conducted by Bäckhed and his colleagues (2004, 2007) showed that adiposity increased in the conventionalized GF mice and was caused by the microbial suppression of *Fiaf* (fasting-induced adipocyte factor or ANGPTL4), an inhibitor of lipoprotein lipase (LPL). LPL promotes release of fatty acids from circulating chylomicrons and very low-density lipoproteins (VLDL), which results in their storage as triglycerides in the adipose tissue. FIAF inhibition of LPL therefore reduces fat storage. One reason why GF mice are protected from obesity is the elevated expression of *Fiaf* in the intestine (Backhed 2004; Backhed *et al.* 2007). A recent study showed that transcription and secretion of FIAF in human colon adenocarcinoma cells is induced by high concentration of SCFA via nuclear receptor peroxisome proliferator activated receptor γ (PPARγ) and butyrate is the strongest activator followed by propionate and acetate (Alex *et al.* 2013), which provides another mechanism through which SCFA prevent adiposity. SCFA have been shown to increase the AMPK activity in liver and muscle tissue, which will activate peroxisome proliferator-activated receptor gamma coactivator (PGC)-1α expression that is important in regulation of cholesterol, lipid, and glucose metabolism (Yamashita *et al.* 2007; Gao *et al.* 2009). SCFA may also influence fatty acid metabolism via GPR43 as adipocytes treated with acetate and propionate exhibited a reduction in lipolytic activity that is not observed in GPR43 deficient mice (Hong *et al.* 2005; Gao *et al.* 2009). Li *et al.* (2014) recently showed that the effect of SCFA in enhancing adipocyte differentiation was not via GPR41 or GPR43 in stromal vascular fraction of porcine subcutaneous fat (Li *et al.* 2014). Thus, further study is needed to identify molecular pathways of SCFA-stimulated adipogenesis.

7.2.4. SCFA Summary

SCFA are one of the most important microbial products and affect a range of host processes-including energy harvest, host-microbe signalling, and gut integrity. Since SCFA have the effect of inhibiting food intake, improving insulin sensitivity, and increasing energy expenditure in the host, they are regarded as healthy microbial products that prevent the development of obesity and diabetes. Even though gastric bypass surgery is still regarded as the most efficient treatment of obesity, SCFA supplementation or probiotic intake can also be considered as a safe treatment alternative to prevent and treat obesity (Yadav *et al.* 2013).

7.3. MICROBIOTA AND BILE ACID METABOLISM

Bile acids are widely known for their role in promoting the absorption of dietary lipids and lipid-soluble nutrients in the intestine. However, more recently bile acids have been identified in metabolic regulation as signalling molecules that regulate a network of lipid, glucose, drug, and energy metabolism (Chiang 2009). Bile acid signaling has also been identified as a mechanism through which microbes regulate host metabolism, as intestinal bacteria modify the structure and abundance of bile acids.

7.3.1. Biosynthesis and Transport of Bile Acids

Bile acids are derived from cholesterol in the liver through a multienzyme process. The full complement of bile acid synthesis requires at least 16 enzymes that catalyze as many as 17 reactions to convert cholesterol into conjugated bile salts (Russell 2003). The immediate products of these reactions are referred to as primary bile acids. In humans, there are two primary bile acids, which are chenodeoxycholic acid (CDCA) and cholic acid (CA). Mice synthesize CA and muricholic acid (MCA), the latter of which is significant in that βMCA is the primary bile acid responsible for activating some signalling pathways discussed below. The first and rate-limiting step of primary bile acids biosynthesis, hydroxylation at the carbon atom C-7, is mediated by the enzyme *CYP7A1*. The final step is conjugation of bile acids with an amino acid. Conjugated bile acids have lower p*Ka* of the terminal acidic group, and therefore they can be freely soluble at acidic pH and resistant to precipitation over a wide range of ionic strengths, calcium concentration, and pH values (Begley *et al.* 2005). The biological consequences are full ionization of conjugated bile acids at physiological pH, reduced ability to cross the cell membrane lipid bilayer, and maintenance of high intraluminal concentration of bile acids, which are essential for promotion and facilitation of fat digestion and absorption (Hofmann and Mysels 1992). More than 97% excreted conjugated bile salts are reabsorbed in the distal ileum and returned to the liver through the hepatic-portal circulation (Macdonald *et al.* 1983). Conjugated bile acids can be absorbed by passive mechanisms along the intestine and active mechanisms mainly located in the terminal ileum and mediated by an efficient transport system (Dietschy 1968). This highly efficient circulation contributes to the accumulation of a pool of recycling bile acids between the liver and the intestine. The process of recycling the bile acid pool is termed enterohepatic circulation. Daily fecal loss of bile acids is only approximately 5% of the bile acid pool (Begley *et al.* 2005). The size of bile acid pool can be maintained at a constant level, approximately 2–4 g in humans (Dawson 2011).

7.3.2. Microbial Involvement in Bile Acid Circulation

After discharge into the small intestine, conjugated bile acids are exposed to indigenous microbes. Gut bacteria modify bile acids via three mechanisms, which include deconjugation, 7α-dehydrogenation, and 7α-dehydroxylation of primary bile acids (Midtvedt 1974). The process results in the formation of secondary bile acids, which increase the chemical diversity of the bile acid pool (Ridlon *et al.* 2006). The primary bile acids cholic and chenodeoxycholic acid are converted into the secondary bile acids deoxycholic acid (3α, 12α-dihydroxy, DCA) and lithocholic acids (3α-hydroxy, LCA), respectively.

Microbes alter both the size of the pool as well as the profile of bile acids. Microbial colonization increases the fecal loss of bile acids (double) (Kellogg and Wostmann 1969), which coincides with an overall reduction in the bile acid pool (Claus *et al.* 2008; Sayin *et al.* 2013). Administration of antibiotics, including streptomycin and penicillin, increases the proportion of taurine-conjugated bile acids (Swann *et al.* 2011). Circulating bile acids under normal conditions are primarily unconjugated whereas GF rats have a circulating bile acid profile dominated by conjugated bile acids, especially taurocholic acid and tauro-β-muricholic acid (more than 93%) (Swann *et al.* 2011). While microbes reduce the bile acid pool size they increase chemical diversity of bile acids, where diversity is greatest in the cecum, colon, and feces (Sayin *et al.* 2013).

Microbial biotransformation of bile acids is mainly attributed to anaerobic bacteria of the genera *Bacteroides, Eubacterium*, and *Clostridium* (Ridlon *et al.* 2006; Nicholson *et al.* 2012). The deconjugation process is performed by various genera such as *Bacteroides, Proteus, Lactobacillus, Clostridium*, and *Streptococci*. The dehydrogenase activity is performed by a wide range of bacteria including species of *E. coli, Pseudomonas, Bacteroides, Eubacterium, Bacillus*, and *Clostridium* (Floch 2002). *Bifidobacterium* spp. were found to excrete bile salt hydrolase (Cholyglycine hydrolase, BSH), an enzyme that catalyzes the hydrolysis of conjugated bile acids into amino acid residues and free bile acids (Jones *et al.* 2008).

7.3.3. Microbial Regulation of Chronic Disease Via Bile Acids Mediation

Obese patients treated orally with vancomycin show reduced fecal secondary bile acids and simultaneous postprandial increase of plasma primary bile acid. Meanwhile, vancomycin treatment decreased peripheral insulin sensitivity. The disappearance of the *Firmicutes* phylum was associated with the deleterious effects on bile acid and glucose metabolism (Vrieze *et al.* 2014). It is proposed that microbes that affect the bile acid pool influence the lipid and glucose metabolism via the Farsenoid X Receptor (FXR) and G-coupled bile acid receptor 1 (TGR5) signalling pathways (Nieuwdorp *et al.* 2014).

7.3.3.1. Farnesoid X Receptor and Metabolism of Bile Acids

It is important that the body can recognize changes in bile acid content so as to maintain an appropriate bile acid pool. The FXR is important in this maintenance as it controls both bile acid biosynthesis and enterohepatic cycling. FXR affects the biosynthesis of bile acids in the liver by repressing the expression of CYP7A1 (rate limiting step in bile acid synthesis) through direct regulation of the short heterodimer partner (SHP). Interruption of this feedback loop (SHP$^{-/-}$ mice) results in increased bile acid pool (Kerr *et al.* 2002). Activation of FXR in the intestine contributes to bile acid regulation through the expression and secretion of fibroblast growth factor-19 (FGF19; FGF15 in rodents). FGF19 is expressed in the ileum and belongs to the fibroblast growth factors family, which governs nutrient metabolism (Beenken and Mohammadi 2009). FGF19 is secreted into the circulation and can suppress expression of CYP7A1 in the liver (Wang *et al.* 2002; Holt *et al.* 2003; Kir *et al.* 2011). FXR-mediated repression of CYP7A1 contributes to the negative feedback regulation of bile acid synthesis (Lefebvre *et al.* 2009), and this feedback mechanism is impacted by microbial metabolism of bile acids. Figures 7.3(a) and 7.3(b) depict the relationship between microbial bile acid deconjugation and FXR signalling.

7.3.3.2. FXR Signalling Pathway and Glucose and Lipid Metabolism

Substantial evidence indicates that the FXR signalling pathway plays an important role in the regulation of glucose metabolism and metabolic disease outcomes. However, due to opposing findings, it is as yet unclear in which direction FXR signalling should be promoted/inhibited to result in a healthy outcome. In rodents, the presence of primary bile acids (specifically tauromuricholic acid or T-MCA) reaching the ileum suppresses the expression of FGF15, resulting in increased bile acid production. Microbes reduce the levels of primary bile acids promoting FGF15 expression in the ileum, resulting in suppressed CYP7A1 transcription and bile acid synthesis in the liver (Sayin *et al.* 2013). One study showed that treatment with a therapeutic antioxidant known as tempol (4-hydroxy-2,2,6,6-tetramethylpiperidine 1-oxyl) caused the accumulation of intestinal primary bile acid T-MCA, which in turn resulted in the inhibition of FXR signalling in the intestine (Li *et al.* 2013). Tempol is a member of nitroxide family and has been used to protect the cell from oxidative injury (Charloux *et al.* 1995), however more recently it has been shown to improve insulin sensitivity, reducing blood glucose and triglyceride levels in obese rats (Banday *et al.* 2005). As obesity is related to inflammation, the use of tempol protected mice treated with a high-fat diet from obesity and insulin resistance. But this wasn't due to a chemical interaction; rather it was due to a remodelling of the intestinal microbiota (Li *et al.* 2013).

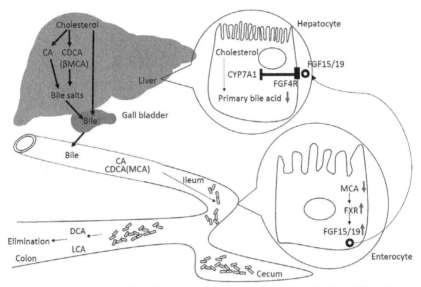

FIGURE 7.3(a). Gut microbiota decrease bile acids synthesis by activating FXR pathway; shown with microbiota. The figure shows the enterohepatic circulation of bile acids and microbial regulation of primary bile acids synthesis. In humans, CA and CDCA are major primary bile acids and CA and MCA are predominate in mice. When secreted by the gall bladder, primary bile acids enter into the intestine. With the presence of gut microbiota, the ratio of MCA/CA isdecreased. Being poor FXR agonists, the reduction in TβMCA leads to the activation of FXR in enterocytes. The increased FGF15/19, a potential suppressor of CYP7A1 in the liver, results in decreased CYP7A1, which inhibits the synthesis of primary bile acid.

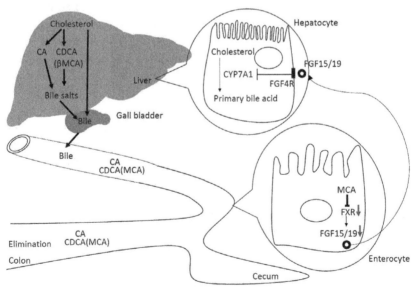

FIGURE 7.3(b). Gut microbiota decrease bile acids synthesis by activating FXR pathway; shown without microbiota. Without gut microbiota, the inhibition of CYP7A1 regulated by FXR pathways is reduced, which leads to stimulation of primary bile acids synthesis.

It has also been reported that FXR deficiency benefits body weight and glucose homeostasis based on diet-induced obesity on FXR$^{-/-}$ mice with obese background (Prawitt et al. 2011). However, FXR$^{-/-}$ mice on a normal background developed severe fatty liver, increased circulating free fatty acids (FFAs), elevated serum glucose levels, and an impaired glucose tolerance (Ma et al. 2006). A fasting-refeeding experiment in FXR$^{-/-}$ mice showed that FXR interfered with glycolytic and lipogenic pathways in the liver through genes involved in hepatic carbohydrate and lipid metabolism (*Lpk*, *Acc-1*, and *Spot14*) (Duran-Sandoval et al. 2005). These studies provide evidence of FXR's role in the dynamic regulation of glucose metabolism, however, due to different animal models (FXR$^{-/-}$ lean/obese), tested ligands, and FXR's different functions in the liver and intestine, contradictory evidence is apparent and a precise mechanism of FXR regulation of glucose has yet to be established (Staels et al. 2010).

7.3.3.3. *G-Protein-Coupled Receptor TGR5 (GPR131, M-BAR of BG37) Signalling Pathways*

Besides FXR, bile acids also act as ligands for TGR5 in the liver and intestine. TGR5 has been recognized to promote glucose homeostasis. Primary and secondary bile acids of lithocholic acid and deoxycholic acid promote glucagon-like peptide-1 (GLP-1) secretion in a murine enteroendocrine cell line STC-1 via the TGR5 pathway (Katsuma et al. 2005). A transgenic mouse model overexpressing TGR5 showed dramatically improved glucose tolerance in mice fed with HFD. The mechanism was associated with a marked increase in secretion of GLP-1 and insulin release after oral glucose administration. Also TGR5 signalling in enteroendocrine L-cells induced intestinal GLP-1 release, leading to the improvement of liver and pancreatic function (Thomas et al. 2009). TGR5 activation and GLP-1 secretion may provide a novel strategy to treat metabolic diseases such as diabetes and nonalcoholic fatty liver disease (Pols et al. 2010). The effects of gut microbiota on glucose homeostasis and insulin sensitivity may contribute to the development of metabolic diseases through regulating the bile acid profile (Duseja and Chawla 2014). Activation of TGR5 in brown adipose tissue and skeletal muscle by bile acids increased energy expenditure and prevented obesity and resistance to insulin in mice. The activation effect mainly induced Type 2 deiodinase (D2) to convert inactive thyroxine into active 3, 5, 3'-triiodothyronine via cAMP signalling pathway, which is known to stimulate mitochondrial oxidative phosphorylation and energy expenditure. The bile acid-TGR5-cAMP-D2 signalling pathway has been identified as the essential machinery for fine-tuning energy homeostasis resulting in improved metabolic control (Watanabe et al. 2006). Gut microbiota may activate TGR5 signalling by increasing hydrophobic bile acids to regulate lipid and glucose homeostasis. As well as its role in regulation of GLP1 and in energy expenditure, TGR5 signalling is anti-inflammatory and can mediate metabolic inflammation, which is discussed below (Yoneno et al. 2013). The relationship between TGR5 signalling and metabolism are shown in Figure 7.4.

7.3.4. Dietary Effects on Microbiota by Affecting Bile Acid Profile

Dietary and antimicrobial compounds can affect the equilibrium of microbiota and bile acid metabolism. For example, high levels of dietary saturated fat promote the growth of the pathobiont *Bilophila wadsworthia*, a response induced by increased taurine-conjugated bile acids. Dietary fats can markedly alter the conditions for microbial assemblage by changing the host bile acid profile, leading to immune imbalance (Devkota et al. 2012). Rats supplemented with different concentrations

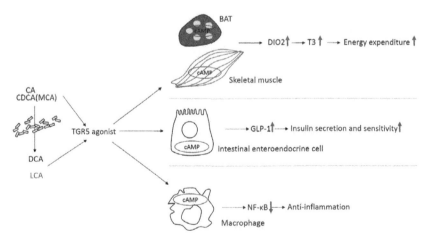

FIGURE 7.4. Bile acid-activated TGR5 signaling pathway associated with gut microbiota in metabolism and inflammation. Though free bile acid and conjugated bile acid are all TGR5 agonists, LCA is the most potent form. Being activated by bile acids, TGR5 stimulates energy expenditure in mice BAT and human skeletal muscle by inducing cAMP/PKA signaling. In the intestine, TGR5 stimulates GLP-1 expression to increase insulin sensitivity, which relates to glucose homeostasis. In macrophage cells, activated TGR5 inhibits NF-κB-dependent proinflammatory cytokine production, which results in anti-inflammation effects.

of cholic acid for 10 days showed altered gut microbiota at the phylum level. Cholic acid reduced microbiota diversity and *Firmicutes* were shown to predominate at the expense of *Bacteroidetes*. Increased input of bile acids resulted in significant inhibition of the *Bacteroidetes* and *Actinobacteria* (Islam *et al.* 2011). These studies demonstrated that bile acids can shape the gut microbiome, and in turn, the microbiota shift can influence the bile acid profile and thereby change host systematic metabolism and health. These studies also reveal an opportunity to modulate the bile acid profile by dietary administration, which can be potentially used in treating metabolic disease.

7.4. GUT MICROBIOTA, TMAO, AND ATHEROSCLEROSIS

Atherosclerosis, with manifestations such as myocardial infarction and stroke, is characterized by an imbalance in lipid metabolism and a maladaptive immune response driven by cholesterol accumulation and inflammation, which includes recruitment of macrophages in the arterial wall (Karlsson *et al.* 2012). Evidence has recently emerged that microbial composition is related to cardiovascular disease (CVD). With the target of discovering unbiased small-molecule metabolic profiles in plasma that predict increased risk for CVD, a downstream product of microbial fermentation, trimethylamine-N-oxide (TMAO), was identified (Wang *et al.* 2011). Further studies have confirmed the association of TMAO levels with major CVD, which is a complex pathophenotype that can be affected by genetic and environmental factors (Loscalzo 2011). The production of TMA and downstream responses are depicted in Figure 7.5.

7.4.1. Dietary Precursors of Trimethylamine-N-Oxide

Choline, phosphatidylcholine (PC), and L-carnitine are the trimethylamine-containing compounds in the diet that can be converted by intestinal microbes to trimethylamine (TMA). TMA is absorbed by colonic cells and converted to trimethylamine-N-oxide (TMAO) by flavin monooxygenase (FMO) enzymes in the liver, or excreted in the urine. TMAO then enters plasma and circulates to reach other cell types such as arterial epithelial cells and macrophages (Zeisel *et al.* 1989; Russell *et al.* 2013).

Choline is a semiessential dietary nutrient in that endogenous synthesis in the liver is insufficient to meet requirements. Choline is found in egg yolk, meat, liver, fish, dairy products, nuts, and soybeans (Zeisel

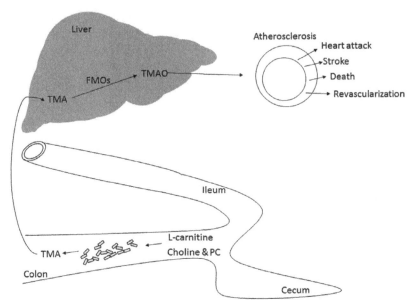

FIGURE 7.5. Gut microbiota metabolism of dietary PC, L-carnitine, and choline and atherosclerosis. L-carnitine and choline are metabolized into TMA by gut commensal microbes, which are metabolized by hepatic enzyme FMOs into TMAO. TMAO influences cholesterol metabolism, which relates to cardiovascular diseases.

2006). The phospholipid PC is the chief dietary source of choline, and is known as a major phospholipid in cell membranes. Besides being a component of eukaryotic cellular membranes, PC is a precursor of signalling molecules, and a primary phospholipid component of circulating VLDL and low-density lipoproteins (LDL). VLDL and LDL are responsible for forward cholesterol transport to the body's tissues away from the liver and elevated levels are associated with adverse health outcomes.

L-carnitine is an abundant nutrient in red meat and has a similar polar head group to choline. Eukaryotes can produce L-carnitine endogenously, but only prokaryotic organisms can catabolize it (Koeth *et al.* 2013). L-carnitine plays an important role in importing activated long-chain fatty acids from the cytosolic compartment into the mitochondria. The complete metabolism of L-carnitine by microorganisms produces glycine, betaine, and acetate. However, some bacteria can convert L-carnitine into TMA (Rebouche and Seim 1998).

7.4.2. TMAO Impacts Host Metabolism

A 3-year study, based on 4,007 patients who were undergoing elective diagnostic cardiac catheterization, showed that increased plasma levels of TMAO was associated with an increased risk of a major adverse CVD. Increasing levels of choline, TMAO, and betaine in plasma were observed with the increased risk of all CVD phenotypes (Tang *et al.* 2013). Atherosclerosis-prone mice ($Apoe^{-/-}$) and atherosclerosis-resistant mice ($Apoe^{+/+}$) confirmed the critical role of FMO in TMAO production, and TMAO levels positively correlated with the size of atheroma. Excessive levels of choline in the diet (1%) given to atherosclerosis-prone mice increased foam cell formation with a concurrent increase in the scavenger receptor CD36 and SRA1 protein of peritoneal macrophages (Wang *et al.* 2011). CD36 and SRA1 are involved in modified LDL uptake by macrophages and intracellular accumulation of cholesterol, thereby contributing to foam cell formation (Silverstein *et al.* 2010). Additionally, increasing the accumulation of cholesterol within cells of the artery wall by TMAO may result from the inhibited removal of cholesterol from peripheral macrophages. Mice consuming a TMAO-containing diet showed 35% decrease in global reverse cholesterol transport than the group fed a normal chow diet. The suppression of reverse cholesterol transport can reduce the rate of cholesterol efflux, thus contributing to atherosclerotic phenotypes. Therefore, the proposed mechanisms through which TMAO promotes atherosclerosis are suppression of reverse cholesterol transport and stimulation of macrophage scavenger receptors and foam cell formation. TMAO can also reduce the expression of key bile acid synthetic enzymes and multiple bile acid transporters in the liver. However, it was not clear whether reduced bile acid pool size contributed to the impairment of reverse cholesterol transport. The relationship between bile acid and TMAO in cholesterol metabolism still requires further study (Koeth *et al.* 2013).

7.4.3. TMA Produced By the Gut Microbiota

Commensal bacteria have an obligate role in the formation of TMAO. Antibiotic treated mice and GF mice generate no increase in plasma TMAO in response to dietary choline, while the levels of plasma TMAO increase readily after microbial colonization (Wang *et al.* 2011). Human subjects treated with antibiotics show almost complete suppression of endogenous TMAO in both plasma and urine after L-carnitine challenge. Upon discontinuation of antibiotics, TMAO in plasma and urine recover with the re-establishment of the intestinal microbiota (Koeth *et al.* 2013).

Choline that escapes absorption in the upper gastrointestinal tract is subject to microbial choline metabolism. Choline is degraded by select bacteria and it has been reported that sulfate-reducing vibrio, such as *Desulfovibrio desulfuricans*, can degrade choline (Hayward and Stadtman 1959; Baker *et al.* 1962; Hagiwara *et al.* 2014). The gene cluster of choline utilization (cut) has been demonstrated to play an essential role in the conversion of choline to TMA by a genetic knockout strategy in sulfate-reducing bacteria *Desulfovibrio alaskensis* G20 strain and heterologous expression in *Escherichia coli* (Craciun and Balskus 2012). While few bacteria have been tested for choline metabolism, bioinformatic analysis identified 89 bacterial genomes in the National Center for Biotechnology Information (NCBI) protein database, which were predicted as choline degraders. In human gut commensals, choline utilization is not evenly distributed. *Cut* gene clusters are mainly found in the *Firmicutes*, *Proteobacteria* and *Actinobacteria* phyla, while they are absent from most Bacteroidetes. Several members of the human intestinal microbiota including *Clostridium, Anaerococcus, Collinsella, Desulfitobacterium, Klebsiella, Escherichia, Providencia, Yokenella*, and *Proteus* spp. and other members may be able to degrade choline to TMA via cut related pathways (Craciun and Balskus 2012). This would suggest that the composition of the microbiota could determine the production of TMA.

7.4.4. Microbiota Alters TMAO Production

Shotgun sequencing of the gut metagenome has identified alterations in microbial community including enrichment of the genus *Collinsella* in patients with symptomatic atherosclerosis, whereas healthy controls were enriched in *Roseburia* and *Eubacterium* (Karlsson *et al.* 2012), supporting an increased production of TMAO. The ability of the microbial population to produce TMA from carnitine is dependent on previous exposure to L-carnitine. Microbial populations adapted to a high animal protein diet are more able to produce TMA from carnitine, indicating that the nature of the microbial population determines the level of TMA. Enriched proportions of the genus *Prevotella* was related to higher plasma TMAO levels while the lower TMAO producers, vegans and vegetarians, tended to have higher population of the genus *Bacteroides* (Koeth *et al.* 2013). It is suggested that the capacity of the microbiome to metabolize L-carnitine to TMAO is owed to the establishment of gut microbiota in response to host dietary habits, as shown by the absence of TMAO elevation in vegan subjects after an L-carnitine challenge test (Backhed 2013). Consistent with the role of microbes in the production of TMAO from dietary PC and the relationship between elevated levels of TMAO with risk of major adverse cardiovascular events (Tang *et al.* 2013), antibiotic treatment completely suppressed the enhancement of atherosclerosis mediated by dietary choline in mice (Wang *et al.* 2011).

Thus far, the exploration of the role of the commensal intestinal microbiota in the well-established link between high levels of red meat consumption and CVD risk has provided new insight of the gut-microbe-diet interaction. Tasks required for the next step in understanding of the significance of the gut microbiota in this context include identification of the specific bacterial sources of TMA, investigating the sequelae of the genetic variation in FMO3 expression, identification of the molecular mechanisms of TMAO-mediated reverse cholesterol transport, and further elucidation of the relationship of the microbiota with another important function of choline and betaine as methyl donors (Loscalzo 2011). While choline requirements must be met, dietary recommendations should limit the intake of PC- and choline-rich foods. Alternatively, strategies can be employed to minimize microbial TMA production in at-risk individuals to reduce the CVD risk in clinical therapy application (Tang *et al.* 2013).

7.4.5. Choline Deficiency and Nonalcoholic Fatty Liver Disease

It has been reported that choline deficiency results in hepatic steatosis and can be reversed by the supplementation of choline (Buchman *et al.* 1995). A study based on 129S6 mice, known to be susceptible to dietary induced obesity, diabetes, and NAFLD, showed that disruptions of choline metabolism contributes to impaired glucose homeostasis and NAFLD. Given that conversion by gut microbiota reduces bioavailability of choline and can result in lower plasma PC levels and high urinary excretion of TMAO, reducing TMA production could also reduce concerns of choline deficiency (Dumas *et al.* 2006). The relationship between choline metabolism and NAFLD shows the essential role of choline. Conversely, the association between TMAO and CVD indicates the extraordinary complexity of the metabolic disease development (Loscalzo 2011).

7.5. METABOLIC INFLAMMATION

While acute inflammation is essential to heal the body after an injury or infection, extensive research over the past decade has made it clear that chronic, low-grade inflammation contributes to the onset of metabolic diseases including obesity, diabetes, CVDs, and nonalcoholic steatohepatitis (NASH) (Hotamisligil 2006; Szabo, Bala et al. 2010; Lumeng and Saltiel 2011; Tarantino 2014). Plasma bacterial endotoxin (e.g., lipopolysaccharide [LPS]) originating from the intestinal microbiota has been implicated in inducing metabolic inflammation resulting in insulin resistance in obesity and type 2 diabetes (Basu et al. 2011; Lassenius et al. 2011). The role of LPS in inducing inflammation and metabolic disease is depicted in Figure 7.6.

LPS is a major component of the outer membrane of Gram-negative bacteria and is released upon microbial death (Rietschel et al. 1996). The lipid A domain, which is highly conserved, is the ligand that activates Toll-like receptor 4 (TLR4) that is present on the cell membrane of myocytes, immune cells, adipocytes, and epithelial cells (Frantz et al. 1999; Devaraj et al. 2008; Vitseva et al. 2008; Ioannidis et al. 2013). Recognition of TLR4 activates downstream proinflammatory responses (Dixon and Darveau 2005) essential for the host to defend against bacterial infection. The concentration of LPS is low in the circulation of healthy individuals; however, it has been shown that obese subjects and diet-induced animal models have increased circulating LPS levels (Cani et al. 2007; Erridge et al. 2007; Pussinen et al. 2011). Importantly, increased adiposity and insulin resistance usually seen in animals fed a high fat diet can also be observed in mice chronically infused with LPS; indicating that LPS triggers adiposity and diabetes via activation of the immune system (Cani et al. 2007).

7.5.1. Consequences of Inflammation on Metabolism

Exposure to LPS results in innate immune activation in many cell types, especially monocytes, dendritic cells, and macrophages. The production and secretion of many proinflammatory cytokines such as tumor necrosis factor α (TNF-α), interleukin 1 (IL-1), and IL-6 have been indicated to play a role in the pathogenesis of chronic inflammation in obesity, insulin resistance, and diabetes (Musso et al. 2011) through their effects on various tissues. Chronic inflammation causes damage to pancreatic beta cells (Hohmeier et al. 2003; Eguchi et al. 2012), hepa-

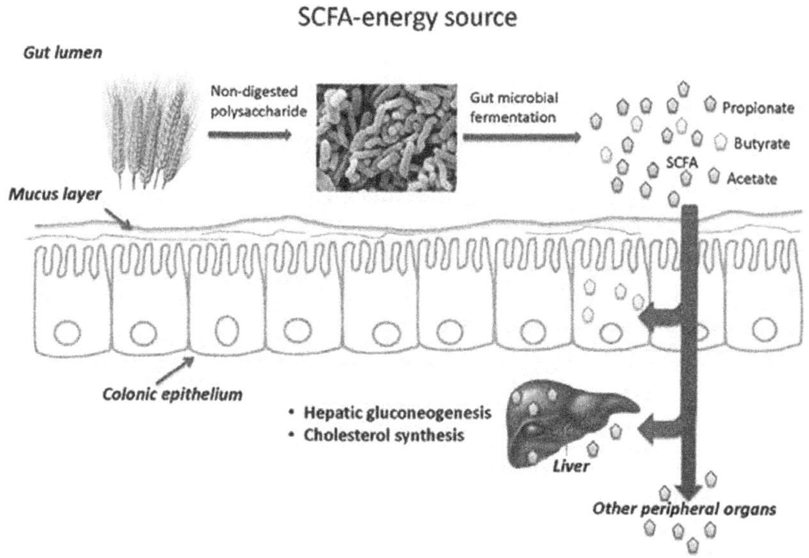

SCFA-energy source

FIGURE 7.6. The role of LPS in inducing inflammation and metabolic diseases. High fat intake increases plasma lipopolysaccharides (LPS). LPS translocates through gut epithelium with chylomicrons and induces local inflammation, which impairs tight-junctions increasing gut permeability. Once circulating in the blood, LPS activates immune cells and induces proinflammatory cytokine production. When plasma LPS reaches peripheral organs, LPS activates TLR4 on the cell membranes and leads to downstream inflammatory responses and immune cell infiltration which impairs insulin signaling and insulin related gene expression in pancreas, adipose tissue, liver and skeletal muscle.

tocytes (Gieling et al. 2009), adipocytes (Hotamisligil 2006), and vascular endothelial cells (Hansson 2005).

Insulin signalling is one of the first processes affected by LPS arriving in the circulation. Insulin signalling is a complex process that involves multiple pathways and cascades of phosphorylation events. Interference with these signalling pathways can alter insulin production or sensitivity and lead to the development of insulin resistance (Hotamisligil 2006). TLR4 is expressed in insulin target tissues, including skeletal muscle, adipose tissue, and liver. Activation of TLR4 by LPS activates proinflammatory kinases such as c-Jun N-terminal kinase (JNK), IκB kinase (IKK), and p38 mitogen-activated protein kinase (MAPK) that target the insulin receptor substrate (IRS) for serine phosphorylation, which in turn impairs insulin signal transduction. The inflammatory kinases also regulate downstream transcription factors, such as NF-κB and interferon regulatory factor (IRF), resulting in increased production of cytokines. The increased cytokines and chemokines induced by TLR4 promote further insulin-desensitization within the target cells and in other cells via paracrine and systemic effects (Kim *et al.* 2014).

The pancreas is at the heart of glucose metabolism, given its role as the sole source of insulin and glucagon. Insulin is secreted in response to nutrients being absorbed into the bloodstream after a meal, particularly glucose. Insulin then instructs cells of the body to take up glucose and as well activates anabolic pathways. The pancreas can suffer directly from inflammation, as well as indirectly as a result of insulin resistance in other tissues. Recent studies have shown that macrophage infiltration in pancreas is triggered by TLRs 2 and 4, which leads to greater expression of inflammatory cytokines, IL-1β, TNF-α, and interferon-γ (IFN-γ) (Ehses *et al.* 2008; Donath *et al.* 2010; Cucak *et al.* 2014; Westwell-Roper *et al.* 2014). Critical events in the progression to diabetes were also observed in chronic inflammation models, including reduced insulin secretion and β cell apoptosis and decreased islet mass, which are due to a pathological activation of the innate immune system by metabolic stress (Donath *et al.* 2010; Wu *et al.* 2013). These all suggest that chronic inflammation triggers insulin resistance and impaired islet function. As a low-grade inflammation activator, LPS has shown effects on β cell dysfunction in a rat model infused with LPS (Hsieh *et al.* 2008). Recently, Amyot *et al.* (2012) demonstrated that LPS could inhibit β cell transcriptional activity in a TLR4-dependent manner and via NF-κB signalling.

The reduced capacity of cells to take up glucose in response to insulin results in elevated blood glucose levels, which in turn activates further insulin secretion causing stress on the pancreas. Skeletal muscle consumes the largest amount of glucose in the body. Probiotic strains of bacteria have been administered to mice and showed indications of greater insulin sensitivity including increased expression of glucose transporter type 4 (GLUT4) and glycogen synthesis related enzymes (Kim *et al.* 2014).

Among metabolic tissues affected by inflammation, adipose was the first described and most studied in metabolic disorders. Adipose tissue acts not only as a storage depot for energy, but also secretes hormones, cytokines, and chemokines that promote inflammation, and conversely adipokines and adiponectin that promote insulin sensitivity (Hotamisligil 2006). Generally, LPS influences the adipose metabolism in four ways. First is the recruitment of macrophages to the adipose tissue, resulting in amplified inflammation (Locati *et al.* 2013). Second is the suppression of insulin signalling in adipose tissue, as discussed above, as well as through activation of phosphoinositide 3-kinase (PI3K)/Akt (Wakayama *et al.* 2014). Third, LPS has been shown to induce lipolysis in human adipocytes via both IKKβ/ NF-κB pathway and protein kinase A (PKA)/hormone-sensitive lipase (HSL) pathways (Grisouard *et al.* 2012). Finally, LPS regulates adipocyte differentiation by downregulating the activity of PPARγ, which is essential to adipogenesis and to maintenance of adipocyte gene expression and function (Luche *et al.* 2013)

The liver plays a key role in distributing nutrients to the rest of the body. This includes the storage and release of glucose and packaging of fatty acids in the form of VLDL and LDL. There are several liver diseases where microbe-induced inflammation plays an important role including NASH and NAFLD. Bacterial translocation has been associated with fat accumulation in the liver and liver fibrosis. Removal of Kuppfer cells, and thus the primary mechanism to induce inflammation in the liver, prevents the development of hepatic steatosis and insulin resistance in response to HFD. The production of TNF-α by Kuppfer cells induced hepatocyte triglyceride accumulation, fatty acid esterification, and reduced fatty acid oxidation and insulin responsiveness (Huang *et al.* 2010). Furthermore, TLR4 deficient mice showed reduced liver injury in a model of NASH (Rivera *et al.* 2007). Inflammation also impacts the ability of macrophages to facilitate cholesterol removal through reverse cholesterol transport (Majdalawieh and Ro 2009). Bacterial overgrowth is one reason for increased bacterial translocation, and thus activation of Kuppfer cells in the liver leading to liver disease (Pardo *et al.* 2000).

7.5.2. LPS, High Fat Diet, and Gut Permeability

HFD has been shown to cause metabolic endotoxemia in animals and humans (Cani *et al.* 2007; Ghanim *et al.* 2009; Pendyala *et al.* 2012), but the underlying molecular mechanisms remain incompletely understood. In healthy humans, postprandial plasma LPS concentration increased significantly compared with the fasting level after a high-fat meal or a mixed meal containing emulsified fat (Erridge *et al.* 2007; Laugerette *et al.* 2011). In order to establish whether plasma LPS levels are based on energy or fat intake, fasting normal subjects were given drinks containing glucose, 100% saturated fat, or orange juice. Elevated LPS levels were only associated with the fat intake (Deopurkar *et al.* 2010). LPS reaches the blood from the intestinal lumen through paracellular and transcellular routes. Impaired epithelial integrity permits increased paracellular transport of LPS. Transcellular transport occurs through receptor-mediated endocytosis (Mani *et al.* 2012). Ghoshal *et al.* (2009) demonstrated that intestinal epithelial cells internalize LPS from the apical surface, which is then transported to the Golgi apparatus where it complexes with chylomicrons. The chylomicron–LPS complex is then secreted into mesenteric lymph and makes its way into the systemic circulation (Ghoshal *et al.* 2009). In the bloodstream, LPS binds to many macromolecules such as albumin, LDL, HDL, and LPS-binding protein. Once combined with LPS-binding protein, LPS can be transferred to membrane-bound or soluble CD14, thereby enabling interactions with TLR4 on cell membranes and downstream inflammatory responses (Beumer *et al.* 2003).

The intestinal epithelium primarily serves as a dynamic barrier, which in the course of its normal function maintains regulated absorption of nutrients and water while excluding potential pathogens. The gut integrity is maintained by commensal bacteria, epithelial intestinal cell factors, and gut hormones (GLP-1, PYY etc.) which could be disrupted when exposed to HFD (Bleau *et al.* 2014). Although evidence to support increased gut permeability in human subjects is minimal, many animal studies have shown increased LPS levels in the lamina propria, mesenteric adipose tissue, and blood resulting from gut barrier dysfunction and bacterial translocation in genetic obese or diet-induced models, reflected by reduced expression of tight junction proteins zona occuldens (ZO)-1, occludin, claudin-1, claudin-3, and junctional adhesion molecule 1 (JAM-1) (Brun *et al.* 2007; Teixeira *et al.* 2012; Bleau *et al.* 2014). In NASH, increased intestinal permeability, increased LPS levels, and tight-junction alterations are also observed (Miele *et al.* 2009; Ilan 2012). Indirect evidence that shows the association between gut permeability and LPS translocation was from probiotic research. *Akkermansia muciniphila* has been shown to restore gut barrier function, and in doing so reduced serum LPS decreasing the effects of diet-induced obesity. Without a change in food intake mice showed improved insulin sensitivity, decreased body weight gain, fat mass development, and fasting hyperglycemia (Everard *et al.* 2013).

7.5.3. LPS Detoxification

To prevent inflammation induced by LPS the body has developed several strategies to detoxify it. The apical brush-border enzyme intestinal alkaline phosphatase (IAP) detoxifies a variety of bacterial components, including LPS, CpG DNA, and flagellin through dephosphorylation (Chen et al. 2010). IAP is produced exclusively in villus-associated enterocytes and secreted into the gut lumen (Alpers et al. 1994). Several in vitro and in vivo studies have shown that IAP is essential in LPS detoxification and preventing bacterial invasion across the gut mucosal barrier. Exogenous IAP not only significantly inhibited LPS-induced inflammatory cytokine production, but also prevented and reversed metabolic syndrome in mice (Goldberg et al. 2008; Kaliannan et al. 2013; Lee et al. 2014). Differences in IAP activities and inflammation between diet-induced obesity prone and resistant rats were also documented (de La Serre et al. 2010). In addition, IAP deficient mice showed increased gut permeability caused by enhanced LPS translocation (Goldberg et al. 2008; Kaliannan et al. 2013) and gained more weight when fed with a HFD (Narisawa et al. 2003). Together these studies demonstrate IAP dephosphorylates LPS preventing inflammation, which would otherwise lead to further damage in other organs that are related to the pathogenesis of metabolic diseases. Interestingly, as a defense protein it has no effect on live bacteria (Chen et al. 2010).

Once in the systemic circulation, LPS can be deactivated or detoxified by immune cells, such as macrophages, or Kupffer cells present in the liver or splenic cells or by binding with acute phase proteins (Satoh et al. 2008; Buttenschoen et al. 2010). Bactericidal/permeability increasing protein (BPI), which is widely expressed in epithelial cells, is capable of binding and neutralizing LPS (Canny et al. 2002). The high affinity of BPI for the lipid A moiety of LPS targets its cytotoxic activity to Gram-negative bacteria by preventing the interaction of lipid A with other proinflammatory molecules (Weiss 2003).

7.6. EARLY LIFE MICROBIOME PROGRAMMING METABOLIC OUTCOMES

The exposure to antibiotics in different situations has led to both reductions and increases in metabolic issues. In many cases where increased intestinal permeability and reduced barrier function have been implicated in metabolic disease, antibiotic treatment is found to be protective against glucose intolerance and insulin resistance. However, there are also studies that have shown disturbances in the microbiota induced by antibiotics can contribute to development of obesity. A recent study showed that low levels of antibiotics during a critical window after birth have a programming effect that results in adverse long-term metabolic outcomes (Cox *et al.* 2014). It is yet unclear how this early life antibiotic treatment increases adipocity later in life as changes were seen in both immunity and hepatic gene expression relevant to adipocity. It was noted, however, that the microbiota after antibiotic withdrawal was similar to controls that had never received antibiotics whereas adipocity phenotypes were maintained, suggesting that the effect was not associated with persistent changes in the microbiota. Epigenetic mechanisms are thus under consideration due to their known role in programming.

7.7 CONCLUSION

Over the past decade it has become clear that gut microbes play an important role in regulating metabolic outcomes of the host. However, the many models used and opposing results indicate that outcomes are dependent on circumstances of an intervention. While there continues to be great excitement over the possibility of regulating microbial populations to promote metabolic health, the complexity of this interaction continues to require extensive study.

7.8. REFERENCES

Al-Lahham, S., H. Roelofsen, F. Rezaee, D. Weening, A. Hoek, R. Vonk *et al.* 2012. Propionic acid affects immune status and metabolism in adipose tissue from overweight subjects. *Eur J Clin Invest* 42(4): 357–364.

Alex, S., K. Lange, T. Amolo, J.S. Grinstead, A.K. Haakonsson, E. Szalowska, *et al.* 2013. Short-Chain Fatty Acids Stimulate Angiopoietin-Like 4 Synthesis in Human Colon Adenocarcinoma Cells by Activating Peroxisome Proliferator-Activated Receptor gamma. *Molecular and Cellular Biology* 33(7): 1303–1316.

Alpers, D.H., A. Mahmood, M. Engle, F. Yamagishi, and K. Deschryverkecskemeti. 1994. The Secretion of Intestinal Alkaline-Phosphatase (Iap) from the Enterocyte. *Journal of Gastroenterology* 29: 63–67.

Amyot, J., M. Semache, M. Ferdaoussi, G. Fontes, and V. Poitout. 2012. Lipopolysaccharides Impair Insulin Gene Expression in Isolated Islets of Langerhans via Toll-Like Receptor-4 and NF-κB Signalling. *Plos One* 7(4).

Arpaia, N., C. Campbell, X. Fan, S. Dikiy, J. van der Veeken, P. deRoos, *et al.* 2013. Metabolites produced by commensal bacteria promote peripheral regulatory T-cell generation. *Nature* 504(7480): 451–455.

Atarashi, K., J. Nishimura, T. Shima, Y. Umesaki, M. Yamamoto, M. Onoue, *et al.* 2008. ATP drives lamina propria T(H)17 cell differentiation. *Nature* 455(7214): 808–U810.

Backhed, F. 2013. Meat-metabolizing bacteria in atherosclerosis. *Nat Med 19*(5): 533–534.

Backhed, F., H. Ding, T. Wang, L.V. Hooper, G.Y. Koh, A. Nagy, *et al.* 2004. The gut microbiota as an environmental factor that regulates fat storage. *Proceedings of the National Academy of Sciences of the United States of America 101*(44): 15718–15723.

Backhed, F., J.K. Manchester, C.F. Semenkovich, and J.I. Gordon. 2007. Mechanisms underlying the resistance to diet-induced obesity in GF mice. *Proceedings of the National Academy of Sciences of the United States of America 104*(3): 979–984.

Baker, F.D., H.R. Papiska, and L.L. Campbell. 1962. Choline fermentation by Desulfovibrio desulfuricans. *J Bacteriol 84*: 973–978.

Banday, A.A., A. Marwaha, L.S. Tallam, and M.F. Lokhandwala. 2005. Tempol reduces oxidative stress, improves insulin sensitivity, decreases renal dopamine D1 receptor hyperphosphorylation, and restores D1 receptor-G-protein coupling and function in obese Zucker rats. *Diabetes 54*(7): 2219–2226.

Barrera, J.G., K.R. Jones, J.P. Herman, D.A. D'Alessio, S.C. Woods, and R.J. Seeley. 2011. Hyperphagia and increased fat accumulation in two models of chronic CNS glucagon-like peptide-1 loss of function. *J Neurosci 31*(10): 3904–3913.

Basu, S., M. Haghiac, P. Surace, J.C. Challier, M. Guerre-Millo, K. Singh, *et al.* 2011. Pregravid obesity associates with increased maternal endotoxemia and metabolic inflammation. *Obesity (Silver Spring) 19*(3): 476–482.

Beenken, A. and M. Mohammadi. 2009. The FGF family: Biology, pathophysiology and therapy. *Nat Rev Drug Discov 8*(3): 235–253.

Begley, M., C.G.M. Gahan, and C. Hill. 2005. The interaction between bacteria and bile. *Fems Microbiology Reviews 29*(4): 625–651.

Bellahcene, M., J.F. O'Dowd, E.T. Wargent, M.S. Zaibi, D.C. Hislop, R.A. Ngala, *et al.* 2013. Male mice that lack the G-protein-coupled receptor GPR41 have low energy expenditure and increased body fat content. *British Journal of Nutrition 109*(10): 1755–1764.

Beumer, C., M. Wulferink, W. Raaben, D. Fiechter, R. Brands, and W. Seinen. 2003. Calf intestinal alkaline phosphatase, a novel therapeutic drug for lipopolysaccharide (LPS)-mediated diseases, attenuates LPS toxicity in mice and piglets. *Journal of Pharmacology and Experimental Therapeutics 307*(2): 737–744.

Bjursell, M., T. Admyre, M. Goransson, A.E. Marley, D.M. Smith, J. Oscarsson, and M. Bohlooly-Y. 2011. Improved glucose control and reduced body fat mass in free fatty acid receptor 2-deficient mice fed a high-fat diet. *American Journal of Physiology-Endocrinology and Metabolism 300*(1): E211–E220.

Bleau, C., A.D. Karelis, D.H. St-Pierre, and L. Lamontagne. 2014. Crosstalk between intestinal microbiota, adipose tissue and skeletal muscle as an early event in systemic low-grade inflammation and the development of obesity and diabetes. *Diabetes Metab Res Rev.*

Brown, A.J., S.M. Goldsworthy, A.A. Barnes, M.M. Eilert, L. Tcheang, D. Daniels, *et al.* 2003. The orphan G protein-coupled receptors GPR41 and GPR43 are activated by propionate and other short chain carboxylic acids. *Journal of Biological Chemistry 278*(13): 11312–11319.

Brun, P., I. Castagliuolo, V. Di Leo, A. Buda, M. Pinzani, G. Palu, and D. Martines. 2007. Increased intestinal permeability in obese mice: New evidence in the pathogenesis of nonalcoholic steatohepatitis. *Am J Physiol Gastrointest Liver Physiol* 292(2): G518–525.

Buchman, A.L., M.D. Dubin, A.A. Moukarzel, D.J. Jenden, M. Roch, K.M. Rice, et al. 1995. Choline Deficiency—A Cause of Hepatic Steatosis during Parenteral-Nutrition That Can Be Reversed with Intravenous Choline Supplementation. *Hepatology* 22(5): 1399–1403.

Buttenschoen, K., P. Radermacher, and H. Bracht. 2010. Endotoxin elimination in sepsis: Physiology and therapeutic application. *Langenbecks Arch Surg* 395(6): 597–605.

Canani, R.B., M.D. Costanzo, L. Leone, G. Bedogni, P. Brambilla, S. Cianfarani, et al. 2011. Epigenetic mechanisms elicited by nutrition in early life. *Nutr Res Rev* 24(2): 198–205.

Cani, P.D., J. Amar, M.A. Iglesias, M. Poggi, C. Knauf, D. Bastelica, et al. 2007. Metabolic endotoxemia initiates obesity and insulin resistance. *Diabetes* 56(7): 1761–1772.

Canny, G., O. Levy, G.T. Furuta, S. Narravula-Alipati, R.B. Sisson, C.N. Serhan, and S.P. Colgan. 2002. Lipid mediator-induced expression of bactericidal/permeability-increasing protein (BPI) in human mucosal epithelia. *Proc Natl Acad Sci USA* 99(6): 3902–3907.

Charloux, C., M. Paul, D. Loisance, and A. Astier. 1995. Inhibition of hydroxyl radical production by lactobionate, adenine, and tempol. *Free Radic Biol Med* 19(5): 699–704.

Chen, K.T., M.S. Malo, A.K. Moss, S. Zeller, P. Johnson, F. Ebrahimi, et al. 2010. Identification of specific targets for the gut mucosal defense factor intestinal alkaline phosphatase. *Am J Physiol Gastrointest Liver Physiol* 299(2): G467–475.

Chervonsky, A.V. 2010. Influence of microbial environment on autoimmunity. *Nature Immunology* 11(1): 28–35.

Chiang, J.Y. 2009. Bile acids: Regulation of synthesis. *J Lipid Res* 50(10): 1955–1966.

Christensen, D.P., M. Dahllof, M. Lundh, D.N. Rasmussen, M.D. Nielsen, N. Billestrup, et al. 2011. Histone Deacetylase (HDAC) Inhibition as a Novel Treatment for Diabetes Mellitus. *Molecular Medicine* 17(5–6): 378–390.

Claus, S.P., T.M. Tsang, Y. Wang, O. Cloarec, E. Skordi, F.P. Martin, et al. 2008. Systemic multicompartmental effects of the gut microbiome on mouse metabolic phenotypes. *Mol Syst Biol* 4: 219.

Cook, S.I. and J.H. Sellin. 1998. Review article: Short chain fatty acids in health and disease. *Alimentary Pharmacology & Therapeutics* 12(6): 499–507.

Cox, L.M., S. Yamanishi, J. Sohn, A.V. Alekseyenko, J.M. Leung, I. Cho, et al. 2014. Altering the intestinal microbiota during a critical developmental window has lasting metabolic consequences. *Cell* 158(4): 705–721.

Craciun, S. and E.P. Balskus. 2012. Microbial conversion of choline to trimethylamine requires a glycyl radical enzyme. *Proceedings of the National Academy of Sciences of the United States of America* 109(52): 21307–21312.

Cucak, H., C. Mayer, M. Tonnesen, L.H. Thomsen, L.G. Grunnet, and A. Rosendahl. 2014. Macrophage contact dependent and independent TLR4 mechanisms induce beta-cell dysfunction and apoptosis in a mouse model of type 2 diabetes. *PLoS One* 9(3): e90685.

Dawson, P.A. 2011. Role of the intestinal bile acid transporters in bile acid and drug disposition. *Handb Exp Pharmacol* (201): 169–203.

de La Serre, C.B., C.L. Ellis, J. Lee, A.L. Hartman, J.C. Rutledge, and H.E. Raybould. 2010. Propensity to high-fat diet-induced obesity in rats is associated with changes in the gut microbiota and gut inflammation. *American Journal of Physiology-Gastrointestinal and Liver Physiology* 299(2): G440–G448.

De Vadder, F., P. Kovatcheva-Datchary, D. Goncalves, J. Vinera, C. Zitoun, A. Duchampt, *et al.* 2014. Microbiota-Generated Metabolites Promote Metabolic Benefits via Gut-Brain Neural Circuits. *Cell* 156(1–2): 84–96.

Deopurkar, R., H. Ghanim, J. Friedman, S. Abuaysheh, C.L. Sia, P. Mohanty, *et al.* 2010. Differential Effects of Cream, Glucose, and Orange Juice on Inflammation, Endotoxin, and the Expression of Toll-Like Receptor-4 and Suppressor of Cytokine Signaling-3. *Diabetes Care* 33(5): 991–997.

Devaraj, S., M.R. Dasu, J. Rockwood, W. Winter, S.C. Griffen, and I. Jialal. 2008. Increased toll-like receptor (TLR)2 and TLR4 expression in monocytes from patients with type 1 diabetes: Further evidence of a proinflammatory state. *Journal of Clinical Endocrinology & Metabolism* 93(2): 578–583.

Devkota, S., Y. Wang, M.W. Musch, V. Leone, H. Fehlner-Peach, A. Nadimpalli, *et al.* 2012. Dietary-fat-induced taurocholic acid promotes pathobiont expansion and colitis in Il10$^{-/-}$ mice. *Nature* 487(7405): 104–108.

Dietschy, J.M. 1968. Mechanisms for the intestinal absorption of bile acids. *J Lipid Res* 9(3): 297–309.

Dixon, D.R. and R.P. Darveau. 2005. Lipopolysaccharide heterogeneity: Innate host responses to bacterial modification of lipid A structure. *Journal of Dental Research* 84(7): 584–595.

Donath, M.Y., M. Boni-Schnetzler, H. Ellingsgaard, P.A. Halban, and J.A. Ehse 2010. Cytokine production by islets in health and diabetes: Cellular origin, regulation and function. *Trends Endocrinol Metab* 21(5): 261–267.

Dumas, M.E., R.H. Barton, A. Toye, O. Cloarec, C. Blancher, A. Rothwell, *et al.* 2006. Metabolic profiling reveals a contribution of gut microbiota to fatty liver phenotype in insulin-resistant mice. *Proceedings of the National Academy of Sciences of the United States of America* 103(33): 12511–12516.

Duran-Sandoval, D., B. Cariou, F. Percevault, N. Hennuyer, A. Grefhorst, T.H. van Dijk, *et al.* 2005. The farnesoid X receptor modulates hepatic carbohydrate metabolism during the fasting-refeeding transition. *J Biol Chem* 280(33): 29971–29979.

Duseja, A. and Y.K. Chawla. 2014. Obesity and NAFLD: The role of bacteria and microbiota. *Clin Liver Dis* 18(1): 59–71.

Eguchi, K., I. Manabe, Y. Oishi-Tanaka, M. Ohsugi, N. Kono, F. Ogata, *et al.* 2012. Saturated fatty acid and TLR signaling link beta cell dysfunction and islet inflammation. *Cell Metab* 15(4): 518–533.

Ehses, J.A., M. Boni-Schnetzler, M. Faulenbach, and M.Y. Donath. 2008. Macrophages, cytokines and beta-cell death in Type 2 diabetes. *Biochem Soc Trans* 36(Pt 3): 340–342.

Erridge, C., T. Attina, C.M. Spickett, and D.J. Webb. 2007. A high-fat meal induces low-grade endotoxemia: Evidence of a novel mechanism of postprandial inflammation. *Am J Clin Nutr* 86(5): 1286–1292.

Everard, A., C. Belzer, L. Geurts, J.P. Ouwerkerk, C. Druart, L.B. Bindels, *et al.* 2013. Cross-talk between Akkermansia muciniphila and intestinal epithelium controls diet-induced obesity. *Proceedings of the National Academy of Sciences of the United States of America* 110(22): 9066–9071.

Fernandes, J., W. Su, S. Rahat-Rozenbloom, T.M.S. Wolever, and E.M. Comelli. 2014. Adiposity, gut microbiota and fecal short chain fatty acids are linked in adult humans. *Nutrition & Diabetes 4.*

Flint, H.J., E.A. Bayer, M.T. Rincon, R. Lamed, and B.A. White. 2008. Polysaccharide utilization by gut bacteria: Potential for new insights from genomic analysis. *Nature Reviews Microbiology* 6(2): 121–131.

Floch, M.H. 2002. Bile salts, intestinal microflora and enterohepatic circulation. *Dig Liver Dis 34* Suppl 2: S54–57.

Frantz, S., L. Kobzik, Y.D. Kim, R. Fukazawa, R. Medzhitov, R.T. Lee, and R.A. Kelly. 1999. Toll4 (TLR4) expression in cardiac myocytes in normal and failing myocardium. *Journal of Clinical Investigation* 104(3): 271–280.

Gao, Z.G., J. Yin, J. Zhang, R.E. Ward, R.J. Martin, M. Lefevre, W.T. Cefalu, and J.P. Ye. 2009. Butyrate Improves Insulin Sensitivity and Increases Energy Expenditure in Mice. *Diabetes* 58(7): 1509–1517.

Ghanim, H., S. Abuaysheh, C.L. Sia, K. Korzeniewski, A. Chaudhuri, J.M. Fernandez-Real, and P. Dandona. 2009. Increase in Plasma Endotoxin Concentrations and the Expression of Toll-like Receptors and Suppressor of Cytokine Signaling-3 in Mononuclear Cells After a High-Fat, High-Carbohydrate Meal Implications for insulin resistance. *Diabetes Care* 32(12): 2281–2287.

Ghoshal, S., J. Witta, J. Zhong, W. de Villiers, and E. Eckhardt. 2009. Chylomicrons promote intestinal absorption of lipopolysaccharides. *Journal of Lipid Research* 50(1): 90–97.

Gieling, R.G., K. Wallace, and Y.P. Han. 2009. Interleukin-1 participates in the progression from liver injury to fibrosis. *American Journal of Physiology-Gastrointestinal and Liver Physiology* 296(6): G1324–G1331.

Goldberg, R.F., W.G. Austen, X.B. Zhang, G. Munene, G. Mostafa, S. Biswas, *et al.* 2008. Intestinal alkaline phosphatase is a gut mucosal defense factor maintained by enteral nutrition. *Proceedings of the National Academy of Sciences of the United States of America* 105(9): 3551–3556.

Grisouard, J., E. Bouillet, K. Timper, T. Radimerski, K. Dembinski, D.M. Frey, *et al.* 2012. Both inflammatory and classical lipolytic pathways are involved in lipopolysaccharide-induced lipolysis in human adipocytes. *Innate Immun* 18(1): 25–34.

Grube, B., P.W. Chong, K.Z. Lau, and H.D. Orzechowski. 2013. A Natural Fiber Complex Reduces Body Weight in the Overweight and Obese: A Double-Blind, Randomized, Placebo-Controlled Study. *Obesity* 21(1): 58–64.

Hagiwara, S., A. Yoshida, Y. Omata, Y. Tsukada, H. Takahashi, H. Kamewada, et al. 2014. Desulfovibrio desulfuricans bacteremia in a patient hospitalized with acute cerebral infarction: Case report and review. *Journal of Infection and Chemotherapy* 20(3–4): 274–277.

Hansson, G.K. 2005. Mechanisms of disease—Inflammation, atherosclerosis, and coronary artery disease. *New England Journal of Medicine* 352(16): 1685–1695.

Hayward, H.R. and T.C. Stadtman. 1959. Anaerobic Degradation of Choline .1. Fermentation of Choline by an Anaerobic, Cytochrome-Producing Bacterium, Vibrio-Cholinicus N Sp. *Journal of Bacteriology* 78(4): 557–561.

Heimann, E., M. Nyman, and E. Degerman. 2014. Propionic acid and butyric acid inhibit lipolysis and de novo lipogenesis and increase insulin-related glucose uptake in primary rat adipocytes. *Adipocyte* 4(1): 1–8.

Hofmann, A.F. and K J. Mysels. 1992. Bile acid solubility and precipitation in vitro and in vivo: The role of conjugation, pH, and Ca^{2+} ions. *J Lipid Res* 33(5): 617–626.

Hohmeier, H.E., V.V. Tran, G. Chen, R. Gasa, and C.B. Newgard. 2003. Inflammatory mechanisms in diabetes: Lessons from the beta-cell. *Int J Obes Relat Metab Disord* 27 Suppl 3: S12–16.

Holt, J.A., G. Luo, A.N. Billin, J. Bisi, Y.Y. McNeill, K.F. Kozarsky, et al. 2003. Definition of a novel growth factor-dependent signal cascade for the suppression of bile acid biosynthesis. *Genes Dev* 17(13): 1581–1591.

Hong, Y.H., Y. Nishimura, D. Hishikawa, H. Tsuzuki, H. Miyahara, C. Gotoh, et al. 2005. Acetate and propionate short chain fatty acids stimulate adipogenesis via GPCR43. *Endocrinology* 146(12): 5092–5099.

Hotamisligil, G.S. 2006. Inflammation and metabolic disorders. Nature 444(7121): 860–867.

Hoverstad, T. and T. Midtvedt. 1986. Short-chain fatty acids in germfree mice and rats. *J Nutr* 116(9): 1772–1776.

Hsieh, P.S., J.Y.H. Chan, J.F. Shyu, Y.T. Chen, and C.H. Loh. 2008. Mild portal endotoxaemia induces subacute hepatic inflammation and pancreatic beta-cell dysfunction in rats. *European Journal of Clinical Investigation* 38(9): 640–648.

Huang, W., A. Metlakunta, N. Dedousis, P.L. Zhang, I. Sipula, J.J. Dube, et al. 2010. Depletion of Liver Kupffer Cells Prevents the Development of Diet-Induced Hepatic Steatosis and Insulin Resistance. *Diabetes* 59(2): 347–357.

Ilan, Y. 2012. Leaky gut and the liver: A role for bacterial translocation in nonalcoholic steatohepatitis. *World J Gastroenterol* 18(21): 2609–2618.

Ioannidis, I., F. Ye, B. McNally, M. Willette, and E. Flano. 2013. Toll-Like Receptor Expression and Induction of Type I and Type III Interferons in Primary Airway Epithelial Cells. *Journal of Virology* 87(6): 3261–3270.

Islam, K.B.M.S., S. Fukiya, M. Hagio, N. Fujii, S. Ishizuka, T. Ooka, et al. 2011. Bile Acid Is a Host Factor That Regulates the Composition of the Cecal Microbiota in Rats. *Gastroenterology* 141(5): 1773–1781.

Jocken, J.W.E., G.H. Goossens, H. Boon, R.R. Mason, Y. Essers, B. Havekes, et al. 2013. Insulin-mediated suppression of lipolysis in adipose tissue and skeletal muscle of obese type 2 diabetic men and men with normal glucose tolerance. *Diabetologia* 56(10): 2255–2265.

Jones, B.V., M. Begley, C. Hill, C.G. Gahan, and J.R. Marchesi. 2008. Functional and comparative metagenomic analysis of bile salt hydrolase activity in the human gut microbiome. *Proc Natl Acad Sci USA 105*(36): 13580–13585.

Kaliannan, K., S.R. Hamarneh, K.P. Economopoulos, S. Nasrin Alam, O. Moaven, P. Patel, *et al.* 2013. Intestinal alkaline phosphatase prevents metabolic syndrome in mice. *Proc Natl Acad Sci USA 110*(17): 7003–7008.

Karaki, S., R. Mitsui, H. Hayashi, I. Kato, H. Sugiya, T. Iwanaga, *et al.* 2006. Short-chain fatty acid receptor, GPR43, is expressed by enteroendocrine cells and mucosal mast cells in rat intestine. *Cell and Tissue Research 324*(3): 353–360.

Karlsson, F.H., F. Fak, I. Nookaew, V. Tremaroli, B. Fagerberg, D. Petranovic, *et al.* 2012. Symptomatic atherosclerosis is associated with an altered gut metagenome. *Nature Communications 3*.

Katsuma, S., A. Hirasawa, and G. Tsujimoto. 2005. Bile acids promote glucagon-like peptide-1 secretion through TGR5 in a murine enteroendocrine cell line STC-1. *Biochemical and Biophysical Research Communications 329*(1): 386–390.

Kellogg, T.F. and B.S. Wostmann 1969. Fecal neutral steroids and bile acids from germ-free rats. *J Lipid Res 10*(5): 495–503.

Kendrick, S.F.W., G. O'Boyle, J. Mann, M. Zeybel, J. Palmer, D.E.J. Jones, and C.P. Day. 2010. Acetate, the Key Modulator of Inflammatory Responses in Acute Alcoholic Hepatitis. *Hepatology 51*(6): 1988–1997.

Kerr, T.A., S. Saeki, M. Schneider, K. Schaefer, S. Berdy, T. Redder, *et al.* 2002. Loss of nuclear receptor SHP impairs but does not eliminate negative feedback regulation of bile acid synthesis. *Dev Cell 2*(6): 713–720.

Kim, S., J.H. Kim, B.O. Park, and Y.S. Kwak. 2014. Perspectives on the therapeutic potential of short-chain fatty acid receptors. *BMB Rep 47*(3): 173–178.

Kimura, I., D. Inoue, T. Maeda, T. Hara, A. Ichimura, S. Miyauchi, *et al.* 2011. Short-chain fatty acids and ketones directly regulate sympathetic nervous system via G protein-coupled receptor 41 (GPR41). *Proc Natl Acad Sci USA 108*(19): 8030–8035.

Kimura, I., K. Ozawa, D. Inoue, T. Imamura, K. Kimura, T. Maeda, *et al.* 2013. The gut microbiota suppresses insulin-mediated fat accumulation via the short-chain fatty acid receptor GPR43. *Nat Commun 4*: 1829.

Kir, S., S.A. Kliewer, and D.J. Mangelsdorf. 2011. Roles of FGF19 in liver metabolism. *Cold Spring Harb Symp Quant Biol 76*: 139–144.

Koeth, R.A., Z.E. Wang, B.S. Levison, J.A. Buffa, E. Org, B.T. Sheehy, *et al.* 2013. Intestinal microbiota metabolism of L-carnitine, a nutrient in red meat, promotes atherosclerosis. *Nature Medicine 19*(5): 576–585.

Lassenius, M.I., K.H. Pietilainen, K. Kaartinen, P.J. Pussinen, J. Syrjanen, C. Forsblom, *et al.* 2011. Bacterial Endotoxin Activity in Human Serum Is Associated With Dyslipidemia, Insulin Resistance, Obesity, and Chronic Inflammation. *Diabetes Care 34*(8): 1809–1815.

Laugerette, F., C. Vors, A. Geloen, M.A. Chauvin, C. Soulage, S. Lambert-Porcheron, *et al.* 2011. Emulsified lipids increase endotoxemia: Possible role in early postprandial low-grade inflammation. *Journal of Nutritional Biochemistry 22*(1): 53–59.

Le Poul, E., C. Loison, S. Struyf, J.Y. Springael, V. Lannoy, M.E. Decobecq, *et al.* 2003.

Functional characterization of human receptors for short chain fatty acids and their role in polymorphonuclear cell activation. *Journal of Biological Chemistry 278*(28): 25481–25489.

Lee, C., J. Chun, S.W. Hwang, S.J. Kang, J.P. Im, and J.S. Kim. 2014. The effect of intestinal alkaline phosphatase on intestinal epithelial cells, macrophages and chronic colitis in mice. *Life Sciences 100*(2): 118–124.

Lefebvre, P., B. Cariou, F. Lien, F. Kuipers, and B. Staels. 2009. Role of bile acids and bile acid receptors in metabolic regulation. *Physiol Rev 89*(1): 147–191.

Li, F., C.T. Jiang, K.W. Krausz, Y.F. Li, I. Albert, H.P. Hao, et al. 2013. Microbiome remodelling leads to inhibition of intestinal farnesoid X receptor signalling and decreased obesity. *Nature Communications 4*.

Li, G.L., W. Yao, and H.L. Jiang. 2014. Short-Chain Fatty Acids Enhance Adipocyte Differentiation in the Stromal Vascular Fraction of Porcine Adipose Tissue. *Journal of Nutrition 144*(12): 1887–1895.

Lin, H.V., A. Frassetto, E.J. Kowalik, A.R. Nawrocki, M.F.M. Lu, J.R. Kosinski, et al. 2012. Butyrate and Propionate Protect against Diet-Induced Obesity and Regulate Gut Hormones via Free Fatty Acid Receptor 3-Independent Mechanisms. *Plos One 7*(4).

Liu, S.M., W.C. Willett, J.E. Manson, F.B. Hu, B. Rosner, and G. Colditz. 2003. Relation between changes in intakes of dietary fiber and grain products and changes in weight and development of obesity among middle-aged women. *American Journal of Clinical Nutrition 78*(5): 920–927.

Locati, M., A. Mantovani, and A. Sica. 2013. Macrophage Activation and Polarization as an Adaptive Component of Innate Immunity. *Development and Function of Myeloid Subsets 120*: 163–184.

Loscalzo, J. 2011. Lipid Metabolism by Gut Microbes and Atherosclerosis. *Circulation Research 109*(2): 127–129.

Louis, P., K.P. Scott, S.H. Duncan, and H.J. Flint. 2007. Understanding the effects of diet on bacterial metabolism in the large intestine. *Journal of Applied Microbiology 102*(5): 1197–1208.

Luche, E., B. Cousin, L. Garidou, M. Serino, A. Waget, C. Barreau, et al. 2013. Metabolic endotoxemia directly increases the proliferation of adipocyte precursors at the onset of metabolic diseases through a CD14-dependent mechanism. *Mol Metab 2*(3): 281–291.

Lumeng, C.N. and A.R. Saltiel. 2011. Inflammatory links between obesity and metabolic disease. *Journal of Clinical Investigation 121*(6): 2111–2117.

Ma, K., P.K. Saha, L. Chan, and D.D. Moore. 2006. Farnesoid X receptor is essential for normal glucose homeostasis. *Journal of Clinical Investigation 116*(4): 1102–1109.

Macdonald, I.A., V.D. Bokkenheuser, J. Winter, A.M. Mclernon, and E.H. Mosbach. 1983. Degradation of Steroids in the Human Gut. *Journal of Lipid Research 24*(6): 675–700.

Majdalawieh, A. and H.S. Ro. 2009. LPS-induced suppression of macrophage cholesterol efflux is mediated by adipocyte enhancer-binding protein 1. *Int J Biochem Cell Biol 41*(7): 1518–1525.

Mani, V., T.E. Weber, L.H. Baumgard, and N.K. Gabler. 2012. Growth and Development Symposium: Endotoxin, Inflammation, and Intestinal Function in Livestock. *Journal of Animal Science* 90(5): 1452–1465.

Manning, S. and R.L. Batterham. 2014. The role of gut hormone peptide YY in energy and glucose homeostasis: Twelve years on. *Annu Rev Physiol* 76: 585–608.

Maslowski, K.M., A.T. Vieira, A. Ng, J. Kranich, F. Sierro, D. Yu, et al. 2009. Regulation of inflammatory responses by gut microbiota and chemoattractant receptor GPR43. *Nature* 461(7268): 1282–U1119.

Meijer, K., P. de Vos, and M.G. Priebe. 2010. Butyrate and other short-chain fatty acids as modulators of immunity: What relevance for health? *Current Opinion in Clinical Nutrition and Metabolic Care* 13(6): 715–721.

Midtvedt, T. 1974. Microbial Bile-Acid Transformation. *American Journal of Clinical Nutrition* 27(11): 1341–1347.

Miele, L., V. Valenza, G. La Torre, M. Montalto, G. Cammarota, R. Ricci, et al. 2009. Increased Intestinal Permeability and Tight Junction Alterations in Nonalcoholic Fatty Liver Disease. *Hepatology* 49(6): 1877–1887.

Musso, G., R. Gambino, and M. Cassader. 2011. Interactions Between Gut Microbiota and Host Metabolism Predisposing to Obesity and Diabetes. *Annual Review of Medicine*, Vol 62, 2011 62: 361–380.

Narisawa, S., L. Huang, A. Iwasaki, H. Hasegawa, D.H. Alpers, and J.L. Millan. 2003. Accelerated fat absorption in intestinal alkaline phosphatase knockout mice. *Mol Cell Biol* 23(21): 7525–7530.

Nicholson, J.K., E. Holmes, J. Kinross, R. Burcelin, G. Gibson, W. Jia, and S. Pettersson. 2012. Host-gut microbiota metabolic interactions. *Science* 336(6086): 1262–1267.

Nielsen, S., Z.K. Guo, C.M. Johnson, D.D. Hensrud, and M.D. Jensen. 2004. Splanchnic lipolysis in human obesity. *Journal of Clinical Investigation* 113(11): 1582–1588.

Nieuwdorp, M., P.W. Gilijamse, N. Pai, and L.M. Kaplan. 2014. Role of the Microbiome in Energy Regulation and Metabolism. *Gastroenterology* 146(6): 1525–1533.

Nohr, M.K., M.H. Pedersen, A. Gille, K.L. Egerod, M.S. Engelstoft, A.S. Husted, et al. 2013. GPR41/FFAR3 and GPR43/FFAR2 as Cosensors for Short-Chain Fatty Acids in Enteroendocrine Cells vs FFAR3 in Enteric Neurons and FFAR2 in Enteric Leukocytes. *Endocrinology* 154(10): 3552–3564.

Pardo, A., R. Bartoli, V. Lorenzo-Zuniga, R. Planas, B. Vinado, J. Riba, et al. 2000. Effect of cisapride on intestinal bacterial overgrowth and bacterial translocation in cirrhosis. Hepatology 31(4): 858–863.

Patil, D.P., D.P. Dhotre, S.G. Chavan, A. Sultan, D.S. Jain, V.B. Lanjekar, et al. 2012. Molecular analysis of gut microbiota in obesity among Indian individuals. *Journal of Biosciences* 37(4): 647–657.

Pendyala, S., J.M. Walker, and P.R. Holt. 2012. A High-Fat Diet Is Associated With Endotoxemia That Originates From the Gut. *Gastroenterology* 142(5): 1100-+.

Pols, T.W.H., J. Auwerx, and K. Schoonjans. 2010. Targeting the TGR5-GLP-1 pathway to combat type 2 diabetes and non-alcoholic fatty liver disease. *Gastroenterologie Clinique Et Biologique* 34(4–5): 270–273.

Prawitt, J., M. Abdelkarim, J.H.M. Stroeve, I. Popescu, H. Duez, V.R. Velagapudi, et *al.* 2011. Farnesoid X Receptor Deficiency Improves Glucose Homeostasis in Mouse Models of Obesity. *Diabetes* 60(7): 1861–1871.

Puddu, A., R. Sanguineti, F. Montecucco, and G.L. Viviani. 2014. Evidence for the Gut Microbiota Short-Chain Fatty Acids as Key Pathophysiological Molecules Improving Diabetes. *Mediators of Inflammation.*

Pussinen, P.J., A.S. Havulinna, M. Lehto, J. Sundvall, and V. Salomaa. 2011. Endotoxemia Is Associated With an Increased Risk of Incident Diabetes. *Diabetes Care* 34(2): 392–397.

Rahat-Rozenbloom, S., J. Fernandes, G.B. Gloor, and T.M. Wolever. 2014. Evidence for greater production of colonic short-chain fatty acids in overweight than lean humans. *Int J Obes (Lond)* 38(12): 1525–1531.

Rebouche, C.J. and H. Seim. 1998. Carnitine metabolism and its regulation in microorganisms and mammals. *Annual Review of Nutrition* 18: 39–61.

Remely, M., E. Aumueller, C. Merold, S. Dworzak, B. Hippe, J. Zanner, *et al.* 2014. Effects of short chain fatty acid producing bacteria on epigenetic regulation of FFAR3 in type 2 diabetes and obesity. *Gene* 537(1): 85–92.

Ridaura, V.K., J.J. Faith, F.E. Rey, J. Cheng, A.E. Duncan, A.L. Kau, *et al.* 2013. Gut microbiota from twins discordant for obesity modulate metabolism in mice. *Science* 341(6150): 1241214.

Ridlon, J.M., D.J. Kang, and P.B. Hylemon. 2006. Bile salt biotransformations by human intestinal bacteria. *Journal of Lipid Research* 47(2): 241–259.

Rietschel, E.T., H. Brade, O. Holst, L. Brade, S. MullerLoennies, U. Mamat, *et al.* 1996. Bacterial endotoxin: Chemical constitution, biological recognition, host response, and immunological detoxification. *Pathology of Septic Shock* 216: 39–81.

Rivera, C.A., P. Adegboyega, N. van Rooijen, A. Tagalicud, M. Allman, and M. Wallace. 2007. Toll-like receptor-4 signaling and Kupffer cells play pivotal roles in the pathogenesis of non-alcoholic steatohepatitis. *J Hepatol* 47(4): 571–579.

Russell, D.W. 2003. The enzymes, regulation, and genetics of bile acid synthesis. *Annual Review of Biochemistry* 72: 137–174.

Russell, W.R., L. Hoyles, H.J. Flint, and M.E. Dumas. 2013. Colonic bacterial metabolites and human health. *Current Opinion in Microbiology* 16(3): 246–254.

Samuel, B.S., A. Shaito, T. Motoike, F E. Rey, F. Backhed, J.K. Manchester, *et al.* 2008. Effects of the gut microbiota on host adiposity are modulated by the short-chain fatty-acid binding G protein-coupled receptor, Gpr41. *Proc Natl Acad Sci U S A* 105(43): 16767–16772.

Satoh, M., S. Ando, T. Shinoda, and M. Yamazaki. 2008. Clearance of bacterial lipopolysaccharides and lipid A by the liver and the role of argininosuccinate synthase. *Innate Immun* 14(1): 51–60.

Sayin, S.I., A. Wahlstrom, J. Felin, S. Jantti, H.U. Marschall, K. Bamberg, *et al.* 2013. Gut Microbiota Regulates Bile Acid Metabolism by Reducing the Levels of Tauro-beta-muricholic Acid, a Naturally Occurring FXR Antagonist. *Cell Metabolism* 17(2): 225–235.

Schwiertz, A., D. Taras, K. Schafer, S. Beijer, N.A. Bos, C. Donus, and P.D. Hardt. 2010. Microbiota and SCFA in lean and overweight healthy subjects. *Obesity (Silver Spring)* 18(1): 190–195.

Sellin, J.H. 1999. SCFAs: The Enigma of Weak Electrolyte Transport in the Colon. *News Physiol Sci 14*: 58–64.

Silverstein, R.L., W. Li, Y.M. Park, and S.O. Rahaman. 2010. Mechanisms of cell signaling by the scavenger receptor CD36: Implications in atherosclerosis and thrombosis. *Trans Am Clin Climatol Assoc 121*: 206–220.

Sina, C., M. Foerster, S. Derer, F.L. Hildebrand, B. Raabe, J. Scheller, *et al.* 2009. G-Protein Coupled Receptor 43 (Gpr43) Is Essential for Neutrophil Recruitment During Intestinal Inflammation. *Gastroenterology 136*(5): A240–A240.

Smith, P.M., M.R. Howitt, N. Panikov, M. Michaud, C.A. Gallini, *et al.* 2013. The microbial metabolites, short-chain fatty acids, regulate colonic Treg cell homeostasis. *Science 341*(6145): 569–573.

Soliman, M.L. and T.A. Rosenberger. 2011. Acetate supplementation increases brain histone acetylation and inhibits histone deacetylase activity and expression. *Molecular and Cellular Biochemistry 352*(1–2): 173–180.

Staels, B., Y. Handelsman, and V. Fonseca. 2010. Bile Acid Sequestrants for Lipid and Glucose Control. *Current Diabetes Reports 10*(1): 70–77.

Swann, J.R., E.J. Want, F.M. Geier, K. Spagou, I.D. Wilson, J.E. Sidaway, *et al.* Systemic gut microbial modulation of bile acid metabolism in host tissue compartments. *Proceedings of the National Academy of Sciences of the United States of America 108*: 4523–4530.

Szabo, G., S. Bala, J. Petrasek, and A. Gattu. 2010. Gut-Liver Axis and Sensing Microbes. *Digestive Diseases 28*(6): 737–744.

Tang, W.H., Z. Wang, B.S. Levison, R.A. Koeth, E.B. Britt, X. Fu, *et al.* 2013. Intestinal microbial metabolism of phosphatidylcholine and cardiovascular risk. *N Engl J Med 368*(17): 1575–1584.

Tarantino, G. 2014. Gut microbiome, obesity-related comorbidities, and low-grade chronic inflammation. *J Clin Endocrinol Metab 99*(7): 2343–2346.

Tazoe, H., Y. Otomo, I. Kaji, R. Tanaka, S.I. Karaki, and A. Kuwahara. 2008. Roles of Short-Chain Fatty Acids Receptors, Gpr41 and Gpr43 on Colonic Functions. *Journal of Physiology and Pharmacology 59*: 251–262.

Teixeira, T.F.S., M.C. Collado, C.L.L.F. Ferreira, J. Bressan, and M.D.G. Peluzio. 2012. Potential mechanisms for the emerging link between obesity and increased intestinal permeability. *Nutrition Research 32*(9): 637–647.

Teixeira, T.F.S., L. Grzeskowiak, S.C.C. Franceschini, J. Bressan, C.L.L.F. Ferreira, and M.C.G. Peluzio. 2013. Higher level of fecal SCFA in women correlates with metabolic syndrome risk factors. *British Journal of Nutrition 109*(5): 914–919.

Thomas, C., A. Gioiello, L. Noriega, A. Strehle, J. Oury, G. Rizzo, *et al.* 2009. TGR5-Mediated Bile Acid Sensing Controls Glucose Homeostasis. *Cell Metabolism 10*(3): 167–177.

Tolhurst, G., H. Heffron, Y.S. Lam, H.E. Parker, A.M. Habib, E. Diakogiannaki, *et al.* 2012. Short-chain fatty acids stimulate glucagon-like peptide-1 secretion via the G-protein-coupled receptor FFAR2. *Diabetes 61*(2): 364–371.

Turnbaugh, P.J., R.E. Ley, M.A. Mahowald, V. Magrini, E.R. Mardis, and J.I. Gordon.

2006. An obesity-associated gut microbiome with increased capacity for energy harvest. *Nature* 444(7122): 1027–1031.

van Hoek, M.J.A. and R.M.H. Merks. 2012. Redox balance is key to explaining full vs. partial switching to low-yield metabolism. *Bmc Systems Biology 6*.

Vernay, M. 1987. Origin and Utilization of Volatile Fatty-Acids and Lactate in the Rabbit—Influence of the Fecal Excretion Pattern. *British Journal of Nutrition* 57(3): 371–381.

Vitseva, O.I., K. Tanriverdi, T.T. Tchkonia, J.L. Kirkland, M.E. McDonnell, C.M. Apovian, *et al.* 2008. Inducible Toll-like receptor and NF-kappaB regulatory pathway expression in human adipose tissue. *Obesity (Silver Spring)* 16(5): 932–937.

Vrieze, A., C. Out, S. Fuentes, L. Jonker, I. Reuling, R.S. Kootte, *et al.* 2014. Impact of oral vancomycin on gut microbiota, bile acid metabolism, and insulin sensitivity. *J Hepatol* 60(4): 824–831.

Vrieze, A., E. Van Nood, F. Holleman, J. Salojarvi, R.S. Kootte, J.F.W.M. Bartelsman, *et al.* 2012. Transfer of Intestinal Microbiota From Lean Donors Increases Insulin Sensitivity in Individuals With Metabolic Syndrome. *Gastroenterology* 143(4): 913–+.

Wakayama, S., A. Haque, N. Koide, Y. Kato, E. Odkhuu, T. Bilegtsaikhan, *et al.* 2014. Lipopolysaccharide impairs insulin sensitivity via activation of phosphoinositide 3-kinase in adipocytes. *Immunopharmacol Immunotoxicol* 36(2): 145–149.

Waldecker, M., T. Kautenburger, H. Daumann, C. Busch, and D. Schrenk. 2008. Inhibition of histone-deacetylase activity by short-chain fatty acids and some polyphenol metabolites formed in the colon. *Journal of Nutritional Biochemistry* 19(9): 587–593.

Wang, L., Y.K. Lee, D. Bundman, Y.Q. Han, S. Thevananther, C.S. Kim, *et al.* 2002. Redundant pathways for negative feedback regulation of bile acid production. *Developmental Cell* 2(6): 721–731.

Wang, Z., E. Klipfell, B.J. Bennett, R. Koeth, B.S. Levison, B. Dugar, *et al.* 2011. Gut flora metabolism of phosphatidylcholine promotes cardiovascular disease. *Nature* 472(7341): 57–63.

Watanabe, M., S.M. Houten, C. Mataki, M.A. Christoffolete, B.W. Kim, H. Sato, *et al.* 2006. Bile acids induce energy expenditure by promoting intracellular thyroid hormone activation. *Nature* 439(7075): 484–489.

Weiss, J. 2003. Bactericidal/permeability-increasing protein (BPI) and lipopolysaccharide-binding protein (LBP): Structure, function and regulation in host defence against Gram-negative bacteria. *Biochem Soc Trans* 31(Pt 4): 785–790.

Westwell-Roper, C., D. Nackiewicz, M. Dan, and J.A. Ehses. 2014. Toll-like receptors and NLRP3 as central regulators of pancreatic islet inflammation in type 2 diabetes. *Immunology and Cell Biology* 92(4): 314–323.

Wolever, T.M.S., P. Spadafora, and H. Eshuis. 1991. Interaction between Colonic Acetate and Propionate in Humans. *American Journal of Clinical Nutrition* 53(3): 681–687.

Wostmann, B.S., C. Larkin, A. Moriarty, and E. Bruckner-Kardoss. 1983. Dietary intake, energy metabolism, and excretory losses of adult male germfree Wistar rats. *Lab Anim Sci* 33(1): 46–50.

Wu, Y., T.T. Wu, J. Wu, L. Zhao, Q. Li, Z. Varghese, *et al.* 2013. Chronic inflammation exacerbates glucose metabolism disorders in C57BL/6J mice fed with high-fat diet. *Journal of Endocrinology* 219(3): 195–204.

Xiong, Y.M., N. Miyamoto, K. Shibata, M.A. Valasek, T. Motoike, R.M. Kedzierski, and M. Yanagisawa. 2004. Short-chain fatty acids stimulate leptin production in adipocytes through the G protein-coupled receptor GPR41. *Proceedings of the National Academy of Sciences of the United States of America* 101(4): 1045–1050.

Yadav, H., J.H. Lee, J. Lloyd, P. Walter, and S.G. Rane. 2013. Beneficial metabolic effects of a probiotic via butyrate-induced GLP-1 hormone secretion. *J Biol Chem* 288(35): 25088–25097.

Yamashita, H., K. Fujisawa, E. Ito, S. Idei, N. Kawaguchi, M. Kimoto, *et al.* 2007. Improvement of obesity and glucose tolerance by acetate in Type 2 diabetic Otsuka Long-Evans Tokushima Fatty (OLETF) rats. *Biosci Biotechnol Biochem* 71(5): 1236–1243.

Yoneno, K., T. Hisamatsu, K. Shimamura, N. Kamada, R. Ichikawa, M.T. Kitazume, *et al.* 2013. TGR5 signalling inhibits the production of pro-inflammatory cytokines by *in vitro* differentiated inflammatory and intestinal macrophages in Crohn's disease. *Immunology* 139(1): 19–29.

Zaibi, M.S., C.J. Stocker, J. O'Dowd, A. Davies, M. Bellahcene, M.A. Cawthorne, *et al.* 2010. Roles of GPR41 and GPR43 in leptin secretory responses of murine adipocytes to short chain fatty acids. *Febs Letters* 584(11): 2381–2386.

Zeisel, S.H. 2006. Choline: Critical role during fetal development and dietary requirements in adults. *Annu Rev Nutr 26*: 229–250.

Zeisel, S.H., K.A. daCosta, M. Youssef, and S. Hensey. 1989. Conversion of dietary choline to trimethylamine and dimethylamine in rats: Dose-response relationship. *J Nutr* 119(5): 800–804.

Zeng, H., D.L. Lazarova, and M. Bordonaro. 2014. Mechanisms linking dietary fiber, gut microbiota and colon cancer prevention. *World J Gastrointest Oncol* 6(2): 41–51.

CHAPTER 8

An Overview of Microbiota-Associated Gastrointestinal Diseases

CLAUDIA HERRERA, M.D., VIRGINIA ROBLES-ALONSO, M.D., and FRANCISCO GUARNER, M.D.

8.1. HOST-MICROBES INTERACTIONS IN THE GASTROINTESTINAL TRACT

CHRONIC microbial colonization that inflicts no evident harm on the host only attracted minor clinical attention during the past century (Guarner 2014). However, animals, including humans, are in permanent association with microbial communities maternally inherited at birth or acquired from the environment. Such communities remained unexplored before the age of molecular techniques because of the difficulties to cultivate and isolate a large majority of the microbial species. Development of novel gene sequencing technologies as well as availability of powerful bioinformatic analysis tools have allowed a dramatic proliferation of research studies over the past few years.

All epithelial surfaces of mammals are colonized by microorganisms, but the gastrointestinal tract harbours the largest microbial burden. The human gastrointestinal tract houses over 10^{14} microbial cells with over 1,000 microbial species, most of them belonging to the domain *Bacteria* (Qin et al. 2010). Hence, the gastrointestinal mucosa is the body's principal site for interaction with the microbial world. The mucosa exhibits a very large surface (considering the villus-crypt structure in an unfolded disposition, the flat extension is comparable to a tennis court) and contains adapted structures and functions for bidirectional communication with microorganisms, including a number of preformed receptors, microbial recognition mechanisms, host-microbe cross-talk pathways, and microbe-specific adaptive responses (Cummings *et al.* 2004; MacDonald *et al.* 2011).

The stomach and duodenum harbour very low numbers of microorganisms adhering to the mucosal surface or in transit, typically less than 10^3 bacteria cells per gram of contents. Acid, bile, and pancreatic secretions kill most ingested microbes, and the phasic propulsive motor activity impedes stable colonisation of the small bowel lumen. There is a progressive increase in numbers of bacteria along the jejunum and ileum, from approximately 10^4 in the jejunum to 10^7 cells per gram of contents at the ileal end. The large intestine is heavily populated by anaerobes with numbers in the region of 10^{12} cells per gram of luminal contents. The colonic transit time is slow (30–70 hours) and microorganisms have the opportunity to proliferate by fermenting available substrates derived from either the diet or endogenous secretions. By far, the colon harbours the largest population of microbial symbionts, which contribute to 60% of solid colonic contents (O'Hara 2006) (Figure 8.1).

Several hundred grams of bacteria living within the intestinal tract certainly affect host homoeostasis. Some resident bacteria in the human gut are associated with toxin formation and pathogenicity when

FIGURE 8.1. The colon is densely colonised by microbial symbionts. The scanning electron micrograph shows clustered groups of bacteria on the mucosal surface of the rat colon (Rattus norvegicus). Bar indicates five microns (Maria Vicario, Vall d'Hebron Research Institute, Spain).

they become dominant, for example, *Clostridium difficile*. Some other resident species are potential pathogens when the integrity of the mucosal barrier is functionally breached. However, the normal interaction between gut bacteria and their host is a symbiotic mutualistic relationship, defined as mutually beneficial for both partners. The host provides a nutrient-rich habitat, and intestinal bacteria confer important benefits on the host's health (Hooper *et al.* 2002). Several beneficial features of gut bacteria are widely recognized, including production of short chain fatty acids, vitamin synthesis, secretion of bacteriocins, and inhibition of pathogens through a multiplicity of mechanisms (Guarner *et al.* 2003; O'Hara 2006). The primary functions of the gut microbiota are ascribed into three categories: metabolic, protective, and trophic functions.

8.1.1. Metabolic Functions

Metabolic functions assist in the fermentation of nondigestible dietary substrates and endogenous mucus. Gene diversity among the microbial community provides a variety of enzymes and biochemical pathways that are distinct from the host's own constitutive resources. Fermentation of carbohydrates is a major source of energy in the colon for bacterial growth and produces short-chain fatty acids (SCFAs) that can be absorbed by the host. This results in salvage of dietary energy, and favours the absorption of ions (Ca, Mg, Fe) in the cecum. Metabolic functions also include the production of vitamins (K, B12, biotin, folic acid, pantothenate) and synthesis of amino acids from ammonia or urea (Metges 2000).

8.1.2. Protective Functions

Protective functions include the barrier effect that prevents invasion by pathogens. The resident bacteria represent a crucial line of resistance to colonization by exogenous microbes or opportunistic bacteria that are present in the gut, but their growth is restricted. The equilibrium between species of resident bacteria provides stability in the microbial population, but use of antibiotics can disrupt the balance (for instance, overgrowth of toxigenic *Clostridium difficile*). The barrier effect is based on the ability of certain bacteria to secrete antimicrobial substances, the bacteriocins, which inhibit the growth of pathogens, and also in the competition for ecological niches.

8.1.3. Trophic Functions

Trophic functions of the gut microbiota include the control of epithelial cell proliferation and differentiation. Epithelial cell turnover is reduced in colonic crypts of germ-free (GF) animals as compared with colonized controls. Cell differentiation is highly influenced by the interaction with resident microorganisms as shown by the expression of a variety of genes in GF animals monoassociated with specific bacterial strains (Hooper and Gordon 2001), and in humans fed with probiotic lactobacilli (van Baarlen *et al.* 2011). Gut bacteria have important trophic effects on mucosal immunocompetent cells and are critical for the development of a healthy immune system. Multiple and diverse interactions between microbes, epithelium, and gut lymphoid tissues are constantly reshaping local and systemic mechanisms of immunity. Commensal microbes play a major role in the induction of regulatory T cells in gut lymphoid follicles (Atarashi *et al.* 2011).

8.2. ANTIBIOTICS AND RISK OF DISEASE

Every day, 10–30 out of 1,000 inhabitants of developed countries consume a defined daily dose of antibiotics as ambulatory patients (Goossens *et al.* 2007). In other words, today an average of 10 million West Europeans and 6 million North Americans, as outpatients, will take a dose of antibiotics. Although most courses of antibiotics result in no immediate, obvious side-effects, there is a concern that antibiotic collateral damage altering the composition of the microbiota will interfere with the functions of this microbial community.

Antibiotic-associated diarrhea is the most commonly recognized complication of antibiotics, and develops in 15–25% of patients on antibiotics. Most episodes of diarrhea induced by antibiotics are mild and self-limited in a few days. However, an increasing number of cases develop more severe forms, including *C. difficile*—associated diarrhea. The antibiotic-induced disturbance of the intestinal microbiota promotes *C. difficile* spore germination within the intestine, vegetative growth, and toxin production, leading to epithelial damage and colitis. Clinical presentation ranges from self-limiting diarrhea to toxic megacolon, fulminant colitis, and death, but even milder cases may suffer recurrent episodes of diarrhea overtime. Antibiotics directed against *C. difficile* can decrease the load of the pathogen and toxin production inducing clinical remission. However, if the microbiota is unable to restore resistance to colonization by *C. difficile*, the patients will develop recurrent episodes of infection (Britton *et al.* 2014). Hence, *C. difficile* infection is the paradigm of how antibiotics can disturb the protective function of the human gut microbiota against pathogens.

There is particular concern with the use of antibiotics during childhood. Antibiotics are among the most prescribed medications during early life. Data from the Centers for Disease Control and Prevention indicate that the average child in the United States receives about three antibiotic treatments in the first 2 years of life and approximately 11 by the age of 10 (Hicks *et al.* 2013). Repeated exposure to antibiotics for the treatment of ear, sinus, and throat infections is common before the age of 3. This is also a period during which the gut microbiota is shaped (Koenig *et al.* 2011). Indeed, from birth to 3 years of age, the composition of the gut community undergoes continuous changes, with a gradual increase in phylogenetic diversity. The introduction of solid meals is associated with an increase in the abundance of Bacteroidetes and a switch from genes facilitating lactate utilization to those linked to carbohydrate utilization, vitamin biosynthesis, and xenobiotic degradation. Superimposed on these patterns of gradual change, the effects of antibiotics result in large shifts in the relative abundance of taxonomic groups. The use of antibiotics induces a decrease in microbial diversity (loss of richness in the ecosystem) and an overgrowth of resistant species, which may even result in an overall increase of microbial load (Panda *et al.* 2014) (Figure 8.2).

Perturbations of the gut microbial ecosystem during its period of development combined with genetic susceptibility may have a long-lasting impact on the immune system leading to disease or predisposition to disease later in life. Indeed, it has been shown that inflammatory bowel diseases (IBD), metabolic disorders (type 2 diabetes, obesity), and atopic diseases are associated with an alteration of the gut community composition.

The incidence and prevalence of childhood IBD is increasing worldwide. A leading hypothesis regarding the pathogenesis is that alterations of the gut microbial community caused by repeated exposure to antibiotics trigger inflammation. Several retrospective and nationwide cohort studies have examined the potential correlation between the use of antibiotics and IBD. Those infants receiving antibiotics before 1 year of age were found to be more likely to be diagnosed with IBD than nonusers (Kronman *et al.* 2012). This association appeared to be strongest in the first 3 months after use and among children with more than 7

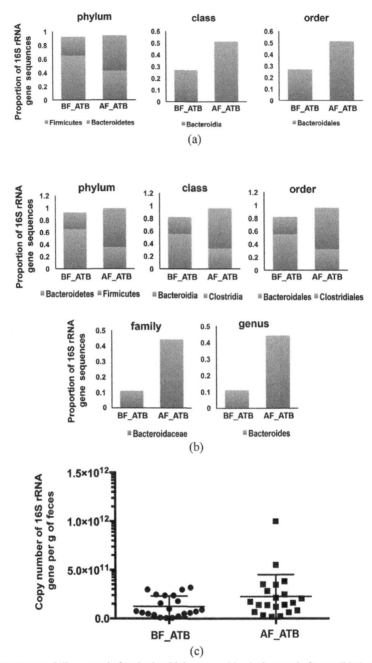

FIGURE 8.2. Differences in fecal microbial communities before and after antibiotic treatment as determined by 16S rRNA gene sequence analysis at various taxonomic levels. (a) Effect of amoxicillin-clavulanic acid. (b) Effect of levofloxacin. Only differences with $q < 0.01$ are shown. BF_ATB = before treatment; AF_ATB = after treatment. (c) Microbial load as assessed by quantitative real-time PCR (qPCR) of the 16S rRNA gene (Panda et al. 2014).

courses of antibiotic treatment (Hviid *et al.* 2011; Virta *et al.* 2012). No definitive link between the type of antibiotic used and IBD was made in any of the studies.

Although it has been demonstrated that human genetics and diet play an important role in determining body weight, it is now widely accepted that the increase in the prevalence of obesity over the past 30 years is also attributable to the alteration of the gut microbial community composition. The demonstration that the obesity phenotype can be transferred to GF recipient mice via microbiome transplantation provided evidence that the gut microbial community contributes to obesity perhaps by increasing caloric recovery from consumed foods. Indeed, obesity has been associated with an alteration of the composition and function of the gut microbial community (Turnbaugh *et al.* 2009). Interestingly, reduced diversity and lower gene counts in the microbial gut community has been associated with increased adiposity, insulin and leptin resistance, and a more pronounced inflammatory phenotype (Le Chatelier *et al.* 2013). These traits are also found after repeated antibiotic treatments. For instance, antibiotic exposure in early life, when host adipocyte populations are developing, has been associated with the development of adiposity in humans (Trasande *et al.* 2013). Since the 1950s, low-dose antibiotics have been widely used as growth promoters in animal husbandry. Experiments using mice have shown that low-dose antibiotics increase fat mass and the percentage of body fat (Cho *et al.* 2012). Coincidentally, the period of accelerated increase in prevalence of obesity in the United States overlaps with both increased dietary caloric intake and antibiotic exposure through food.

In conclusion, antibiotics are powerful medicines to fight against pathogens and cure infectious diseases. However, despite the well-documented resilience of the gut microbiota, treatment with these drugs may be associated with persistent changes in microbial composition and with potential long-term consequences for host immunity and metabolic activities. Many of these unintended consequences come about from the use of antibiotics in early life, during microbial community acquisition, a period which in turn is involved in the education of the host's immune system.

8.3. THE GUT MICROBIOTA IN INFLAMMATORY BOWEL DISEASE

IBD comprise two clinical entities, namely Crohn's disease (CD) and ulcerative colitis (UC). Both are chronic relapsing diseases, where environmental, genetic, and microbial factors are implicated in the pathogenesis of inflammatory lesions in the gut wall. Incidence and prevalence of IBD are increasing in different regions around the world, indicating its emergence as a global disease (Molodecky *et al.* 2012). Noteworthy, increasing incidence is temporally associated with socio-economic development, modern lifestyle, and changes in the microbial environment of humans (hygiene) suggesting that the aetiology is linked with altered microbial colonization of the gut. It is postulated that in IBD, mucosal lesions are generated by an excessive immune response to the microbiota that colonize the intestine. It is proposed that an altered microbial colonization would trigger or at least contribute to the uncontrolled immune response in genetically predisposed individuals.

Reported changes in the gut microbiota of patients with UC include reduction in phylogenetic diversity, intraindividual variability of the microbial composition over time (temporal instability of the ecosystem), and over- or under-representation of certain microbial species. Studies performed in twin pairs allow discriminating between genetic imprinting and environmental factors. In a recent study, 16S rDNA genetic sequences were analysed in sigmoid biopsies from twin pairs discordant for UC. Results showed less bacterial diversity together with more *Actinobacteria* and *Proteobacteria* and less *Bacteroidetes* (attributable to species from the *Prevotellaceae* family) at phylum level in the pairs affected by UC as compared with the healthy siblings. Interestingly, the nonaffected twins were characterized by lower diversity than unrelated healthy controls; this finding reflects some degree of heritability of gut microbial communities (Lepage *et al.* 2011). By contrast, another study carried out on 40 twin pairs with at least one sibling affected by IBD, did not find differences between UC-affected and healthy twins (Willing *et al.* 2010). The fact that in this study a majority of patients were on long-term remission may explain the difference and suggest restoration of the microbial composition during periods of remission. The potential effects of the drugs used as maintenance therapy (or during previous phases of the disease) on the microbial changes are not known.

Recent research underscores the concept that the imbalance in the gut microbiome results not only from shifts in the dominant phyla but also from changes in low-abundance species, which seem to play critical functions in the ecosystem. For instance, bacterial species with anti-inflammatory properties such as *Faecalibacterium prausnitzii* (belonging to the *Clostridium leptum* group) are depleted both in active UC and in remission (Varela *et al.* 2013). A study investigating *F. prausnitzii* by quantitative real time PCR found significant reductions in abundance not only in UC patients but also in their healthy relatives as compared to unrelated healthy controls. Moreover, the recovery of the *F. prausnitzii* population after relapse was associated with maintenance of clinical remission. Some other species have been shown to be overrepresented in UC. Both *E. coli* (Sokol *et al.* 2006) and *Fusobacterium varium* (Ohkusa *et al.* 2002) show higher abundance in fecal samples from UC patients than controls. Notably, experimental work indicates that rectal infusion of *Fusobacterium varium* induces colitis in the mouse (Ohkusa *et al.* 2003).

Dietary factors are powerful modulators of the gut microbiota and there is experimental evidence that saturated fat may contribute to intestinal inflammation through changes in the microbiota. A diet rich on saturated milk-derived fat produces an enrichment of taurocholate in bile and promotes overgrowth of the sulphite-reducing bacteria *Bilophila wadsworthia*, increasing the incidence of colitis in genetically IBD-susceptible IL-10 knock-out mice (Devkota *et al.* 2012). Sulphite- and sulphate-reducing bacteria produce hydrogen sulphide that is a highly toxic moiety for the colonic epithelium (Roediger *et al.* 1997; Lennon *et al.* 2014).

As in UC, gut microbiota changes in CD patients are characterized by a fall in phylogenetic diversity at the expense of the *Firmicutes* phylum (Manichanh *et al.* 2006; Kang *et al.* 2010), as well as temporal instability. In the twin study by Willing *et al.* (2010) the microbial compositions of individuals with CD differed from those of healthy individuals, and could by stratified by Crohn's phenotype. When the ileum is involved *Faecalibacterium* and *Roseburia* genera (belonging to Firmicutes phylum) were depleted, whereas *Enterobacteriaceae* and *Ruminococcus gnavus* were overrepresented. Interestingly, changes in microbial composition have been shown to have prognostic value: a reduction of *Faecalibacterium prausnitzii* abundance in ileal mucosal samples is associated with a higher risk of postoperative recurrence of ileal disease (Sokol *et al.* 2008).

The potential role of drugs on the microbiome changes in IBD patients was not established by previous investigations. However, a recent study recruited 447 children with new-onset CD and collected fecal and mucosal biopsy samples prior to treatment. The study confirmed that the disease is strongly associated with a drop in species diversity,

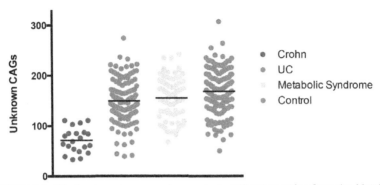

FIGURE 8.3. Whole genome sequencing analysis of fecal samples from the MetaHIT cohort of European individuals detected a high percentage of nonannotated genes (unknown taxonomy and functionality). By searching for coabundance gene groups (CAGs) among the 396 samples of the cohort, 741 gene-rich groups (containing more than 700 genes, median 1,700) were identified leading to the detection of previously unknown genetic entities. Interestingly, it was reported a dramatic drop of such unknown species in fecal samples from 21 Crohn's disease patients, studied while on clinical remission (Nielsen et al. 2014).

increased abundance of species from *Enterobacteriaceae*, *Pasteurellacaea*, *Veillonellaceae*, and *Fusobacteriaceae*, and decreased abundance of *Erysipelotrichales*, *Bacteroidales*, and *Clostridiales*. It was found that the microbial changes observed in tissue samples were not reflected in stool samples collected at the same time, emphasizing the need to examine tissue biopsies along with stool samples (Gevers *et al.* 2014).

Presence of adherent-invasive *Escherichia coli* (AIEC) strains is frequent in patients with CD. These strains survive and replicate extensively within macrophages without inducing host cell death, and their high replication rates induce the secretion of large amounts of tumour necrosis factor alpha (Martinez-Medina *et al.* 2009).

A recent publication proposes a novel bioinformatic approach to classify "orphan" genes, which cannot be assigned to a taxonomic order. The new strategy searches groups of genes with coabundance in a series of samples from different individuals, considering that such gene clusters reveal the presence of a distinct biological entity (eukaryotic cell, bacteria, phage, virus, etc.). From a cohort of 396 fecal samples, more than 7,000 of such coabundance gene groups (CAGs) were identified leading to the detection of hundreds of genetic entities previously unknown. Most CAGs were small in size, less than 700 genes, whereas 741 CAGs were gene rich (more than 700 genes, median 1,700), representative of a number of prokaryotic or eukaryotic cells, where only a subset of 227 CAGs were identifiable. Figure 8.3 shows abundance of large CAGs of unknown taxonomy in individuals of the European cohort. A dramatic drop of such unknown species was found in fecal samples from 21 CD patients (Nielsen *et al.* 2014). This approach opens a new opportunity to identify species depleted in CD.

8.4. THE GUT MICROBIOTA IN FUNCTIONAL BOWEL DISORDERS

Functional gastrointestinal disorders encompass a number of digestive conditions where patients refer abdominal symptoms in the absence of detectable cause by conventional diagnostic methods. Among them, several types are defined based on the presumable region of the gut originating the symptoms and their pattern. The most common disorder among those is the so-called Irritable Bowel Syndrome (IBS), which is defined by abdominal pain or discomfort associated to altered bowel habit (Longstreth 2005). IBS affects 10–20% of the population in the developed world (Gwee 2005), has a profound impact on quality of life, and imposes a tremendous socioeconomic burden ($26.4 billion estimated annual cost in the United States). Specific clinical diagnostic criteria for IBS have been developed by consensus (Longstreth *et al.* 2006), but IBS is a heterogeneous condition. Because the expression capability and the symptom repertoire of the gut are very limited, the same symptom pattern can be produced by various pathophysiological mechanisms.

A complex net of feedback reflex mechanisms controls gut homeostasis and, normally, the digestive process should evolve unperceived. However, under some circumstances, afferent pathways can be activated inducing conscious perception of symptoms. Some data indicate that IBS symptoms can be produced by three mechanisms: (1) abnormal gut contents (i.e., excess of gas), (2) altered gut motor responses, and (3) gut hypersensitivity. Some subsets of IBS patients complain of excessive gas production that may interfere with daily activities. The amount of gas production can be clinically determined (Azpiroz *et al.* 2007). Other subsets of IBS patients have delayed motor responses of the gut. Motility and motor reflexes in the gut can be evaluated by intestinal manometry or by wireless capsule endoscopy (Malagelada *et al.* 2012). Finally, some IBS patients have visceral hypersensitivity and perceive symptoms in response to stimuli that are well tolerated by healthy subjects.

Despite its large incidence and social impact, the aetiology of IBS remains poorly understood. These sensory-reflex dysfunctions have been associated with miscellaneous causes, including low-grade inflammation, bacterial gastroenteritis, psychological disturbances, and stress. Gut microbes and their metabolic products are likely to be involved in altered gastrointestinal function given their demonstrated ability to influence intestinal permeability and immune function, activity in the enteric nervous system, pain modulation systems, and the brain (Mayer et al. 2014).

The examination of plausible differences in the gut microbiome in IBS has been for long a challenging task due to the huge complexity of this ecosystem, as well as by the heterogeneity of IBS patients. Some studies have demonstrated that subtypes of IBS patients, selected by clinical criteria, such as stool form, present instability of the microbiota at three months intervals (Matto et al. 2005). Furthermore, a considerable number of studies (22 studies comprising 827 subjects) suggest that gut microbiome composition in IBS patients exhibits some differences as compared to healthy subjects. Significant shifts in microbial community composition have been reported in IBS patients with different subtypes of IBS (IBS- diarrhea, IBS-Constipation, IBS-Mixed), ages (paediatric versus adult), or compartments (mucosa versus stool samples) (Mayer et al. 2014). However, there is a lack of consensus on the wide range of gut microbial differences between IBS subjects and healthy controls and the specific microbial changes that may be corre-

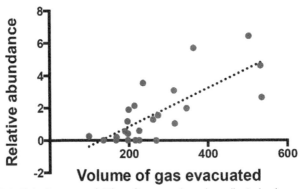

FIGURE 8.4. Colonic gas production after a meal can be collected using a rectal balloon catheter connected via a line without leaks to a barostat, and the volume is continuously recorded for 6 hours (Manichanh et al. 2014). In subjects who consulted for excessive passage of gas per anus, and not in controls, fecal abundance of Bilophila wadsworthia was found positively correlated with the volume of gas evacuated ($r = 0.64$).

lated to disease outcome. Some general trends are decreased diversity in small bowel microbiota with increased abundance of Gram-negative organisms in IBS. Based on analysis of fecal samples, regardless of analytical methodology used, a number of studies reported decreased relative abundance of the genera *Bifidobacterium* and *Lactobacillus*, and increased *Firmicutes:Bacteroidetes* ratios at the phylum level (Simren *et al.* 2013; Mayer *et al.* 2014).

The colonic microbiota contributes to both the production and consumption of intestinal gas, and in patients with excess flatulence increased abundance of *Bacteroides fragilis* and *Bilophila wadsworthia* has been reported (Manichanh *et al.* 2014). *Bacteroides fragilis* is an opportunistic human pathogen, causing infections of the peritoneal cavity, and appendicitis via abscess formation. *Bilophila wadsworthia* is a sulphate-reducing species producing hydrogen sulphide gas which has irritant effects on the colonic epithelium. It was observed that the volume of intestinal gas correlated with abundance of the *Bilophila* species in patients with excess flatulence but not in controls, suggesting a major contribution of this species to the production of intestinal gas (Manichanh *et al.* 2014) (Figure 8.4).

There is conflicting evidence regarding alterations in the production of microbial metabolites in patients with IBS, and on the beneficial effects of gut microbial manipulations with prebiotics, probiotics, and antibiotics in some IBS patients.

8.5. THE GUT MICROBIOTA IN LIVER DISEASES

There is continuous bidirectional communication between the gut and the liver through bile, hormones, inflammatory mediators, and products of digestion and absorption. Thus, compositional structure and metabolic functions of the intestinal microbiota can be expected to have both direct and indirect effects on liver function and physiology. Alteration in gut microbiota composition or "dysbiosis" has been demonstrated in several acute and chronic liver diseases. The dysbiosis is characterized by changes in the *Bacteroides:Firmicutes* ratio and an increase in potentially pathogenic taxa such as *Enterobacteriaceae* in diseased states. Such alterations of the gut microbiota associated with liver diseases will be briefly reviewed here.

Nonalcoholic fatty liver disease (NAFLD) is the hepatic manifestation of the metabolic syndrome, and one of the most common liver diseases in the Western world. Nonalcoholic steatohepatitis (NASH) is an

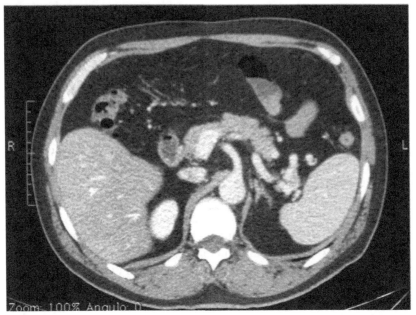

FIGURE 8.5. Nonalcoholic fatty liver disease is often associated with abdominal obesity, which is a known risk factor for insulin resistance and cardiovascular disease. This CT scan of an individual with NAFLD shows accumulation of intra-abdominal fat packed into the peritoneal cavity in between the bowel loops as opposed to subcutaneous fat, which is found underneath the skin. This central form of obesity is associated with dysbiosis, altered intestinal permeability, increased microbial translocation, and systemic low-grade inflammation.

advanced stage of NAFLD and features steatosis, necroinflammation, and fibrosis. Animal studies have shown that the gut microbiota can contribute to all the histological components of NAFLD. Dietary factors and changes in diet are determinants of the composition of the microbiome (Wu *et al.* 2011). NAFLD patients are often obese and insulin resistant (Figure 8.5). Obesity is accompanied by an intestinal metagenome that has an increased capacity to collect energy from the host diet (Backhed *et al.* 2004; Turnbaugh *et al.* 2009). Bacterial enzymes aid in the digestion of otherwise indigestible dietary polysaccharides and extraction of calories from them (Turnbaugh *et al.* 2006). Enteric bacteria can also suppress the synthesis and secretion of "fasting-induced adipocyte factor" from epithelial cells in the small intestine, resulting in increased activity of lipoprotein lipase and increased accumulation of triglycerides in the liver (Bäckhed *et al.* 2004; Bäckhed *et al.* 2007). Several human studies in the field of obesity have found differences in the gut microbiota between obese and lean individuals. It was suggested that obese humans harbour considerably fewer *Bacteroidetes* and more Firmicutes than lean controls, although this has not been a consistent finding (Ley *et al.* 2006; Le Chatelier *et al.* 2013; Zhu *et al.* 2013). An increase of *Actinobacteria* in obese individuals has also been reported (Turnbaugh *et al.* 2009). A recent study that characterized the gut microbiota of NASH, obese, and healthy children and adolescents, found a significant decrease of Firmicutes in NASH and obese patients. There was also an increased abundance of *Escherichia* species in NASH microbiomes (Zhu *et al.* 2013). In anaerobic conditions, *Escherichia* species take the mixed-acid fermentation pathway, a major product of which is ethanol, which may be a factor in driving the disease progression to NASH (Clark 1989). Another study in adults with biopsy-proven NASH or simple steatosis and living liver donors as healthy controls found that patients with NASH had a lower percentage of *Bacteroidetes* and higher fecal *C. coccoides* compared to those with simple steatosis and healthy controls (Mouzaki *et al.* 2013).

Alterations in choline appear to influence the development of NAFLD. The gut microbiota play a role in choline metabolism by catalysing the first step in its conversion to dimethylamine and trimethylamine. High-fat diets favour the promotion of microbial species that convert dietary choline into methylamines, reducing circulating plasma levels of phosphatidylcholine (Dumas *et al.* 2006). Phosphatidylcholine is necessary for the assembly and secretion of very-low-density lipoproteins (VLDL), and microbiota-induced choline deficiency results in triglyceride accumulation in hepatocytes (Jiang *et al.* 2005).

Alcohol abuse is one of the leading causes of chronic liver disease. There is limited data of the effect of alcohol on the gut microbiota and the role of the microbiome in alcoholic liver disease. An animal model of alcoholic liver disease in mice observed an altered gut microbiota with significant reductions in *Lactobacillus, Pediococcus, Leuconostoc,* and *Lactococcus* and an increase in the number of *Verrucomicrobia,* and *Bacteroidetes,* such as *Bacteroidales, Bacteroides,* and *Porphyromonadaceae* in the alcohol treated group (Yan *et al.* 2011).

Mutlu *et al.* (2009) evaluated the effect of daily alcohol administration in rats. There were alterations in mucosa-associated microbiota after 10 weeks (Mutlu *et al.* 2009). The same group characterized the mucosa-associated colonic microbiota in patients with alcoholic cirrhosis and alcoholic patients without liver disease and healthy controls (Mutlu *et al.* 2012). They found that the proportion of *Bacteroidaceae* was lower in alcoholic patients than in nonalcoholic individuals. It seems, based on clinical and preclinical studies, that excessive alcohol intake is accompanied with dysbiosis.

Liver fibrosis may result in end-stage liver disease or cirrhosis, which eventually disrupts the metabolic functions of the liver. The development of cirrhosis in these patients is the major determinant of morbidity and mortality (Poynard et al. 2000). Several studies have assessed the taxonomic composition of the intestinal microbiota in patients with cirrhosis (Chen et al. 2011; Islam et al. 2011; Bajaj et al. 2012; Wu et al. 2012; Xu et al. 2012). A common feature of cirrhosis is an increase of potentially pathogenic bacteria, accompanied by reduced proportions of beneficial bacteria. A study that characterized fecal microbiome communities in 36 patients with liver cirrhosis and 24 healthy controls found the proportion of phylum *Bacteroidetes* significantly reduced in cirrhosis patients and an increase in *Proteobacteria* and *Fusobacteria* compared with controls. Patients with alcoholic cirrhosis had significant increases in the *Prevotellaceae* family compared with patients with hepatitis B–related cirrhosis or healthy individuals. The authors also observed a positive correlation between Child-Turcotte-Pugh (CTP) score and *Streptococcaceae*, whereas *Lachnospiraceae* decreased significantly in patients with cirrhosis and correlated negatively with CTP score (Chen et al. 2011).

The gut microbiota has been implicated in the major complications of liver cirrhosis, including hepatic encephalopathy (HE), spontaneous bacterial peritonitis, and variceal bleeding. HE is a broad term reflecting a spectrum of neuropsychiatric abnormalities observed in patients with liver dysfunction (Bajaj 2010). It is divided into two primary components: overt HE, which can be diagnosed clinically through a constellation of signs and symptoms, and minimal HE, which requires specialized diagnostic testing. It seems that ammonia generated by the enteric microbiota is the critical driver of this pathologic process. Studies of microbiota have shown no difference in stool microbiota between cirrhosis patients with or without HE, but mucosal microbiomes differ by having lower *Roseburia* and higher levels of *Enterococcus, Veillonella, Megasphaera,* and *Burkholderia* abundance. These altered microbiomes are associated with poor cognition, endotoxaemia, and inflammation in hepatic encephalopathy patients compared to cirrhosis patients without hepatic encephalopathy (Bajaj et al. 2012a, 2012b).

Bacterial translocation occurs in healthy individuals and is important for immune system development, but also can be harmful. In patients with cirrhosis, bacterial translocation induces inflammation and hemodynamic derangement (Bellot *et al.* 2013). Infections such as spontaneous bacterial peritonitis and bacteraemia develop in patients with end-stage liver disease, caused by migration of intestinal bacteria into the peritoneal cavity or circulation. Up to 80% of infections are caused by Gram-negative bacilli, especially *Escherichia coli*, a microbe that belongs to the *Enterobacteriaceae* family, which in turn, as mentioned above, is increased in the microbiota of patients with cirrhosis (Chen *et al.* 2011, Bajaj *et al.* 2012, Kakiyama *et al.* 2013).

Bile acids mediate communication between the liver and intestine. Conjugated bile acids are absorbed in the terminal ileum to return to the liver. Intestinal bacteria in the large intestine generate secondary bile acids by deconjugation and dehydroxylation. Recent studies have evaluated the bile acid pool as a modulator of the composition of the gut microbiome and vice versa. Animal model studies have demonstrated that increased bile acid levels in the colon select against the *Bacteroidetes* and *Actinobacteria* and favour the *Firmicutes* (Islam *et al.* 2011). Human gut bacteria carrying out bile acid 7α-dehydroxylation have been shown to belong to the genus *Clostridium*, which are Gram-positive, anaerobic, spore-forming members of the *Firmicutes* (Ridlon *et al.* 2010). Kakiyama *et al.* (2013) analyzed fecal microbiota and fecal bile acids in cirrhotic patients and found a significant decrease in total bile acids in feces of patients with advanced cirrhosis. These patients also had a higher *Enterobacteriaceae* abundance but lower *Lachonospiraceae, Ruminococcaceae,* and *Blautia* abundance (Kakiyama *et al.* 2013). These data suggest that dysbiosis might be occurring in patients with cirrhosis in part due to low bile acid input into the gut. Changes in serum bile acids also have been reported in experimental models of NASH and alcoholic liver disease (Tanaka *et al.* 2012; Xie *et al.* 2013).

In conclusion, there appears to be an association between intestinal dysbiosis and liver disease, and there is evidence for a more fundamental role in the aetiology of certain liver conditions, such as NAFLD and NASH. Changes in the gut microbiome appear to cause liver disease mostly in animal models and a few clinical studies. The gut microbiota can influence complications of cirrhosis, such as HE. Future studies should assess microbial gene expression, proteins, and metabolites, and focus on longitudinal study of patient cohorts.

8.6. REFERENCES

Atarashi, K., T. Tanoue, T. Shima, A. Imaoka, T. Kuwahara, Y. Momose, *et al.* 2011.

Induction of colonic regulatory T cells by indigenous Clostridium species. *Science* *331*(6015): 337–341.

Azpiroz, F., M. Bouin, M. Camilleri, E.A. Mayer, P. Poitras, J. Serra, and R.C. Spiller. 2007. Mechanisms of hypersensitivity in IBS and functional disorders. *Neurogastroenterol Motil 19*(1 Suppl): 62–88.

Bäckhed, F., H. Ding, T. Wang, L.V. Hooper, G.Y. Koh, A. Nagy, *et al.* 2004. The gut microbiota as an environmental factor that regulates fat storage. *Proc Natl Acad Sci USA 101*(44): 15718–15723.

Bäckhed, F., J.K. Manchester, C.F. Semenkovich, and J.I. Gordon. 2007. Mechanisms underlying the resistance to diet-induced obesity in germ-free mice. *Proc Natl Acad Sci USA 104*(3): 979–984.

Bajaj, J.S. 2010. Review article: the modern management of hepatic encephalopathy. *Aliment Pharmacol Ther 31*(5): 537–547.

Bajaj, J.S., P.B. Hylemon, J.M. Ridlon, D.M. Heuman, K. Daita, M.B. White, *et al.* 2012. Colonic mucosal microbiome differs from stool microbiome in cirrhosis and hepatic encephalopathy and is linked to cognition and inflammation. *Am J Physiol Gastrointest Liver Physiol 303*(6): G675–685.

Bajaj, J.S., J.M. Ridlon, P.B. Hylemon, L.R. Thacker, D.M. Heuman, S. Smith, *et al.* 2012. Linkage of gut microbiome with cognition in hepatic encephalopathy. *Am J Physiol Gastrointest Liver Physiol 302*(1): G168–175.

Bellot, P., R. Frances, and J. Such. 2013. Pathological bacterial translocation in cirrhosis: Pathophysiology, diagnosis and clinical implications. *Liver Int 33*(1): 31–39.

Britton, R.A. and V.B. Young. 2014. Role of the intestinal microbiota in resistance to colonization by *Clostridium difficile*. *Gastroenterology 146*(6): 1547–1553.

Chen, Y., F. Yang, H. Lu, B. Wang, Y. Chen, D. Lei, *et al.* 2011. Characterization of fecal microbial communities in patients with liver cirrhosis. *Hepatology 54*(2): 562–572.

Cho, I., S. Yamanishi, L. Cox, B.A. Methe, J. Zavadil, K. Li, *et al.* 2012. Antibiotics in early life alter the murine colonic microbiome and adiposity. *Nature 488*(7413): 621–626.

Clark, D.P. 1989. The fermentation pathways of *Escherichia coli*. *FEMS Microbiol Rev 5*(3): 223–234.

Cummings, J.H., J.M. Antoine, F. Azpiroz, R. Bourdet-Sicard, P. Brandtzaeg, P.C. Calder, *et al.* 2004. PASSCLAIM—gut health and immunity. *Eur J Nutr 43* Suppl 2: II118–II173.

Devkota, S., Y. Wang, M.W. Musch, V. Leone, H. Fehlner-Peach, A. Nadimpalli, *et al.* 2012. Dietary-fat-induced taurocholic acid promotes pathobiont expansion and colitis in Il10$^{-/-}$ mice. *Nature 487*(7405): 104–108.

Dumas, M.E., R.H. Barton, A. Toye, O. Cloarec, C. Blancher, A. Rothwell, *et al.* 2006. Metabolic profiling reveals a contribution of gut microbiota to fatty liver phenotype in insulin-resistant mice. *Proc Natl Acad Sci USA 103*(33): 12511–12516.

Frank, D.N., A.L. St Amand, R.A. Feldman, E.C. Boedeker, N. Harpaz, and N.R. Pace 2007. Molecular-phylogenetic characterization of microbial community imbalances in human inflammatory bowel diseases. *Proc Natl Acad Sci USA 104*(34): 13780–13785.

Gevers, D., S. Kugathasan, L.A. Denson, Y. Vazquez-Baeza, W. Van Treuren, B. Ren, *et al.* 2014. The treatment-naive microbiome in new-onset Crohn's disease. *Cell Host Microbe* 15(3): 382–392.

Goossens, H., M. Ferech, S. Coenen, P. Stephens, and G. European. 2007. Comparison of outpatient systemic antibacterial use in 2004 in the United States and 27 European countries. *Clin Infect Dis* 44(8): 1091–1095.

Guarner, F. 2014. Decade in review-gut microbiota: The gut microbiota era marches on. *Nat Rev Gastroenterol Hepatol.*

Guarner, F. and J.R. Malagelada. 2003. Gut flora in health and disease. *Lancet* 361(9356): 512–519.

Gwee, K.A. 2005. Irritable bowel syndrome in developing countries—A disorder of civilization or colonization? *Neurogastroenterol Motil* 17(3): 317–324.

Hicks, L.A., T.H. Taylor, Jr., and R.J. Hunkler. 2013. U.S. outpatient antibiotic prescribing, 2010. *N Engl J Med* 368(15): 1461–1462.

Hooper, L.V. and J.I. Gordon. 2001. Commensal host-bacterial relationships in the gut. *Science* 292(5519): 1115–1118.

Hooper, L. V., T. Midtvedt, and J.I. Gordon. 2002. How host-microbial interactions shape the nutrient environment of the mammalian intestine. *Annu Rev Nutr* 22: 283–307.

Hviid, A., H. Svanstrom, and M. Frisch. 2011. Antibiotic use and inflammatory bowel diseases in childhood. *Gut* 60(1): 49–54.

Islam, K.B., S. Fukiya, M. Hagio, N. Fujii, S. Ishizuka, T. Ooka, *et al.* 2011. Bile acid is a host factor that regulates the composition of the cecal microbiota in rats. *Gastroenterology* 141(5): 1773–1781.

Jiang, X.C., Z. Li, R. Liu, X.P. Yang, M. Pan, L. Lagrost, *et al.* 2005. Phospholipid transfer protein deficiency impairs apolipoprotein-B secretion from hepatocytes by stimulating a proteolytic pathway through a relative deficiency of vitamin E and an increase in intracellular oxidants. *J Biol Chem* 280(18): 18336–18340.

Kakiyama, G., W.M. Pandak, P.M. Gillevet, P.B. Hylemon, D.M. Heuman, K. Daita, *et al.* 2013. Modulation of the fecal bile acid profile by gut microbiota in cirrhosis. *J Hepatol* 58(5): 949–955.

Kang, S., S.E. Denman, M. Morrison, Z. Yu, J. Dore, M. Leclerc, and C.S. McSweeney. 2010. Dysbiosis of fecal microbiota in Crohn's disease patients as revealed by a custom phylogenetic microarray. *Inflamm Bowel Dis* 16(12): 2034–2042.

Koenig, J.E., A. Spor, N. Scalfone, A.D. Fricker, J. Stombaugh, R. Knight, *et al.* 2011. Succession of microbial consortia in the developing infant gut microbiome. *Proc Natl Acad Sci USA* 108 Suppl 1: 4578–4585.

Kronman, M.P., T.E. Zaoutis, K. Haynes, R. Feng, and S.E. Coffin. 2012. Antibiotic exposure and IBD development among children: A population-based cohort study. *Pediatrics* 130(4): e794–803.

Le Chatelier, E., T. Nielsen, J. Qin, E. Prifti, F. Hildebrand, G. Falony, *et al.* 2013. Richness of human gut microbiome correlates with metabolic markers. *Nature* 500(7464): 541–546.

Lennon, G., A. Balfe, N. Bambury, A. Lavelle, A. Maguire, N.G. Docherty, *et al.* 2014. Correlations between colonic crypt mucin chemotype, inflammatory grade and Desulfovibrio species in ulcerative colitis. *Colorectal Dis* 16(5): O161–169.

Lepage, P., R. Hasler, M.E. Spehlmann, A. Rehman, A. Zvirbliene, A. Begun, et al. 2011. Twin study indicates loss of interaction between microbiota and mucosa of patients with ulcerative colitis. *Gastroenterology 141*(1): 227–236.

Ley, R.E., P.J. Turnbaugh, S. Klein, and J.I. Gordon. 2006. Microbial ecology: human gut microbes associated with obesity. *Nature 444*(7122): 1022–1023.

Longstreth, G.F. 2005. Definition and classification of irritable bowel syndrome: current consensus and controversies. *Gastroenterol Clin North Am 34*(2): 173–187.

Longstreth, G.F., W.G. Thompson, W.D. Chey, L.A. Houghton, F. Mearin, and R.C. Spiller. 2006. Functional bowel disorders. *Gastroenterology 130*(5): 1480–1491.

MacDonald, T.T., I. Monteleone, M.C. Fantini, and G. Monteleone. 2011. Regulation of homeostasis and inflammation in the intestine. *Gastroenterology 140*(6): 1768–1775.

Malagelada, C., F. De Lorio, S. Segui, S. Mendez, M. Drozdzal, J. Vitria, et al. 2012. Functional gut disorders or disordered gut function? Small bowel dysmotility evidenced by an original technique. *Neurogastroenterol Motil 24*(3): 223–228, e104–225.

Manichanh, C., A. Eck, E. Varela, J. Roca, J.C. Clemente, A. Gonzalez, et al. 2014. Anal gas evacuation and colonic microbiota in patients with flatulence: Effect of diet. *Gut 63*(3): 401–408.

Manichanh, C., L. Rigottier-Gois, E. Bonnaud, K. Gloux, E. Pelletier, L. Frangeul, et al. 2006. Reduced diversity of fecal microbiota in Crohn's disease revealed by a metagenomic approach. *Gut 55*(2): 205–211.

Martinez-Medina, M., X. Aldeguer, M. Lopez-Siles, F. Gonzalez-Huix, C. Lopez-Oliu, G. Dahbi, et al. 2009. Molecular diversity of *Escherichia coli* in the human gut: New ecological evidence supporting the role of adherent-invasive *E. coli* (AIEC) in Crohn's disease. *Inflamm Bowel Dis 15*(6): 872–882.

Matto, J., L. Maunuksela, K. Kajander, A. Palva, R. Korpela, A. Kassinen, and M. Saarela. 2005. Composition and temporal stability of gastrointestinal microbiota in irritable bowel syndrome—A longitudinal study in IBS and control subjects. *FEMS Immunol Med Microbiol 43*(2): 213–222.

Mayer, C., J. Call, A. Albiach-Serrano, E. Visalberghi, G. Sabbatini, and A. Seed. 2014. Abstract knowledge in the broken-string problem: Evidence from nonhuman primates and pre-schoolers. *PLoS One 9*(10): e108597.

Mayer, E.A., T. Savidge, and R.J. Shulman. 2014. Brain-gut microbiome interactions and functional bowel disorders. *Gastroenterology 146*(6): 1500–1512.

Metges, C.C. 2000. Contribution of microbial amino acids to amino acid homeostasis of the host. *J Nutr 130*(7): 1857S–1864S.

Molodecky, N.A., I.S. Soon, D.M. Rabi, W.A. Ghali, M. Ferris, G. Chernoff, et al. 2012. Increasing incidence and prevalence of the inflammatory bowel diseases with time, based on systematic review. *Gastroenterology 142*(1): 46–54 e42; quiz e30.

Mouzaki, M., E.M. Comelli, B.M. Arendt, J. Bonengel, S.K. Fung, S.E. Fischer, et al. 2013. Intestinal microbiota in patients with nonalcoholic fatty liver disease. *Hepatology 58*(1): 120–127.

Mutlu, E., A. Keshavarzian, P. Engen, C.B. Forsyth, M. Sikaroodi, and P. Gillevet. 2009. Intestinal dysbiosis: A possible mechanism of alcohol-induced endotoxemia and alcoholic steatohepatitis in rats. *Alcohol Clin Exp Res 33*(10): 1836–1846.

Mutlu, E.A., P.M. Gillevet, H. Rangwala, M. Sikaroodi, A. Naqvi, P.A. Engen, et al. 2012. Colonic microbiome is altered in alcoholism. *Am J Physiol Gastrointest Liver Physiol 302*(9): G966–978.

Nielsen, H.B., M. Almeida, A.S. Juncker, S. Rasmussen, J. Li, S. Sunagawa, et al. 2014. Identification and assembly of genomes and genetic elements in complex metagenomic samples without using reference genomes. *Nat Biotechnol 32*(8): 822–828.

O'Hara, A.M. 2006. The gut flora as a forgotten organ. *EMBO Rep 7*(7): 688–693.

Ohkusa, T., I. Okayasu, T. Ogihara, K. Morita, M. Ogawa, and N. Sato. 2003. Induction of experimental ulcerative colitis by *Fusobacterium varium* isolated from colonic mucosa of patients with ulcerative colitis. *Gut 52*(1): 79–83.

Ohkusa, T., N. Sato, T. Ogihara, K. Morita, M. Ogawa, and I. Okayasu. 2002. Fusobacterium varium localized in the colonic mucosa of patients with ulcerative colitis stimulates species-specific antibody. *J Gastroenterol Hepatol 17*(8): 849–853.

Panda, S., I. El khader, F. Casellas, J. Lopez Vivancos, M. Garcia Cors, A. Santiago, et al. 2014. Short-term effect of antibiotics on human gut microbiota. *PLoS One 9*(4): e95476.

Poynard, T., V. Ratziu, Y. Benhamou, P. Opolon, P. Cacoub, and P. Bedossa. 2000. Natural history of HCV infection. *Baillieres Best Pract Res Clin Gastroenterol 14*(2): 211–228.

Qin, J., R. Li, J. Raes, M. Arumugam, K.S. Burgdorf, C. Manichanh, et al. 2010. A human gut microbial gene catalogue established by metagenomic sequencing. *Nature 464*(7285): 59–65.

Ridlon, J.M., D.J. Kang, and P.B. Hylemon. 2010. Isolation and characterization of a bile acid inducible 7α-dehydroxylating operon in *Clostridium hylemonae* TN271. Anaerobe 16(2): 137–146.

Roediger, W.E., J. Moore, and W. Babidge. 1997. Colonic sulfide in pathogenesis and treatment of ulcerative colitis. *Dig Dis Sci 42*(8): 1571–1579.

Simren, M., G. Barbara, H.J. Flint, B.M. Spiegel, R.C. Spiller, S. Vanner, et al. 2013. Intestinal microbiota in functional bowel disorders: A Rome foundation report. *Gut 62*(1): 159–176.

Sokol, H., P. Lepage, P. Seksik, J. Dore, and P. Marteau. 2006. Temperature gradient gel electrophoresis of fecal 16S rRNA reveals active *Escherichia coli* in the microbiota of patients with ulcerative colitis. *J Clin Microbiol 44*(9): 3172–3177.

Sokol, H., B. Pigneur, L. Watterlot, O. Lakhdari, L.G. Bermudez-Humaran, J.J. Gratadoux, et al. 2008. *Fecalibacterium prausnitzii* is an anti-inflammatory commensal bacterium identified by gut microbiota analysis of Crohn disease patients. *Proc Natl Acad Sci USA 105*(43): 16731–16736.

Tanaka, N., T. Matsubara, K.W. Krausz, A.D. Patterson, and F.J. Gonzalez. 2012. Disruption of phospholipid and bile acid homeostasis in mice with nonalcoholic steatohepatitis. *Hepatology 56*(1): 118–129.

Trasande, L., J. Blustein, M. Liu, E. Corwin, L.M. Cox, and M.J. Blaser. 2013. Infant antibiotic exposures and early-life body mass. *Int J Obes (Lond) 37*(1): 16–23.

Turnbaugh, P.J., M. Hamady, T. Yatsunenko, B.L. Cantarel, A. Duncan, R.E. Ley, et al. 2009. A core gut microbiome in obese and lean twins. *Nature 457*(7228): 480–484.

Turnbaugh, P.J., R.E. Ley, M.A. Mahowald, V. Magrini, E.R. Mardis, and J.I. Gordon. 2006. An obesity-associated gut microbiome with increased capacity for energy harvest. *Nature* 444(7122): 1027–1031.

van Baarlen, P., F. Troost, C. van der Meer, G. Hooiveld, M. Boekschoten, R.J. Brummer, and M. Kleerebezem. 2011. Human mucosal *in vivo* transcriptome responses to three *lactobacilli* indicate how probiotics may modulate human cellular pathways. *Proc Natl Acad Sci USA 108* Suppl 1: 4562–4569.

Varela, E., Manichanh, C., Gallart, M., Torrejon, A., Borruel, N., Casellas, F., *et al.* 2013. Colonisation by *Faecalibacterium prausnitzii* and maintenance of clinical remission in patients with ulcerative colitis. *Alimentary pharmacology & therapeutics. Jul;* 38(2):151–61.

Virta, L., A. Auvinen, H. Helenius, P. Huovinen, and K.L. Kolho. 2012. Association of repeated exposure to antibiotics with the development of pediatric Crohn's disease —A nationwide, register-based finnish case-control study. *Am J Epidemiol* 175(8): 775–784.

Willing, B.P., J. Dicksved, J. Halfvarson, A.F. Andersson, M. Lucio, Z. Zheng, *et al.* 2010. A pyrosequencing study in twins shows that gastrointestinal microbial profiles vary with inflammatory bowel disease phenotypes. *Gastroenterology* 139(6): 1844–1854 e1841.

Wu, G.D., J. Chen, C. Hoffmann, K. Bittinger, Y.Y. Chen, S.A. Keilbaugh, *et al.* 2011. Linking long-term dietary patterns with gut microbial enterotypes. *Science* 334(6052): 105–108.

Wu, Z.W., H.F. Lu, J. Wu, J. Zuo, P. Chen, J.F. Sheng, *et al.* 2012. Assessment of the fecal *lactobacilli* population in patients with hepatitis B virus-related decompensated cirrhosis and hepatitis B cirrhosis treated with liver transplant. *Microb Ecol* 63(4): 929–937.

Xie, G., W. Zhong, H. Li, Q. Li, Y. Qiu, X. Zheng, *et al.* 2013. Alteration of bile acid metabolism in the rat induced by chronic ethanol consumption. *FASEB J* 27(9): 3583–3593.

Xu, M., B. Wang, Y. Fu, Y. Chen, F. Yang, H. Lu, *et al.* 2012. Changes of fecal Bifidobacterium species in adult patients with hepatitis B virus-induced chronic liver disease. *Microb Ecol* 63(2): 304–313.

Yan, A.W., D.E. Fouts, J. Brandl, P. Starkel, M. Torralba, E. Schott, *et al.* 2011. Enteric dysbiosis associated with a mouse model of alcoholic liver disease. *Hepatology* 53(1): 96–105.

Zhu, L., S.S. Baker, C. Gill, W. Liu, R. Alkhouri, R.D. Baker, and S.R. Gill. 2013. Characterization of gut microbiomes in nonalcoholic steatohepatitis (NASH) patients: A connection between endogenous alcohol and NASH. *Hepatology* 57(2): 601–609.

CHAPTER 9

An Overview of Microbiota-Associated Epigenetic Disorders

DAWN D. KINGSBURY and HOLLY H. GANZ

9.1. INTRODUCTION

9.1.1. What Is Epigenetics?

EACH individual's characteristics, denoted as a phenotype, arise from interactions between his or her genome, the functional component of genetic material, known as the epigenome, and the environment. The term epigenetics, a blending of the words "epigenesis" and "genetics", was first used by Conrad Waddington (Waddlngton 1942). Epigenetics is the study of factors that lead to the functional uniqueness of a phenotype manifested through the expression of certain genes (turned on) and the repression of others (turned off). The resulting distinctly altered cellular function is heritably modified, notably without any change to the genomic DNA sequence.

Within each cell, genomic DNA is tightly wrapped around proteins called histones, which form the superstructure known as chromatin. The resulting chromatin is further folded into chromosomes (Figure 9.1). Folding helps to compact the DNA for storage in the cell nucleus and it is used to control which genes are expressed. When the cell is stimulated, signals are sent to the nucleus to not only express but also to regulate gene function. In response to cellular signals, gene regulatory proteins change epigenetic tags associated with the DNA, making certain genes more or less available for transcription. Most simply put, these post-transcriptional modifications are orchestrated by a population of protein complexes that can incorporate (write), remove (erase),

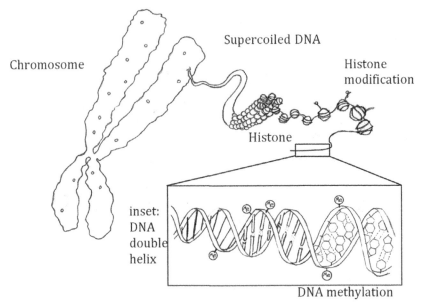

FIGURE 9.1. Chromosome structure consists of tightly coiled DNA wrapped around histones. Addition of various functional chemical groups, e.g., methyl (CH_3) to histones and DNA, are two common epigenetic modifications. Drawing by Seana K. Davidson.

and/or bind (read) these codes. In reality, the interactions are extremely complex, and cutting edge technologies are identifying crosstalk and downstream effects of combinatorial modifications.

Epigenetic mechanisms include modifications of chromatin structure through covalent modification of histones. Lysine residues on the histone proteins may be modified by acetylation, methylation, formylation, proprionylation, butyrylation, crotonylation, malonylation, succinylation, 5-hydroxylation, ADP-ribosylation, sumoylation, and/or ubiquitination. Arginine residues may be modified by methylation and citrullination. Serine and threonine residues may be modified by acetylation, phosphorylation, and glycosylation. Tyrosine residues may be modified by acetylation, phosphorylation, and hydroxylation. Histidine residues may be modified by phosphorylation. Most DNA contains methylated bases that are introduced after DNA replication by enzymes known as methyltransferases. Broadly, hypermethylation silences genes, and hypomethylation activates genes with some exceptions. Direct modification of DNA can also occur by hydroxymethylation, formylation, and carboxylation of its cytosine residues (Rothbart and Strahl 2014).

More recently, regulatory noncoding RNAs (ncRNA) were identified as important epigenetic regulators of gene transcription and translation (Esteller 2011; Consortium 2012). ncRNAs are functional RNA molecules that are transcribed from DNA but are not translated into protein, and instead have roles in heterochromatin formation, histone modification, DNA methylation targeting, and gene silencing. They are divided in two main groups based on size. The short ncRNAs are less than 30 nucleotides, and the long ncRNAs are greater than 200 nucleotides. The short ncRNAs come in many classes: microRNAs (miRNA), short interfering RNAs (siRNA), transcription initiation RNAs, splice site RNAs, and piwi-interacting RNAs (piRNA) (Collins *et al.* 2010; Morris and Mattick 2014). In-between in length and currently being investigated are promoter-associated RNA and enhancer RNA types (Kaikkonen *et al.* 2011; Camp *et al.* 2014). These epigenetic mechanisms form a complex spatial and temporal hierarchy to regulate gene expression contextually. Again, while these epigenetic factors do not involve any changes to the primary DNA sequence, the epigenetic tags on the genome may be copied from one generation to the next, and can result in long-term changes in phenotype, including cells' potential to proliferate and differentiate.

9.1.2. Microbially-Produced Metabolites Can Affect Gene Expression and Health

Metabolites produced by the microbiome can further affect the epigenome and gene expression in the host, potentially contributing to the development of a number of important health disorders, including cancer, autism, diabetes, arthritis, and metabolic syndrome. The intestinal microbiome "crosslinks" the environment, especially the diet, with the host genome by producing low molecular weight compounds that affect epigenomic mechanisms and gene expression (Shenderov 2012). An overview of potential epigenetic mechanisms of various microbial metabolites, physiological effects, and associated bacterial groups is provided in Table 9.1.

Microbial metabolism produces a number of short-chain fatty acids (SCFAs) that are bioactive. Acetate, propionate, and butyrate are SCFAs produced from the breakdown of fiber and oligosaccharides by gut microbiota that have profound effects on gut health (Tremaroli and Bäckhed 2012). *Firmicutes* are among the most important butyrate producers in the gut microbiome, particularly *Faecalibacterium prausnit-*

TABLE 9.1. Epigenetic Mechanisms of Microbial Metabolites and Their Physiological Effects.

Metabolite	Epigenetic Mechanism	Physiological Effects	Bacterial Groups Associated with Metabolic Pathway
SCFA: Butyrate, Acetate propionate, formate, valerate, caproate	Histone modifications	Cancer prevention and therapy; anti-inflammatory	Firmicutes, particularly *Faecalibacterium prausnitzii* and *Eubacterium rectale/Roseburia* spp
Betaine, Choline-ethanolamine, TMA	DNA methylation	Fetal brain development and function, Increased atherosclerosis risk	*E. coli*
AA: tryptophan*, indole, indole-3-acetate, tryptamine	AhR-DRE transcription	Reduced colonic inflammation	Bacilli (phylum Firmicutes) *Lactobacillus* spp
Hydrogen and H$_2$S	Direct free radical associated DNA damage; DNA hypomethylation	Impaired NAD regeneration/cytochrome oxidase, suppressed butyrate utilization, decreased mucus synthesis; pneumatosis intestinalis	*Desulfovibrioaceae* (phylum Proteobacteria)
Bile acids: deoxycholic acid	DNA hypomethylation, B-catentin signaling	Cancer progression	Unknown; an antibiotic-induced change
Alcohols: 1,2 propanediol, acetaldehyde	Decreased folate (methyl donor)	Increased intracellular invasion and persistence	Adherent-invasive *Escherichia coli* (Crohns Dz)
Neuropeptides: GABA, serotonin*, 4-ethylphenylsulfate	Increased GABA receptor expression	Gut motility, protection from depression, anxiety	Bifidobacteria, *Lactobacillus* spp; particularly *rhamnosus*, *Bacteroides fragilis* decreases

zii (in clostridial cluster IV) and *Eubacterium rectale/Roseburia* spp (in clostridial cluster XIVa) (Preshaw *et al.* 2012).

Butyrate and acetate provide a protective function for the intestinal mucosal surface through reduced inflammation, and they act as histone deacetylase (HDAC) inhibitors in the epigenetic control of host cell responses (Licciardi *et al.* 2010). Butyrate is also essential for determining the role of histone acetylation in chromatin structure and function and is the most effective SCFA in stimulating or repressing the expression of specific genes (Davie 2003). HDAC inhibitors such as butyrate are known anticancer agents, reducing the growth of certain cancers by enhancing the expression of certain genes involved in cell-cycle arrest and apoptosis (Avivi-Green *et al.* 2001; Acharya *et al.* 2005; Hofmanová et al 2005; Licciardi et al 2010). There is keen interest in HDAC inhibitors and their effects on epigenetic mechanisms governing a number of chronic illnesses (Inaba and Amano 2010; Esteller 2011; Lightfoot *et al.* 2012). Understanding the role of the microbiome in epigenetic mechanisms of illness is doubly important because it may assist in the development of risk assessment and early detection diagnostic tools for markers that are amendable to therapies that manipulate microbial community structure and/or gene function.

Tryptophan is an essential, aromatic amino acid obtained by the diet and microbiota (Young 1994). Ninety percent of the neurotransmitter serotonin (5-hydroxytryptamine) is synthesized in the gut from tryptophan. Along with its roles in digestion and metabolism, serotonin has broad body-wide effects, regulating behavior, pain perception, response to stress, as well as aspects of cardiovascular, respiratory, and urogenital function (Berger *et al.* 2009). Tryptophan is synthesized from chorismate by several phyla (*Proteobacteria, Actinobacteria,* and *Firmicutes*) of the microbiota (Xie *et al.* 2003). Studies utilizing germ-free (GF) mice and antibiotic treated rats confirm the key role for bacteria in tryptophan metabolism (Wikoff *et al.* 2009; Zheng *et al.* 2011; Marcobal *et al.* 2013).

The plot thickens when the microbial catabolism of tryptophan yields indoles such as indole-3-aldehyde, indole-3-acetate, indirubin, and tryptamine. These are all ligands for the aryl hydrocarbon receptor (AhR) (Zelante *et al.*; Jin *et al.* 2014). This cell-surface protein can activate transcription through a specific signal transduction pathway.

The aryl hydrocarbon receptor has been primarily described in the context of environmental toxins and xenobiotic ligands (namely halogenated aromatic hydrocarbons, e.g., dioxin, publicized with Viktor

Yushchenko's poisoning in 2004), plant flavonoids, and endogenously formed kynurenine (from tryptophan metabolism). AhR-ligand interactions come in three flavors: agonist, weak agonist/antagonist, or selective AhR modulators (SAhRM). Classic agonists bind AhR, which travels to the nucleus to dimerize with AhR nuclear translocator (ARNT) and occupies dioxin-responsive elements (DRE) that lie upstream from transcriptional start sites of target genes. Staying in the cytoplasm, SAhRMs bind and promote non-DRE activated protein-protein interactions with no translocation to the nucleus for DRE-mediated transcription. Classically AhR's main function is described as bioactivation of carcinogens by the metabolism of xenobiotics. AhR also regulates biochemical pathways involving immunity—both inflammatory and anti-inflammatory effects, fatty acid and cholesterol synthesis, energy metabolism, cell cycle, and epithelial barrier integrity (Murray *et al.* 2014).

In head and neck squamous cell carcinoma cells, AhR mediates increased expression of growth factors (e.g., amphiregulin, epiregulin, platelet derived growth factor A, vascular endothelial growth factor A, and fibroblast growth factor 9), along with the proinflammatory cytokine IL-6 (DiNatale *et al.* 2011; John *et al.* 2014). Both factors are associated with an aggressive phenotype. Other AhR proposed roles in tumor invasiveness and metastasis include initiation of epithelial-mesenchymal transition, decreases in cell adhesion, increases in cell motility, and decreases in tumor immune surveillance. As an aside, AhR activity can influence many signaling pathways and can be influenced by many particular indices (e.g., nutrient availability, cell cycle status, redox state, hypoxia, cell-cell contact, and cytokine level). Response of AhR ligands and subsequent signaling is circumstantial. For example, while increased nuclear AhR presence is a negative prognosticator for prostatic carcinoma progression, AhR is a positive prognosticator in hormone-dependent breast cancer due to the mutual antagonism between AhR and the estrogen receptor (Murray *et al.* 2014).

Metabolites aside, certain bacteria also promote the expression of AhR. For example, lipopolysaccharide (LPS) stimulation of Toll-like Receptor 4 (TLR4) increases AhR expression by the binding of the transcription factor complex RelA/p50 to the NF-κB (nuclear factor kappa-light-chain-enhancer of activated B cells) binding site in AhR gene promoter (Vogel *et al.* 2014). Enhancement of AhR can then increase the sensitivity to toxins and other ligands leading to a more robust response to potentially harmful exposures. As to whether this response also increases inflammation, there is evidence to suggest the promotion through TLR4 may be anti-inflammatory in nature yet this has not been proven in humans.

9.2. MICROBIAL INFLUENCE ON HUMAN HEALTH

Exploring the effects of microbes on the various systems of health is best achieved in the same way as a physical exam of a patient (Figure 9.2). The outcomes may follow one of two paths: inflammatory or neoplastic. Regarding the latter, epigenetic processes have been profoundly linked to cancer development. Viruses being obligate intracellular parasites have developed numerous ways of hijacking cell processes. While viruses, especially herpesviruses and bacteriophages, are members of the human microbiota, the epigenetic manipulations by oncogenic viruses are not included here. Instead this chapter focuses on bacteria, fungi, and archaea.

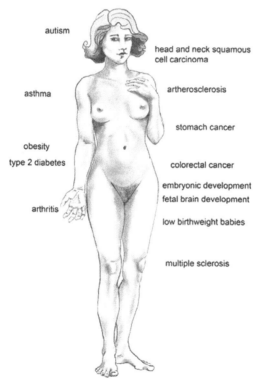

FIGURE 9.2. This chapter explores the body systems for various medical conditions where the human microbiome may induce epigenetic changes contributing to disease. Drawing by Seana K. Davidson.

9.2.1. Weight and body Condition

As described in previous chapters, the gut microbiome modulates absorption, storage, and energy harvest from the diet with consequences for host obesity (Bäckhed et al. 2004, 2007; Martin et al. 2007; Turnbaugh et al. 2008). Thus, gut microbes are an obvious potential therapeutic target in the prevention and reduction of obesity. Microbially-produced SCFAs regulate the expression of host genes promoting the deposition of absorbed energy in adipocytes (Bäckhed et al. 2004, 2007; Dumas et al. 2006; Martin et al. 2007). For example, host energy storage is regulated by the gut microbiome through host signaling pathways; for exmaple, microbial suppression of the fasting-induced adipose factor (Fiaf), a lipoprotein lipase inhibitor that promotes adiposity (Bäckhed et al. 2004).

The gut microbiota may also play active roles in the development of insulin resistance (Dumas et al. 2006), which, in addition to obesity, is linked to a number of other important health problems, including type 2 diabetes mellitus, hypertension, hyperlipidaemia, artherosclerosis, and polycystic ovarian disease (Saltiel and Kahn 2001). Epigenetic marks arising in host DNA from early nutrition may induce long-term changes in gene expression that can lead to the development of some of these disorders. For example, in a study of effects of the periconceptional diet in female sheep, Sinclair et al. (2007) restricted specific B vitamins (i.e., B12 and folate) and methionine and found that this dietary intervention generated widespread epigenetic alterations to DNA methylation in offspring and the adult offspring were heavier and fatter, had altered immune responses to antigenic challenge, were insulin-resistant, and had elevated blood pressure. In addition to DNA methylation, small noncoding RNAs or microRNAs also regulate gene expression and may play important roles in obesity, insulin resistance, and related disorders (Canani et al. 2011).

9.2.2. Head and Neck

The oral cavity has a well-documented resident microbial population (Dewhirst et al. 2010). As described elsewhere, many members of the oral microbiome are beneficial or may act as opportunistic pathogens. Some more pathogenic bacteria in the oral cavity are associated with periodontal disease (e.g., *Porphyromonas gingivalis*) and oral squamous cell carcinoma, (e.g., *Capnocytophaga gingivalis, Prevotella melaninogenica* and *Streptococcus mitis* [Mager *et al.* 2005; Hooper *et al.* 2009]). The accepted pathogenesis for oral disease features inflammation that arises when bacterial plaque forms on teeth. Over time, the chronic inflammation induced by this bacterial biofilm results in periodontal disease. When inflammation progresses, the epithelial barrier is compromised by ulceration. Microbes are allowed entry into underlying connective tissue where they can cause local effects and into blood capillaries where they can spread via the oral hematogenous route to cause distant effects.

Here is an example of epigenetic mechanisms of disease being associated with the oral microbiome locally. Later in the urogenital section of this chapter, a distance effect on preterm delivery and low birth-weight babies is linked to the oral microbiome. Epidemiologic studies support the clinical observation that the prevalence of head and neck squamous cell carcinoma (HNSCC) is linked to dental hygiene (Mager *et al.* 2005; Rosenquist *et al.* 2005; Guha *et al.* 2007). This leads to the hypothesis that the oral flora is influencing epithelial homeostasis, likely through the regulation of inflammation.

To identify possible mechanisms, Bebek *et al.* (2012) retrospectively analyzed patients with HNSCC. They intended to determine whether tumor versus adjacent normal tissue would have differing and specific microbiomes. Any relationship would then correspond to altered methylation of inflammatory genes. Indeed, the authors identified increased promoter hypermethylation of *IL8, RARB, TGFBR2*, and *MDR1* in the tumors compared with matched normal tissue by screening for inflammation and HNSCC-associated genes (Bebek *et al.* 2012). *IL8* codes for an inflammatory chemokine that recruits neutrophils. *RARB* encodes the retinoic acid receptor beta that is a nuclear receptor limiting cell growth by regulating gene expression. *TGFBR2* is a tumor suppressor gene encoding the transmembrane protein kinase transforming growth factor, beta receptor II. *MDR1* (multidrug resistance protein 1 or permeability glycoprotein or CD243) is an efflux pump responsible for expunging xenobiotics from normal tissue and providing some tumors broad chemo-resistance. Characterization of bacterial community composition using 16S rRNA sequencing revealed differences between the microbiome communities of tumor tissue compared to adjacent normal tissue. Multivariate analysis of covariance verifies that these specific microbiota, notably Enterobacteriaceae and *Tenericutes*, are associated with hypermethylation of the promoter for MDR1. These authors suggest that such microbiome-linked epigenetic findings may translate diagnostically into using DNA methylation profiles for prediction or early detection of HNSCC and/or therapeutically, using demethylation agents and/or specialized probiotics as adjunct therapies (Bebek *et al.* 2012).

9.2.3. Peripheral Lymph Nodes/Lymphatic/Immune System

Chapter 6 is entirely devoted to the role of the microbiota on the immune system. Briefly, the gut microbiome also has epigenetic links to lymphoma and leukemia. *Helicobacter pylori, Campylobacter jejuni, Chlamydia psittaci,* and *Borrelia burgdorferi* infections were associated with lymphoma of the stomach, small intestine, ocular adnexa, and skin, respectively. Gastric lymphomas have more methylated genes. The aberrant CpG methylation of specific genes including the tumor suppressors *p16 cyclin-dependent kinase inhibitor 2A, KIP2* cyclin-dependent kinase inhibitor 1C, *DAPK* death-associated protein kinase, *H-cadherin* (*HCAD*) cell-adhesion molecule, *MINT31* (a promoter locus regulating calcium channels involved with the tumor suppressor p53), and the DNA methyltransferase *MGMT* are consistently associated with *H. pylori* infection and lost after eradication therapy (Oka *et al.* 2011).

The microRNA miR-155, considered an onco-miRNA, is upregulated in many tumors including various B cell, Hodgkin's, and Burkitt's lymphomas. miR-155 negatively regulates the tumor suppressor gene *SOCS1*, causing activation of STAT3 via the JAK signaling cascade. The oncoprotein STAT3 increases cell survival, proliferation, and metastasis. Also, miR-155 suppresses PU.1 levels in Hodgkin's lymphoma. PU.1 is a transcription factor essential for late differentiation of B cells. The toll like receptor TLR4 detects lipopolysaccharide on Gram-negative bacteria. miR-155 is up-regulated by TLR4 in a NF-kB dependent mechanism (Aalaei-Andabili *et al.* 2013).

9.2.4. Heart and Lungs/Cardiovascular and Respiratory System

9.2.4.1. Cardiovascular Disease

Future therapies for cardiovascular disease may include probiotics that affect choline metabolism as well as dietary interventions. Recent studies link microbial metabolism of dietary choline and phosphatidylcholine (Koeth *et al.* 2013) and betaine and trimethylamine N-oxide (TMAO) (Arumugam *et al.* 2011) to atherosclerosis and cardiovascular pathogenesis in humans. Microbial metabolism of L-carnitine, a trimethylamine that is abundant in red meat, produces TMAO and accelerates artherosclerosis in mice (Koeth *et al.* 2013). Choline and betaine are involved in the methylation cycle: for example, betaine (trimethylglycine) serves as an alternate methyl source for remethylating homocysteine to methionine via betaine:homocysteine methyltransferase. Increasing levels of betaine by increasing S-adenosylmethionine and the S-adenosylmethionine: S-adenosylhomocysteine ratio could increase methylation potential and suppress gene expression. Choline and betaine deficiencies may also produce epigenetic changes in genes linked to atherosclerosis (Dong *et al.* 2002; Zaina *et al.* 2005). Alternatively, a positive correlation between total choline in blood and cardiovascular disease was identified by Danne *et al.* (2007) and LeLeiko *et al.* (2009). Additional studies are needed to determine how altered genomic methylation affects gene expression in vascular cells, such as studies of the methylated genome with different levels of choline and betaine in mice that are prone and resistant to atherosclerosis (Loscalzo 2011).

9.2.4.2. Respiratory

Whilst the respiratory tract has its own resident microbiota, most of the current literature has elucidated the role of distant microbial populations in the oral cavity and the intestine in disorders such as atherosclerosis. Bacteria can affect epigenetic mechanisms governing the development of asthma. In a laboratory study with mice, Brand *et al.* (2011) show that maternal exposure to the bacterium *Acinetobacter lwoffii* F78 during gestation prevented the development of an asthmatic phenotype in the offspring. Furthermore, the authors found that the interferon-gamma (*IFNG*) promoter of CD41 T cells in the progeny provide significant protection against loss of histone 4 (H4) acetylation, which is closely associated with IFN-g expression.

9.2.5. Gastrointestinal/Digestive System

9.2.5.1. Helicobacter Pylori-Associated Gastric Cancer

Helicobacter pylori is the only bacterium classified by the World Health Organization as a class 1 carcinogen, so few would question its classification as a pathogen (IARC 1994). However, *H. pylori* has coexisted as part of the human gastric microbiota for approximately the last hundred thousand years (Moodley *et al.* 2012) and still currently infects half the world's population (Hunt *et al.* 2011). Moreover, the absence of *Helicobacter* has been shown to be risk factor for certain diseases such as asthma, tuberculosis, and esophageal cancer (Chen and Blaser 2007, 2008; Blaser *et al.* 2008; Islami and Kamangar 2008; Perry *et al.* 2010; Whiteman *et al.* 2010). Only a minority of *H. pylori* infected individuals develop disease, which manifests as either gastric or duodenal ulceration or neoplasia. *H. pylori*-associated cancer presents as either gastric adenocarcinoma or mucosa-associated lymphoid tissue lymphoma (Pereira and Medeiros 2014; Plummer *et al.* 2014). While the majority of early gastric MALT lymphoma cases go into remission upon *H. pylori* eradication therapy, gastric cancer is the third leading cause of cancer death in men (the fifth in women) worldwide (Jemal *et al.* 2011). Cancer risk is a summation of the polymorphic nature of the human host/patient and his or her bacterial population, taking into account other environmental risk factor exposures.

Intimate contact with the gastric epithelium allows *H. pylori* to directly inject effectors into host cells to affect epigenetic changes. The Cag pathogenicity island encodes a syringe like pilus structure, type IV secretion system that can translocate the virulence factor cytotoxin-associated gene A (*CagA*). A target of *CagA*, Scr-homology protein tyrosine phosphatase (SHP) mediates host (Ras)-ERK (MAP-kinase) signaling, activating the nuclear factor κB and β-catenin pathways, which lead to changes in cell proliferation, adhesion, and shape. Cell polarity is altered resulting in intestinal metaplasia. Cell survival, even in the face of DNA damage, occurs secondary to *CagA* mediated proteasomal degradation of tumor suppressor p53 and enhanced tumor necrosis factor receptor-associated factor 6 (TRAF6)-mediated Lys 63-linked ubiquitination of transforming growth factor-β-activated kinase 1 (TAK1) (Wang *et al.* 2014a).

Bacterial nuclear targeting proteins (e.g., urease subunit A, outer membrane protein Omp18 and 50S ribosomal protein L20 delivered via outer membrane vesicles) have molecular masses greater than 40 kDa and cannot passively diffuse through the nuclear pore complex. They have specific amino acid sequences, referred to as nuclear localization signals (NLSs), and are recognized by nuclear transporter receptors (importins) to facilitate selective transport into the host nuclei. Lee *et al.* (2012) identified 22 *H. pylori* proteins targeting the nuclei of cultured COS-seven African green monkey kidney cells.

Robust gene silencing can occur via regional methylation, specifically in cytosine-phosphate-guanine dinucleotide rich areas (CpG islands) of gene promoter regions. The mechanisms may include an overexpression of endogenous DNA methyltransferases (DNMTs), injected bacterial DNMTs, a reduction of DNA demethylation activity (Zhao and Bu 2012). Hypermethylation is also commonly observed in inflammation-associated neoplasia. Inflammatory cells such as neutrophils produce hypochlorous (HOCl) and hypobromous (HOBr) acid, which can react with DNA to produce 5-chlorocytosine and 5-bromocytosine respectively. These inflammation-damaged 5-halocytosines cannot be distinguished from 5-methylcytosine by the enzyme DNMT-1 or methyl-binding proteins. Consequently, de novo DNA methylation may occur (Kundu and Surh 2008). The pathological result of hypermethylation is down regulation of tumor suppressor genes.

In gastric cancer the tumor suppressor genes, runt-related transcription factor3 (RUNX3), p16 also known as cyclin-dependent kinase inhibitor 2A (*CDKN2A*), the gene encoding a calcium-dependent cell-to-cell adhesion glycoprotein (*CDH1*), and *MLH1*—a DNA mismatch repair gene (microsatellite instability results from its suppression) are all inactivated more frequently by methylation than by mutation. Individuals infected with *H. pylori* have higher methylation levels than those uninfected. Decreases in methylation can be tracked with eradication of *Helicobacter*. Furthermore, methylation levels correlate with gastric cancer risk and severity (Ushijima and Hattori 2012).

Hypomethylation usually corresponds with gene activation. Global hypomethylation has been used as a cancer marker. However, recent investigations in gastric cancer did not reveal consistent global hypomethylation. Interrogating specific normally methylated repetitive elements, Alu repeats, were found to be hypomethylated in *H. pylori* infected and gastric cancer biopsies (Yoshida *et al.* 2011). Hypomethylation of Alu elements, which are distributed throughout the genome, can influence heteropyknosis—the differential condensation of various regions of the chromosomes; this may create chromosomal instability by chromatin decondensation and homologous recombination.

Epithelial-derived cancer cells often overexpress *MUC1*. *MUC1* is a cell surface-associated mucin containing a cytoplasmic domain, which regulates intracellular signaling and gene transcription. *H. pylori*'s CagA interacts with *MUC1* to increase gastric cancer cell proliferation. *H. pylori* upregulates MUC1 expression in gastric epithelial cell culture via CpG hypomethylation and signal transducer and activator of transcription 3 (STAT3) binding of the *MUC1* promoter (Guang *et al.* 2014).

STAT3 also binds the promoter of the transmembrane protein with epidermal growth factor and two follistatin motifs 2 gene (*TMEFF2*) albeit repressively. Important in human gastric carcinogenesis, tumor progression, and survival correlate with decreased TMEFF2 levels, which are down regulated by a mechanism other than hypermethylation. *H. pylori* infection (*in vitro* and *in vivo*) decreases *TMEFF2* expression. Further complicating the schema, bidirectional regulation exists, whereby TMEFF2 suppresses the phosphorylation of STAT3 and TMEFF2-induced down regulation of phosphorylated STAT3 depends on SHP1 (Sun *et al.* 2014).

MiRNAs are also being investigated for the role they play in initiation and progression of neoplastic *H. pylori* disease. Briefly, here are two examples. *H. pylori*'s CagA is important in suppressing miR-210. Epigenetic silencing of miR-210 occurs by hypermethylation of the miR-210 gene. Increased methylation of *miR-210* is found in *H. pylori*-positive human gastric biopsies and is associated with inflammation in the absence of hypoxia. miR-210 targets the oncogene STMN1 (also known as stathmin1, oncoprotein18, metablastin) (and *DIMT1* (demethyladenosine transferase 1). Inhibition of *miR-210* expression augments cell proliferation by activating *STMN1* (Kiga *et al.* 2014).

The oncogenic miRNAs miR-142-5p and miR-155 are overexpressed in MALT gastric lymphoma. Clinical disease course and *H. pylori* eradication treatment response are associated with miRNA expression levels. These miRNAs suppress *TP53INP1*, a proapoptotic stress-induced p53 target gene. Suppression could lead to the inhibition of apoptosis, accelerated proliferation, and tumor progression (Saito *et al.* 2012).

In conclusion, these findings widely support gastric cancer as an epigenetic disease directly influenced by a member of the gastric microbiota. Elucidation of the molecular mechanisms controlling epigenetic influences provides a glimpse at the spider web-like complexity of control over a cell's fate and gives us targets for ever improving prognostic markers, diagnostic and therapeutic tools.

9.2.5.2. Colorectal Cancer

Composition and activity of microbiota of the colon are thought to alter risk for the development of colorectal cancer (CRC). For example, the bacteria Bacteroides vulgatus and *Bacteroides stercoris* were linked to an increased risk for colorectal cancer, whereas *Lactobacillus acidophilus*, *Lactobacillus* S06, and *Eubacterium aerofaciens* species were associated with a reduced risk (Moore and Moore 1995; Lightfoot *et al.* 2013). Dietary consumption of red meat and fat are associated with an increased risk of CRC, which could arise from dietary effects on the composition and metabolic activity of the microbiota (Hambly *et al.* 1997; Davis and Milner 2009; Muegge *et al.* 2011). Experimental studies with mice have found that intestinal microbiota play a pivotal role in inflammation-induced CRC (Hu *et al.* 2013). Microbes may modulate the development of CRC through the production of SCFAs and other metabolites that produce epigenetic modifications that inactivate tumor suppressor genes (Lightfoot *et al.* 2013). Dietary fibers and microbially-produced butyrate may play a protective role against CRC by a modulation of a canonical Wnt signaling pathway that is constitutively activated in many CRCs (Bordonaro *et al.* 2008; Yang *et al.* 2013). Additional potential epigenetic mechanisms mediated by the intestinal microbiota may include inhibition of histone deacetylases by SCFAs in colonic epithelial cells (Waby *et al.* 2010; Kilner *et al.* 2012) and regulation of miRNA expression through DNA methylation and histone modifications (Bandres *et al.* 2009; Mazeh *et al.* 2013).

9.2.6. Urogenital/Reproductive/Endocrine Systems

Vaginal microbiota can affect fetal development, imbalance in maternal gut, and long term health of offspring through epigenetic mechanisms (Shenderov 2012). Furthermore, maternal and neonatal diet may interact with genes controlling lipid and carbohydrate metabolism, inducing alterations in epigenetic regulations that have long term effects on the development of a number of chronic illnesses later in life, such as metabolic syndrome, insulin resistance, type 2 diabetes, obesity, dyslipidaemia, hypertension, and cardiovascular disease (Canani *et al.* 2011). As described in previous chapters, microbial inoculation of the gastrointestinal tract of a newborn infant likely comes from a variety of sources, including vaginal, fecal, and skin-associated microbes from the mother, milk, and the environment (Song *et al.* 2013). In addition to nutrients, mammalian milk supplies bioactive components that affect the composition of the microbiota inhabiting the infant gastrointestinal tract (Ward *et al.* 2007). Adult health may reflect the development and function of these intestinal microbiota early in life, mediated by long lasting, epigenetic mechanisms (Canani *et al.* 2011).

Though controversy exists in the current literature (Vettore *et al.* 2008; Fogacci *et al.* 2011), epidemiological and interventional studies have bolstered the anecdotal clinical observations linking poor oral health to low birth weight infants (Lopez *et al.* 2005; Corbella *et al.* 2012). Evidence-based medicine is offering explanations for the high degree of heterogeneity amongst studies. Meta-analyses are identifying cofounding variables (da Rosa *et al.* 2012). Subcategorizing these large cohorts by dental disease severity (Dasanayake 2013) or by dental treatment outcome (responders versus nonresponders) (Jeffcoat *et al.* 2011) has elucidated statistically significant differences between patient subsets and pregnancy complication prevalence. Lastly genetic predisposition (e.g., a specific single nucleotide polymorphism in the inflammatory response gene encoding the prostaglandin E receptor *PTGER3*) is shown to be associated with the outcome of periodontal therapy and preterm birth (Jeffcoat *et al.* 2014).

Turning toward the microbial side, the finding of microbial DNA most resembling the oral microbiome within the placenta is most interesting (Aagaard *et al.* 2014). Certainly in the face of pregnancy complications, oral pathogens such as *Porhyromonas gingivalis* have been implicated in human (Barak *et al.* 2007; León *et al.* 2007; Katz *et al.* 2009) and animal studies (Lin *et al.* 2003; Boggess *et al.* 2005; Bélanger *et al.* 2008). Curiously, demonstration of transplacental migration via oral infection in rodents has been problematic (Arce *et al.* 2009). The molecular mechanisms by which *P. gingivalis* inhibits cellular proliferation and induces cellular death are being worked out in human trophoblast cell line [HTR-8/SVneo]. Fundamentally, a message originating on the cell surface, proceeds through a chain of proteins that generally communicate by adding phosphate groups to adjacent partners. The cascade proceeds all the way to the cell nucleus where transcription factors change gene expression and alter translation of RNA into protein. Cell division signaling commonly occurs through the Ras-Raf-MEK-ERK (MAPK/ERK) pathway. ERK 1/2 plays a dual role in controlling proliferation and cell death. When *P. gingivalis* invades trophoblasts FAS receptor (also known as APO-1, CD95, TNFRSF6) the most studied member of the death receptor family is induced. The tumor suppressor protein p53, which is responsible for regulating the cell cycle and preventing genomic mutation, accumulates in its active, phosphorylated form, and enhances miRNA maturation (Suzuki *et al.* 2009). These findings suggest that *P. gingivalis* may be inducing DNA damage. Indeed, Ataxia telangiectasia and Rad3 related kinase (ATR, FRP1) an essential regulator of DNA damage checkpoints, as well as its downstream signaling molecule CHEK2 are activated. Examining the checks and balances/feedback loops, p53 degradation related protein MDM2 (mouse double minute 2 homolog, E3 ubiquitin-protein ligase) is not induced. The *P. gingivalis* virulence factor, gingipains (trypsin-like cysteine proteinases) are degrading the negative regulator of p53 MDM2. The end result is trophoblast cell cycle arrest in G_1 and apoptosis—a form of programmed cell death. The regulation of apoptosis is associated with Ets1 activation. Ets1 is a transcription factor that can act to activate and/or repress genes by binding to DNA or other proteins. Further investigations are needed to translate these in vitro findings into the complexity of live models and clinical medicine (Inaba *et al.* 2012). For instance, finding microbial DNA in the placentas of healthy mothers may lead to hypotheses that lean more towards a homeostatic role such as inducing tolerance whilst the other end of the health disease continuum demonstrating a pathologic role inducing preterm delivery and low birth weight infants (Egija Zaura, unpublished).

9.2.7. Orthopedic (Skeletal)/Muscular/Neurologic (Nervous) Systems

9.2.7.1. Multiple Sclerosis

Independently, both epigenetic mechanisms (Mastronardi *et al.* 2006; Casaccia-Bonnefil *et al.* 2008) and the microbiome (Ochoa-Repáraz *et al.* 2011; Wang *et al.* 2014b) have been identified as playing important roles in the manifestation of multiple sclerosis.

The developed world has a high burden of allergic and autoimmune disease. The hygiene hypothesis is one theory that helps explain the increasing global distribution and temporal changes in the incidence of these pathologies. This hypothesis proposes that a critical factor for an individual at risk for allergic or autoimmune disease may be the lack of previously common infectious exposures (Leibowitz *et al.* 1966; Strachan 1989). Extending "lack of infection" to include commensal microbes lacking from altered diet or other factors is not a far stretch.

A prominent theme amongst investigations into allergic and autoimmune disease has been a relative deficiency of regulatory T cell (T_{reg}) activity (Wing and Sakaguchi 2006; Brusko *et al.* 2008). Functionally, these cells control inflammation and allow peripheral tolerance to ubiquitous antigens, for example, self, commensal and environmental/allergens. T_{reg} cells have been shown to control effector T cells by acting directly on dendritic cells, by inducing cell death of activated cells, by modulating the cytokine milieu, by release of adenosine nucleosides and by microRNA laden exosomes. The latter two providing epigenetic cell nonautonomous mechanisms (Vignali *et al.* 2008; Bopp *et al.* 2007; Chaudhry *et al.* 2011; Kastenmuller *et al.* 2011; Chatila and Williams 2014; Okoye *et al.* 2014).

Classic characterization of regulatory T cells includes the cell surface markers CD4 and CD25 (interleukin-2 receptor α-chain), as well as the transcription factor Foxp3 (forkhead box P3). However, many accounts attest to suppressive populations lacking Foxp3, as well as FoxP3+ populations serving as proinflammatory effectors given a specific cytokine milieu (Anderson *et al.* 2007; Dominguez-Villar *et al.* 2011; Huber *et al.* 2011; Komatsu *et al.* 2014).

Furthermore, distinct mechanisms of suppression depend on the context and location of the inflammatory response being controlled. For example, the T_{reg} cells that are found to be controlling specific subsets of effector cells (e.g., Th1, Th2, Th17, Tfh) have similar transcription factor upregulation (Figure 9.3). At present, these matched transcription factors are hypothesized to allow migration and survival within these select inflammatory sites, rather than having a directly suppressive effect or leading to differentiation into an effector phenotype.

The demonstrated plasticity and heterogeneity of T_{reg} cells necessitates better determination of the stability and function of suppressive cells, perhaps this is best achieved via their epigenome. For example, T_{reg} cells have specific demethylated regions including T_{reg} function-associated genes (e.g., *Foxp3* intron 1 and *Ctla4* exon 2). These genes are constitutively enhanced. In contrast, FoxP3 dependent genes are repressed by FoxP3 which is an activity seen only after T cell receptor (TCR) stimulation (e.g., CD3 and LAT [Linker for Activation of T cells]) (Morikawa and Sakaguchi 2014).

CTLA4 (cytotoxic T-lymphocyte-associated protein 4, CD152) is an inhibitory protein receptor. The primary intracellular mechanism of inhibition is CTLA4's ability to bind protein phosphatases, specifically protein phosphatase 2 (PP2A) and Src homology 2-containing phosphotyrosine phosphatase (SHP-2), which results in dephosphorylation and inactivation of TCR signaling. A cell extrinsic mechanism of CTLA4 suppressive activity involves binding CD80 and CD86 on antigen presenting cells, thereby competing for binding with the co-stimulatory molecule CD28 (Oderup *et al.* 2006; Qureshi *et al.* 2011).

A recent study elucidates a mechanism by which a commensal microbe alters T_{reg} cells (Round and Mazmanian 2010). This brings us

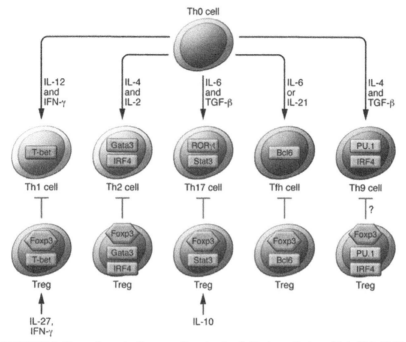

FIGURE 9.3. T_{reg} cells controlling specific subsets of effector cells (e.g., Th1, Th2, Th17, Tfh) have similar transcription factor upregulation to their paired effectors. These factors may have roles in getting the T_{reg} cells to those locations and facilitating survival in those environments particular to their paired effectors, rather than determining suppressive or effector functions for the T_{reg} cells (Chaudhy and Rudensky 2013).

again to the concept that the mucosal immune system is in constant homeostatic surveillance of the gut microbiota. The innate immune system is able to distinguish microbes from self by utilizing pattern recognition receptors such as toll like receptors (TLRs), which recognize evolutionarily conserved molecular patterns on microbes (PAMPs) (Janeway 1989). Refer to Chapter 5 on the microbiome and the immune system for further details.

Polysaccharide A (PSA) is produced by the intestinal pathobiont *Bacteroides fragilis*. PSA has been shown to modulate inflammation by binding to TLR2 on regulatory T lymphocytes, activating them and suppressing inflammation (Round and Mazmanian 2008). Using the murine experimental autoimmune encephalomyelitis model for multiple sclerosis, PSA controls neuroinflammation via a TLR2/CD39 dependent process (Wang *et al.* 2014b). Independently, Lal *et al.* (2011) has shown how TLR2 acts epigenetically on foxp3 in regulatory T cells in the response to *Staphlococcus aureus*-derived peptidoglycan stimulation via MyD88 (myeloid differentiation primary response gene 88) signaling. Specifically, Interferon regulatory factor 1 (IRF1) binds to the foxp3 promoter (and intronic enhancer) which negatively regulates foxp3 expression (Lal *et al.* 2011). Differences in heterodimerization of TLR2 with TLR1 in the case of Gram-negative bacteria and with TLR6 in the case of Gram-positive bacteria help explain differences of TLR2 response to different bacteria, but further work dissecting the epigenetic mechanisms is needed.

CD39 is a transmembrane ectonucleotidase, which inhibits inflammation by removing the phosphates from ATP (adenosine triphosphate) and ADP (diphosphate). In a murine macrophage cell line (RAW 264.7) the expression of *CD39* was dependent on the level of the cellular second messenger cyclic AMP (adenosine monophosphate). Signaling occurred across multiple pathways (PKA, PI3K, AKT, and ERK dependent). The transcription factors cAMP response element binding protein (CREB) and activating transcription factor 2 (ATF2) bind to the promoter of *CD39* to induce its expression (Liao *et al.* 2010). T_{reg}s have been shown to mediate suppressive effects by delivering cAMP to effector cells via gap junctions (Bopp *et al.* 2007).

The plasticity and complexity of T_{reg} cells necessitates a quick response to many environmental clues, which is best facilitated by epigenetic change. As more research is conducted, many immune-mediated and allergic diseases may come to be classified as epigenetic disorders. The influence of the microbiota on these epigenetic changes will follow closely.

Jhangi *et al.* (2014) preliminarily demonstrated that Archaea may be increased in multiple sclerosis (MS) patients. Initially, 61 MS patients and 43 healthy controls of a larger patient cohort had their fecal microbiota characterized by 454 and Illumina 16S sequencing. A 6.6-fold higher mean relative abundance of *Methanobrevibacteriaceae* was found in the MS patients. The cell wall and lipid membranes of this organism are highly immunogenic. Interestingly, two anti-inflammatory bacteria, *Butyricimonas*, a genus in the *Bacteroides* phylum, and *Lachnospiraceae*, a family in the *Firmicutes* phylum, were found in higher numbers in healthy and treated controls, respectively. Both of these bacteria are butyrate producers. Butyrate is known to reduce inflammatory signaling (Jhangi *et al.* 2014).

Additional studies are needed to address whether epigenetic changes caused by the microbiome affects the risk of developing multiple sclerosis (Huynh and Casaccia 2013).

9.2.7.2. Autism

Autism spectrum disorder collectively refers to a heterogeneous group of neurodevelopmental disorders, characterized by restricted, repetitive behaviors as well as impairments in social interaction and communication. A potential role for the microbiota was suspected when antibiotics caused a temporary improvement from autistic symptoms in a small group of children were being treated for chronic gastrointestinal signs (Sandler *et al.* 2000). Subsequently, autistic children have been found to have higher *Clostridium histolyticum* in their fecal flora. These bacteria could produce neurotoxins that result in gastrointestinal and systemic effects (Parracho *et al.* 2005). Autistic children have lower gastrointestinal microbial diversity. With lower abundance of versatile carbohydrate-degrading fermenters such as *Prevotella, Coprococcus*, and unclassified *Veillonellaceae* (Kang *et al.* 2013). Another study again revealed increases in *Firmicutes* (Clostridia) and *Proteobacteria* with decreases in *Bacteroidetes*. This dysbiosis was concurrent with carbohydrate maldigestion and malabsorption (Williams *et al.* 2011). Alterations in metabolism lead to shifts in microbial metabolites. Fecal total SCFAs are reported to be higher in children with autism (Wang *et al.* 2012). A rat model of autism shows the adverse effects of the SCFA, propionic acid on brain tissue (MacFabe *et al.* 2007). Mechanisms of propionic acid involve mitochondrial dysfunction and oxidative stress (Clark-Taylor and Clark-Taylor 2004, MacFabe 2012). Similar pathology is shown in autism patients (Chauhan and Chauhan 2006). Many epigenetic alterations in autism have been explored. For example, Rett syndrome is caused by mutations in *MECP2* an X-linked gene that encodes methyl CpG binding protein 2. *MECP2* binds to methylated DNA and acts as a transcriptional regulator. A recent study links human maternal infection with autism risk, similar to animal model work (Lee *et al.* 2015). Significant interactive effects were seen between maternal infection during pregnancy and copy number variants, another type of genome modification (Mazina *et al.* 2015). Genome-wide methylome maps are determining the genes with altered methylation. Autism candidate genes (e.g., *CNTNAP2, GABRB3, MACROD2*, and *NRXN1*) have been found to be significantly enriched in partial methylated domains in neuronal cells and placental tissue. Dietary folate serves as methyl donor and can profoundly alter DNA methylation. Deficiencies in very early pregnancy have been implicated in autism (LaSalle 2013). The large intestinal microbes provide a large store of available folate (Kim *et al.* 2004; Aufreiter *et al.* 2009). An integrated approach interfacing microbial, epigenetic, genetic, and environmental factors will be key to solving the complex etiology of autism and providing optimal diagnostics and treatments.

9.3. CONCLUSIONS

We are only beginning to characterize some of the potential roles played by human-associated microbiota in disease etiology, particularly via epigenetic mechanisms. Increased understanding of the mechanisms involved will help us to develop new approaches to treat these important diseases.

9.4. REFERENCES

Aagaard, K., J. Ma, K.M. Antony, R. Ganu, J. Petrosino, and J. Versalovic. 2014. The Placenta Harbors a Unique Microbiome. *Science Translational Medicine* 6:237ra265.

Aalaei-Andabili, S.H., M. Fabbri, and N. Rezaei. 2013. Reciprocal effects of Toll-like receptors and miRNAs on biological processes in human health and disease: A systematic review. *Immunotherapy* 5:1127–1142.

Acharya, M.R., A. Sparreboom, J. Venitz, and W.D. Figg. 2005. Rational development of histone deacetylase inhibitors as anticancer agents: A review. *Molecular pharmacology* 68:917–932.

Anderson, C.F., M. Oukka, V.J. Kuchroo, and D. Sacks. 2007. $CD4^+$ $CD25^-$ $Foxp3^-$Th1 cells are the source of IL-10-mediated immune suppression in chronic cutaneous leishmaniasis. *The Journal of experimental medicine* 204:285–297.

Arce, R.M., S.P. Barros, B. Wacker, B. Peters, K. Moss, and S. Offenbacher. 2009. Increased TLR4 Expression in Murine Placentas after Oral Infection with Periodontal Pathogens. *Placenta* 30:156–162.

Arumugam, M., J. Raes, E. Pelletier, D. Le Paslier, T. Yamada, D.R. Mende, *et al.* 2011. Enterotypes of the human gut microbiome. *Nature* (London).

Aufreiter, S., J.F. Gregory, C.M. Pfeiffer, Z. Fazili, Y.I. Kim, N. Marcon, *et al.* 2009. Folate is absorbed across the colon of adults: Evidence from cecal infusion of 13C-labeled [6S]-5-formyltetrahydrofolic acid. *The American Journal of Clinical Nutrition* 90:116–123.

Avivi-Green, C., S. Polak-Charcon, Z. Madar, and B. Schwartz. 2001. Apoptosis cascade proteins are regulated in vivo by high intracolonic butyrate concentration: Correlation with colon cancer inhibition. *Oncology Research Featuring Preclinical and Clinical Cancer Therapeutics* 12:83–95.

Bäckhed, F., H. Ding, T. Wang, L.V. Hooper, G.Y. Koh, A. Nagy, *et al.* 2004. The gut microbiota as an environmental factor that regulates fat storage. *Proceedings of the National Academy of Sciences of the United States of America* 101:15718–15723.

Bäckhed, F., J.K. Manchester, C.F. Semenkovich, and J.I. Gordon. 2007. Mechanisms underlying the resistance to diet-induced obesity in germ-free mice. *Proceedings of the National Academy of Sciences 104*:979–984.

Bandres, E., X. Agirre, N. Bitarte, N. Ramirez, R. Zarate, J. Roman-Gomez, *et al.* 2009. Epigenetic regulation of microRNA expression in colorectal cancer. *International Journal of Cancer 125*:2737–2743.

Barak, S., O. Oettinger-Barak, E.E. Machtei, H. Sprecher, and G. Ohel. 2007. Evidence of Periopathogenic Microorganisms in Placentas of Women With Preeclampsia. *Journal of periodontology 78*:670–676.

Bebek, G., K.L. Bennett, P. Funchain, R. Campbell, R. Seth, J. Scharpf, *et al.* 2012. Microbiomic subprofiles and MDR1 promoter methylation in head and neck squamous cell carcinoma. *Human molecular genetics 21*:1557–1565.

Bélanger, M., L. Reyes, K. von Deneen, M.K. Reinhard, A. Progulske-Fox, and M.B. Brown. 2008. Colonization of maternal and fetal tissues by *Porphyromonas gingivalis* is strain-dependent in a rodent animal model. *American Journal of Obstetrics and Gynecology 199*:86.e81–86.e87.

Berger, M., J.A. Gray, and B.L. Roth. 2009. The expanded biology of serotonin. *Annual review of medicine 60*:355–366.

Blaser, M.J., Y. Chen, and J. Reibman. 2008. Does *Helicobacter pylori* protect against asthma and allergy? *Gut 57*:561–567.

Boggess, K.A., P.N. Madianos, J.S. Preisser, K.J. Moise Jr., and S. Offenbacher. 2005. Chronic maternal and fetal *Porphyromonas gingivalis* exposure during pregnancy in rabbits. *American Journal of Obstetrics and Gynecology 192*:554–557.

Bopp, T., C. Becker, M. Klein, S. Klein-Heßling, A. Palmetshofer, E. Serfling, *et al.* 2007. Cyclic adenosine monophosphate is a key component of regulatory T cell–mediated suppression. *The Journal of experimental medicine 204*:1303–1310.

Bordonaro, M., D.L. Lazarova, and A.C. Sartorelli. 2008. Butyrate and Wnt signaling. *Cell Cycle 7*:1178–1183.

Brand, S., R. Teich, T. Dicke, H. Harb, A.Ö. Yildirim, J. Tost, *et al.* 2011. Epigenetic regulation in murine offspring as a novel mechanism for transmaternal asthma protection induced by microbes. *Journal of Allergy and Clinical Immunology 128*:618–625. e617.

Brusko, T.M., A.L. Putnam, and J.A. Bluestone. 2008. Human regulatory T cells: Role in autoimmune disease and therapeutic opportunities. *Immunological Reviews 223*:371–390.

Camp, J.G., C.L. Frank, C.R. Lickwar, H. Guturu, T. Rube, A.M. Wenger, *et al.* 2014. Microbiota modulate transcription in the intestinal epithelium without remodeling the accessible chromatin landscape. *Genome research 24*:1504–1516.

Canani, R.B., M. Di Costanzo, L. Leone, G. Bedogni, P. Brambilla, S. Cianfarani, *et al.* 2011. Epigenetic mechanisms elicited by nutrition in early life. *Nutrition research reviews 24*:198–205.

Casaccia-Bonnefil, P., G. Pandozy, and F. Mastronardi. 2008. Evaluating epigenetic landmarks in the brain of multiple sclerosis patients: A contribution to the current debate on disease pathogenesis. *Progress in Neurobiology 86*:406–416.

Chatila, T.A. and C.B. Williams. 2014. Regulatory T Cells: Exosomes Deliver Tolerance. *Immunity 41*:3–5.

Chaudhry, A., R.M. Samstein, P. Treuting, Y. Liang, Marina C. Pils, *et al.* 2011. Interleukin-10 Signaling in Regulatory T Cells Is Required for Suppression of Th17 Cell-Mediated Inflammation. *Immunity 34*:566–578.

Chauhan, A. and V. Chauhan. 2006. Oxidative stress in autism. *Pathophysiology 13*:171–181.

Chen, Y. and M.J. Blaser. 2007. Inverse associations of *Helicobacter pylori* with asthma and allergy. *Archives of internal medicine 167*:821–827.

Chen, Y. and M.J. Blaser. 2008. *Helicobacter pylori* colonization is inversely associated with childhood asthma. *Journal of infectious diseases 198*:553–560.

Clark-Taylor, T. and B.E. Clark-Taylor. 2004. Is autism a disorder of fatty acid metabolism? Possible dysfunction of mitochondrial β-oxidation by long chain acyl-CoA dehydrogenase. *Medical hypotheses 62*:970–975.

Collins, L.J., X.S. Chen, and B. Schonfeld. 2010. The epigenetics of non-coding RNA. Handbook of Epigenetics: The New Molecular and Medical Genetics. New York: Elsevier:49–61.

Consortium, E.P. 2012. An integrated encyclopedia of DNA elements in the human genome. *Nature 489*:57–74.

Corbella, S., S. Taschieri, L. Francetti, F. De Siena, and M. Del Fabbro. 2012. Periodontal disease as a risk factor for adverse pregnancy outcomes: A systematic review and meta-analysis of case–control studies. *Odontology 100*:232–240.

da Rosa, M.I., P.D.S. Pires, L.R. Medeiros, M.I. Edelweiss, and J. Martínez-Mesa. 2012. Periodontal disease treatment and risk of preterm birth: A systematic review and meta-analysis Tratamento de doença periodontal e risco de parto prematuro: Revisão sistemática e metanálise. *Cad. Saúde Pública 28*:1823–1833.

Danne, O., C. Lueders, C. Storm, U. Frei, and M. Möckel. 2007. Whole blood choline and plasma choline in acute coronary syndromes: Prognostic and pathophysiological implications. *Clinica chimica acta 383*:103–109.

Dasanayake, A.P. 2013. Scaling and Root Planing Is Effective in Reducing Preterm Birth Only in High-Risk Groups. *Journal of Evidence Based Dental Practice 13*:42–44.

Davie, J.R. 2003. Inhibition of histone deacetylase activity by butyrate. *The Journal of nutrition 133*:2485S–2493S.

Davis, C.D. and J.A. Milner. 2009. Gastrointestinal microflora, food components and colon cancer prevention. *The Journal of nutritional biochemistry 20*:743–752.

Dewhirst, F.E., T. Chen, J. Izard, B.J. Paster, A.C.R. Tanner, W.H. Yu, *et al.* 2010. The Human Oral Microbiome. *Journal of Bacteriology 192*:5002–5017.

DiNatale, B.C., J.C. Schroeder, and G.H. Perdew. 2011. Ah receptor antagonism inhibits constitutive and cytokine inducible IL6 production in head and neck tumor cell lines. *Molecular Carcinogenesis 50*:173–183.

Dominguez-Villar, M., C.M. Baecher-Allan, and D.A. Hafler. 2011. Identification of T helper type 1-like, Foxp3+ regulatory T cells in human autoimmune disease. *Nature medicine 17*:673–675.

Dong, C., W. Yoon, and P.J. Goldschmidt-Clermont. 2002. DNA methylation and atherosclerosis. *The Journal of nutrition 132*:2406S–2409S.

Dumas, M.E., R.H. Barton, A. Toye, O. Cloarec, C. Blancher, A. Rothwell, *et al.* 2006. Metabolic profiling reveals a contribution of gut microbiota to fatty liver phenotype in insulin-resistant mice. *Proceedings of the National Academy of Sciences* *103*:12511–12516.

Esteller, M. 2011. Non-coding RNAs in human disease. *Nature Reviews Genetics* *12*:861–874.

Fogacci, M.F., M.V. Vettore, and A.T.T. Leão. 2011. The effect of periodontal therapy on preterm low birth weight: A meta-analysis. *Obstetrics & Gynecology 117*:153–165.

Guang, W., S.J. Czinn, T.G. Blanchard, K.C. Kim, and E.P. Lillehoj. 2014. Genetic regulation of MUC1 expression by *Helicobacter pylori* in gastric cancer cells. *Biochemical and Biophysical Research Communications 445*:145–150.

Guha, N., P. Boffetta, V. Wünsch Filho, J.E. Neto, O. Shangina, D. Zaridze, *et al.* 2007. Oral health and risk of squamous cell carcinoma of the head and neck and esophagus: Results of two multicentric case-control studies. *American journal of epidemiology 166*:1159–1173.

Hambly, R.J., C.J. Rumney, J.M. Fletcher, P.J. Rijken, and I.R. Rowland. 1997. Effects of high-nand low-risk diets on gut microflora-associated biomarkers of colon cancer in human flora-associated rats.

Hofmanová, R.J., A. Vaculová, A. Lojek, and A. Kozubík. 2005. Interaction of polyunsaturated fatty acids and sodium butyrate during apoptosis in HT-29 human colon adenocarcinoma cells. *European journal of nutrition 44*:40–51.

Hooper, S.J., M.J. Wilson, and S.J. Crean. 2009. Exploring the link between microorganisms and oral cancer: A systematic review of the literature. *Head & neck 31*:1228–1239.

Hu, B., E. Elinav, S. Huber, T. Strowig, L. Hao, A. Hafemann, *et al.* 2013. Microbiota-induced activation of epithelial IL-6 signaling links inflammasome-driven inflammation with transmissible cancer. *Proceedings of the National Academy of Sciences 110*:9862–9867.

Huber, S., N. Gagliani, E. Esplugues, W. O'Connor Jr., F.J. Huber, A. Chaudhry, *et al.* 2011. Th17 cells express interleukin-10 receptor and are controlled by Foxp3[-] and Foxp3[+] regulatory CD4[+] T cells in an interleukin-10 dependent manner. *Immunity 34*:554.

Hunt, R.H., S.D. Xiao, F. Megraud, R. Leon-Barua, F. Bazzoli, S. Van der Merwe, *et al.* 2011. Helicobacter pylori in developing countries. World Gastroenterology Organisation Global Guideline. *J Gastrointestin Liver Dis 20*:299–304.

Huynh, J.L. and P. Casaccia. 2013. Epigenetic mechanisms in multiple sclerosis: Implications for pathogenesis and treatment. *The Lancet Neurology 12*:195–206.

IARC, W.G.O.T.E.O.C.R.T.H. 1994. Schistosomes, liver flukes and *Helicobacter pylori*. *IARC Monogr Eval Carcinog Risks Hum 61*:1–241.

Inaba, H. and A. Amano. 2010. Roles of oral bacteria in cardiovascular diseases-from molecular mechanisms to clinical cases: Implication of periodontal diseases in development of systemic diseases. *Journal of pharmacological sciences 113*:103–109.

Inaba, H., M. Kuboniwa, H. Sugita, R.J. Lamont, and A. Amano. 2012. Identification of Signaling Pathways Mediating Cell Cycle Arrest and Apoptosis Induced by Porphyromonas gingivalis in Human Trophoblasts. *Infection and Immunity 80*:2847–2857.

Islami, F. and F. Kamangar. 2008. *Helicobacter pylori* and esophageal cancer risk: A meta-analysis. *Cancer Prevention Research 1*:329–338.

Janeway, C.A. 1989. Approaching the asymptote? Evolution and revolution in immunology. Pages 1–13 in Cold Spring Harbor symposia on quantitative biology. Cold Spring Harbor Laboratory Press.

Jeffcoat, M., S. Parry, M. Sammel, B. Clothier, A. Catlin, and G. Macones. 2011. Periodontal infection and preterm birth: Successful periodontal therapy reduces the risk of preterm birth. *BJOG: An International Journal of Obstetrics & Gynaecology 118*:250–256.

Jeffcoat, M.K., R.L. Jeffcoat, N. Tanna, and S.H. Parry. 2014. Association of a Common Genetic Factor, PTGER3, With Outcome of Periodontal Therapy and Preterm Birth. *Journal of periodontology 85*:446–454.

Jemal, A., F. Bray, M.M. Center, J. Ferlay, E. Ward, and D. Forman. 2011. Global cancer statistics. *CA: A Cancer Journal for Clinicians 61*:69–90.

Jhangi, S., R. Gandhi, B. Glanz, S. Cook, P. Nejad, D. Ward, *et al.* 2014. Increased Archaea Species and Changes with Therapy in Gut Microbiome of Multiple Sclerosis Subjects (S24.001). *Neurology 82*:S24.001.

Jin, U.H., S.O. Lee, G. Sridharan, K. Lee, L.A. Davidson, A. Jayaraman, *et al.* 2014. Microbiome-Derived Tryptophan Metabolites and Their Aryl Hydrocarbon Receptor-Dependent Agonist and Antagonist Activities. *Molecular pharmacology 85*:777–788.

John, K., T.S. Lahoti, K. Wagner, J.M. Hughes, and G.H. Perdew. 2014. The Ah receptor regulates growth factor expression in head and neck squamous cell carcinoma cell lines. *Molecular Carcinogenesis 53*:765–776.

Kaikkonen, M.U., M.T.Y. Lam, and C.K. Glass. 2011. Non-coding RNAs as regulators of gene expression and epigenetics. *Cardiovascular research 90*:430–440.

Kang, D.W., J.G. Park, Z.E. Ilhan, G. Wallstrom, J. LaBaer, J.B. Adams, and R. Krajmalnik-Brown. 2013. Reduced incidence of Prevotella and other fermenters in intestinal microflora of autistic children. *PLoS ONE 8*:e68322.

Kastenmuller, W., G. Gasteiger, N. Subramanian, T. Sparwasser, D.H. Busch, Y. Belkaid, *et al.* 2011. Regulatory T Cells Selectively Control CD8+ T Cell Effector Pool Size via IL-2 Restriction. *The Journal of Immunology 187*:3186–3197.

Katz, J., N. Chegini, K.T. Shiverick, and R.J. Lamont. 2009. Localization of P. gingivalis in Preterm Delivery Placenta. *Journal of Dental Research 88*:575–578.

Kiga, K., H. Mimuro, M. Suzuki, A. Shinozaki-Ushiku, T. Kobayashi, T. Sanada, *et al.* 2014. Epigenetic silencing of miR-210 increases the proliferation of gastric epithelium during chronic *Helicobacter pylori* infection. *Nat Commun 5*.

Kilner, J., J.S. Waby, J. Chowdry, A.Q. Khan, J. Noirel, P.C. Wright, *et al.* 2012. A proteomic analysis of differential cellular responses to the short-chain fatty acids butyrate, valerate and propionate in colon epithelial cancer cells. *Molecular BioSystems 8*:1146–1156.

Kim, T.H., J. Yang, P.B. Darling, and D.L. O'Connor. 2004. A large pool of available folate exists in the large intestine of human infants and piglets. *The Journal of nutrition 134*:1389–1394.

Koeth, R.A., Z. Wang, B.S. Levison, J.A. Buffa, B.T. Sheehy, E.B. Britt, et al. 2013. Intestinal microbiota metabolism of l-carnitine, a nutrient in red meat, promotes atherosclerosis. *Nature medicine* 19:576–585.

Komatsu, N., K. Okamoto, S. Sawa, T. Nakashima, M. Oh-hora, T. Kodama, et al. 2014. Pathogenic conversion of Foxp3+ T cells into TH17 cells in autoimmune arthritis. Nat Med 20:62–68.

Kundu, J.K. and Y.J. Surh. 2008. Inflammation: Gearing the journey to cancer. *Mutation research* 659:15–30.

Lal, G., N. Yin, J. Xu, M. Lin, S. Schroppel, Y. Ding, et al. 2011. Distinct Inflammatory Signals Have Physiologically Divergent Effects on Epigenetic Regulation of Foxp3 Expression and Treg Function. *American Journal of Transplantation* 11:203–214.

LaSalle, J.M. 2013. Epigenomic strategies at the interface of genetic and environmental risk factors for autism. *Journal of human genetics* 58:396–401.

Lee, B.K., C. Magnusson, R.M. Gardner, Å. Blomström, C.J. Newschaffer, I. Burstyn, et al. 2015. Maternal hospitalization with infection during pregnancy and risk of autism spectrum disorders. *Brain, behavior, and immunity* 44:100–105.

Lee, J.H., S.H. Jun, S.C. Baik, D.R. Kim, J.Y. Park, Y.S. Lee, et al. 2012. Prediction and screening of nuclear targeting proteins with nuclear localization signals in *Helicobacter pylori*. *Journal of Microbiological Methods* 91:490–496.

Leibowitz, U., A. Antonovsky, J.M. Medalie, H.A. Smith, L. Halpern, and M. Alter. 1966. Epidemiological study of multiple sclerosis in Israel. II. Multiple sclerosis and level of sanitation. *Journal of neurology, neurosurgery, and psychiatry* 29:60.

LeLeiko, R.M., C.S. Vaccari, S. Sola, N. Merchant, S.H. Nagamia, M. Thoenes, and B.V. Khan. 2009. Usefulness of Elevations in Serum Choline and Free F_2-Isoprostane to Predict 30-Day Cardiovascular Outcomes in Patients With Acute Coronary Syndrome. *The American journal of cardiology* 104:638–643.

León, R., N. Silva, A. Ovalle, A. Chaparro, A. Ahumada, M. Gajardo, et al. 2007. Detection of Porphyromonas gingivalis in the amniotic fluid in pregnant women with a diagnosis of threatened premature labor. *Journal of periodontology* 78:1249–1255.

Liao, H., M.C. Hyman, A.E. Baek, K. Fukase, and D.J. Pinsky. 2010. cAMP/CREB-mediated Transcriptional Regulation of Ectonucleoside Triphosphate Diphosphohydrolase 1 (CD39) Expression. *Journal of Biological Chemistry* 285:14791–14805.

Licciardi, P.V., S.S. Wong, M.L. Tang, and T.C. Karagiannis. 2010. Epigenome targeting by probiotic metabolites. *Gut Pathog* 2:24.

Lightfoot, Y.L., T. Yang, B. Sahay, and M. Mohamadzadeh. 2012. Targeting aberrant colon cancer-specific DNA methylation with lipoteichoic acid-deficient Lactobacillus acidophilus. *Gut microbes* 4:84–88.

Lightfoot, Y.L., T. Yang, B. Sahay, and M. Mohamadzadeh. 2013. Targeting aberrant colon cancer-specific DNA methylation with lipoteichoic acid-deficient Lactobacillus acidophilus. *Gut microbes* 4:84–88.

Lin, D., M.A. Smith, J. Elter, C. Champagne, C.L. Downey, J. Beck, and S. Offenbacher. 2003. Porphyromonas gingivalis Infection in Pregnant Mice Is Associated with Placental Dissemination, an Increase in the Placental Th1/Th2 Cytokine Ratio, and Fetal Growth Restriction. *Infection and Immunity* 71:5163–5168.

Lopez, N.J., I. Da Silva, J. Ipinza, and J. Gutiérrez. 2005. Periodontal therapy reduces the rate of preterm low birth weight in women with pregnancy-associated gingivitis. *Journal of periodontology* 76:2144–2153.

Loscalzo, J. 2011. Lipid metabolism by gut microbes and atherosclerosis. *Circulation research* 109:127–129.

MacFabe, D.F. 2012. Short-chain fatty acid fermentation products of the gut microbiome: Implications in autism spectrum disorders. *Microbial ecology in health and disease 23*.

MacFabe, D.F., D.P. Cain, K. Rodriguez-Capote, A.E. Franklin, J.E. Hoffman, F. Boon, et al. 2007. Neurobiological effects of intraventricular propionic acid in rats: Possible role of short chain fatty acids on the pathogenesis and characteristics of autism spectrum disorders. *Behavioural brain research* 176:149–169.

Mager, D.L., A.D. Haffajee, P.M. Devlin, C.M. Norris, M.R. Posner, and J.M. Goodson. 2005. The salivary microbiota as a diagnostic indicator of oral cancer: A descriptive, non-randomized study of cancer-free and oral squamous cell carcinoma subjects. *Journal of translational medicine* 3:27.

Marcobal, A., P.C. Kashyap, T.A. Nelson, P.A. Aronov, M.S. Donia, A. Spormann, et al. 2013. A metabolomic view of how the human gut microbiota impacts the host metabolome using humanized and gnotobiotic mice. *The ISME journal* 7:1933–1943.

Martin, F.P.J., M.E. Dumas, Y. Wang, C. Legido-Quigley, I.K. Yap, H. Tang, et al. 2007. A top-down systems biology view of microbiome-mammalian metabolic interactions in a mouse model. *Molecular systems biology 3*.

Mastronardi, F.G., D.D. Wood, J. Mei, R. Raijmakers, V. Tseveleki, H.M. Dosch, et al. 2006. Increased citrullination of histone H3 in multiple sclerosis brain and animal models of demyelination: A role for tumor necrosis factor-induced peptidylarginine deiminase 4 translocation. *The Journal of neuroscience* 26:11387–11396.

Mazeh, H., I. Mizrahi, N. Ilyayev, D. Halle, B.L. Brücher, A. Bilchik, et al. 2013. The diagnostic and prognostic role of microRNA in colorectal cancer-a comprehensive review. *Journal of Cancer* 4:281.

Mazina, V., J. Gerdts, S. Trinh, K. Ankenman, T. Ward, M.Y. Dennis, et al. 2015. Epigenetics of Autism-related Impairment: Copy Number Variation and Maternal Infection. *Journal of Developmental & Behavioral Pediatrics* 36:61–67.

Moodley, Y., B. Linz, R.P. Bond, M. Nieuwoudt, H. Soodyall, C.M. Schlebusch, et al. 2012. Age of the Association between *Helicobacter pylori* and Man. *PLoS Pathog* 8:e1002693.

Moore, W. and L.H. Moore. 1995. Intestinal floras of populations that have a high risk of colon cancer. *Applied and environmental microbiology* 61:3202–3207.

Morikawa, H. and S. Sakaguchi. 2014. Genetic and epigenetic basis of T_{reg} cell development and function: From a FoxP3-centered view to an epigenome-defined view of natural T_{reg} cells. *Immunological Reviews* 259:192–205.

Morris, K.V. and J.S. Mattick. 2014. The rise of regulatory RNA. *Nat Rev Genet* 15:423–437.

Muegge, B.D., J. Kuczynski, D. Knights, J.C. Clemente, A. González, L. Fontana, et al. 2011. Diet drives convergence in gut microbiome functions across mammalian phylogeny and within humans. *Science* 332:970–974.

Murray, I.A., A.D. Patterson, and G.H. Perdew. 2014. Aryl hydrocarbon receptor ligands in cancer: Friend and foe. *Nat Rev Cancer 14*:801–814.

Ochoa-Repáraz, J., D.W. Mielcarz, S. Begum-Haque, and L.H. Kasper. 2011. Gut, bugs, and brain: Role of commensal bacteria in the control of central nervous system disease. *Annals of neurology 69*:240–247.

Oderup, C., L. Cederbom, A. Makowska, C.M. Cilio, and F. Ivars. 2006. Cytotoxic T lymphocyte antigen-4-dependent down-modulation of costimulatory molecules on dendritic cells in CD4+ CD25+ regulatory T-cell-mediated suppression. *Immunology 118*:240–249.

Oka, T., H. Sato, M. Ouchida, A. Utsunomiya, and T. Yoshino. 2011. Cumulative epigenetic abnormalities in host genes with viral and microbial infection during initiation and progression of malignant lymphoma/leukemia. *Cancers 3*:568–581.

Okoye, I.S., S.M. Coomes, V.S. Pelly, S. Czieso, V. Papayannopoulos, T. Tolmachova, *et al.* 2014. MicroRNA-containing T-regulatory-cell-derived exosomes suppress pathogenic T helper 1 cells. *Immunity 41*:89–103.

Parracho, H.M.R.T., M.O. Bingham, G.R. Gibson, and A.L. McCartney. 2005. Differences between the gut microflora of children with autistic spectrum disorders and that of healthy children. *Journal of medical microbiology 54*:987–991.

Pereira, M.I. and J.A. Medeiros. 2014. Role of *Helicobacter pylori* in gastric mucosa-associated lymphoid tissue lymphomas. *World journal of gastroenterology: WJG 20*:684.

Perry, S., B.C. de Jong, J.V. Solnick, M.D.l.L. Sanchez, S. Yang, P.L. Lin, *et al.* 2010. Infection with *Helicobacter pylori* Is Associated with Protection against Tuberculosis. *PLoS ONE 5*:e8804.

Plummer, M., S. Franceschi, J. Vignat, D. Forman, and C. de Martel. 2014. Global burden of gastric cancer attributable to pylori. *International Journal of Cancer*:n/a-n/a.

Preshaw, P.M., A.L. Alba, D. Herrera, S. Jepsen, A. Konstantinidis, K. Makrilakis, and R. Taylor. 2012. Periodontitis and diabetes: A two-way relationship. *Diabetologia 55*:21–31.

Qureshi, O.S., Y. Zheng, K. Nakamura, K. Attridge, C. Manzotti, E.M. Schmidt, *et al.* 2011. Trans-endocytosis of CD80 and CD86: A molecular basis for the cell-extrinsic function of CTLA-4. *Science 332*:600–603.

Rosenquist, K., J. Wennerberg, E.B. Schildt, A. Bladström, B. Göran Hansson, and G. Andersson. 2005. Oral status, oral infections and some lifestyle factors as risk factors for oral and oropharyngeal squamous cell carcinoma. A population-based case-control study in southern Sweden. *Acta Oto-laryngologica 125*:1327–1336.

Rothbart, S.B. and B.D. Strahl. 2014. Interpreting the language of histone and DNA modifications. *Biochimica et Biophysica Acta (BBA)—Gene Regulatory Mechanisms 1839*:627–643.

Round, J.L. and S.K. Mazmanian. 2010. Inducible Foxp3+ regulatory T-cell development by a commensal bacterium of the intestinal microbiota. *Proceedings of the National Academy of Sciences of the United States of America 107*:12204–12209.

Saito, Y., H. Suzuki, H. Tsugawa, H. Imaeda, J. Matsuzaki, K. Hirata, *et al.* 2012. Overexpression of miR-142-5p and miR-155 in gastric mucosa-associated lymphoid tissue (MALT) lymphoma resistant to *Helicobacter pylori* eradication. *PLoS ONE* 7:e47396.

Saltiel, A.R. and C.R. Kahn. 2001. Insulin signalling and the regulation of glucose and lipid metabolism. *Nature* 414:799–806.

Sandler, R.H., S.M. Finegold, E.R. Bolte, C.P. Buchanan, A.P. Maxwell, M.L. Väisänen, *et al.* 2000. Short-term benefit from oral vancomycin treatment of regressive-onset autism. *Journal of Child Neurology* 15:429–435.

Shenderov, B.A. 2012. Gut indigenous microbiota and epigenetics. *Microbial ecology in health and disease* 23.

Sinclair, K.D., C. Allegrucci, R. Singh, D.S. Gardner, S. Sebastian, J. Bispham, *et al.* 2007. DNA methylation, insulin resistance, and blood pressure in offspring determined by maternal periconceptional B vitamin and methionine status. *Proceedings of the National Academy of Sciences* 104:19351–19356.

Song, S.J., M.G. Dominguez-Bello, and R. Knight. 2013. How delivery mode and feeding can shape the bacterial community in the infant gut. *CMAJ: Canadian Medical Association Journal* 185:373.

Strachan, D.P. 1989. Hay fever, hygiene, and household size. *BMJ: British Medical Journal* 299:1259.

Sun, T.T., J.Y. Tang, W. Du, H.J. Zhao, G. Zhao, S.L. Yang, *et al.* 2014. Bidirectional regulation between TMEFF2 and STAT3 may contribute to *Helicobacter pylori*-associated gastric carcinogenesis. *International Journal of Cancer*:n/a-n/a.

Suzuki, H.I., K. Yamagata, K. Sugimoto, T. Iwamoto, S. Kato, and K. Miyazono. 2009. Modulation of microRNA processing by p53. *Nature* 460:529–533.

Tremaroli, V. and F. Bäckhed. 2012. Functional interactions between the gut microbiota and host metabolism. *Nature* 489:242–249.

Turnbaugh, P.J., F. Bäckhed, L. Fulton, and J.I. Gordon. 2008. Diet-induced obesity is linked to marked but reversible alterations in the mouse distal gut microbiome. *Cell host & microbe* 3:213–223.

Ushijima, T. and N. Hattori. 2012. Molecular Pathways: Involvement of *Helicobacter pylori*–Triggered Inflammation in the Formation of an Epigenetic Field Defect, and Its Usefulness as Cancer Risk and Exposure Markers. *Clinical Cancer Research* 18:923–929.

Vettore, M.V., A.T. Leao, M.D.C. Leal, M. Feres, and A. Sheiham. 2008. The relationship between periodontal disease and preterm low birthweight: Clinical and microbiological results. *Journal of periodontal research* 43:615–626.

Vignali, D.A.A., L.W. Collison, and C.J. Workman. 2008. How regulatory T cells work. *Nature Reviews Immunology* 8:523–532.

Vogel, C.F.A., E.M. Khan, P.S.C. Leung, M.E. Gershwin, W.L.W. Chang, D. Wu, *et al.* 2014. Cross-talk between Aryl Hydrocarbon Receptor and the Inflammatory Response: A Role for Nuclear Factor-κB. *Journal of Biological Chemistry* 289:1866–1875.

Waby, J.S., H. Chirakkal, C. Yu, G.J. Griffiths, R. Benson, C.D. Bingle, and B.M. Corfe. 2010. Sp1 acetylation is associated with loss of DNA binding at promoters associated with cell cycle arrest and cell death in a colon cell line. *Mol Cancer* 9:275.

Waddlngton, C.H. 1942. The Pupal Contraction as an Epigenetic Crisis in Drosophila. Pages 181–188 in *Proceedings of the Zoological Society of London*. Wiley Online Library.

Wang, F., W. Meng, B. Wang, and L. Qiao. 2014a. Helicobacter pylori-induced gastric inflammation and gastric cancer. *Cancer Letters 345*:196–202.

Wang, L., C.T. Christophersen, M.J. Sorich, J.P. Gerber, M.T. Angley, and M.A. Conlon. 2012. Elevated fecal short chain fatty acid and ammonia concentrations in children with autism spectrum disorder. *Digestive diseases and sciences 57*:2096–2102.

Wang, Y., K.M. Telesford, J. Ochoa-Repáraz, S. Haque-Begum, M. Christy, E.J. Kasper, et al. 2014b. An intestinal commensal symbiosis factor controls neuroinflammation via TLR2-mediated CD39 signalling. *Nat Commun 5*.

Ward, R.E., M. Niñonuevo, D.A. Mills, C.B. Lebrilla, and J.B. German. 2007. In vitro fermentability of human milk oligosaccharides by several strains of bifidobacteria. *Molecular nutrition & food research 51*:1398–1405.

Whiteman, D.C., P. Parmar, P. Fahey, S.P. Moore, M. Stark, Z.Z. Zhao, et al. 2010. Association of *Helicobacter pylori* infection with reduced risk for esophageal cancer is independent of environmental and genetic modifiers. *Gastroenterology 139*:73–83.

Wikoff, W.R., A.T. Anfora, J. Liu, P.G. Schultz, S.A. Lesley, E.C. Peters, and G. Siuzdak. 2009. Metabolomics analysis reveals large effects of gut microflora on mammalian blood metabolites. *Proceedings of the National Academy of Sciences 106*:3698–3703.

Williams, B.L., M. Hornig, T. Buie, M.L. Bauman, M.C. Paik, I. Wick, et al. 2011. Impaired carbohydrate digestion and transport and mucosal dysbiosis in the intestines of children with autism and gastrointestinal disturbances. *PLoS ONE 6*:e24585.

Wing, K. and S. Sakaguchi. 2006. Regulatory T cells as potential immunotherapy in allergy. *Current opinion in allergy and clinical immunology 6*:482–488.

Xie, G., N.O. Keyhani, C.A. Bonner, and R.A. Jensen. 2003. Ancient Origin of the Tryptophan Operon and the Dynamics of Evolutionary Change. *Microbiology and Molecular Biology Reviews 67*:303–342.

Yang, T., J.L. Owen, Y.L. Lightfoot, M.P. Kladde, and M. Mohamadzadeh. 2013. Microbiota impact on the epigenetic regulation of colorectal cancer. *Trends in Molecular medicine 19*:714–725.

Yoshida, T., S. Yamashita, T. Takamura-Enya, T. Niwa, T. Ando, S. Enomoto, et al. 2011. Alu and Sat? hypomethylation in *Helicobacter pylori*-infected gastric mucosae. *International Journal of Cancer 128*:33–39.

Young, V.R. 1994. Adult amino acid requirements: The case for a major revision in current recommendations. *The Journal of Nutrition 124*:1517S–1523S.

Zaina, S., M.W. Lindholm, and G. Lund. 2005. Nutrition and aberrant DNA methylation patterns in atherosclerosis: More than just hyperhomocysteinemia? *The Journal of nutrition 135*:5–8.

Zelante, T., Rossana G. Iannitti, C. Cunha, A. De Luca, G. Giovannini, G. Pieraccini, et al. Tryptophan Catabolites from Microbiota Engage Aryl Hydrocarbon Receptor and Balance Mucosal Reactivity via Interleukin-22. *Immunity 39*:372–385.

Zhao, C. and X. Bu. 2012. Promoter methylation of tumor-related genes in gastric carcinogenesis. *Histology and Histopathology* 27:1271.

Zheng, X., G. Xie, A. Zhao, L. Zhao, C. Yao, N.H.L. Chiu, *et al.* 2011. The footprints of gut microbial–mammalian co-metabolism. *Journal of Proteome Research* 10:5512–5522.

CHAPTER 10

Fecal Microbiota Transplantation in Gastrointestinal Disease

DR. ROWENA ALMEIDA and DR. ELAINE O. PETROF

10.1. INTRODUCTION

THE human microbiome refers to the microbial population of bacteria, viruses, archaea, protozoans, and fungi that are present within the human body. It is estimated to contain 10–100 trillion microbial cells, of which the intestinal microbiota is the most diverse and densely populated. The gut harbors at least 10^{14} bacteria and archaea, composed of approximately 1,100 prevalent species, with at least 160 species per individual (Qin et al. 2010). It is now appreciated that the microbial environment is integral for the maintenance of human health and the pathogenesis of disease. Use of germ-free (GF) animal models and advances in metagenomics and bio computational analysis have revolutionized the study of the microbiota by generating a comprehensive catalog of the intestinal microbial composition. This knowledge is now being translated in diverse clinical settings, of which fecal microbiota transplantation (FMT) and intestinal microbiota therapeutics will be the focus of this chapter.

10.2. BASIC CONCEPTS IN INTESTINAL MICROBIOME FUNCTION

The gastrointestinal microbiota contains 10-fold more bacteria than the total number of human cells. This complex ecosystem with its diversity and richness plays a fundamental role in the regulation of host physiology such that increased bacterial diversity has been found to be associated with fewer comorbidities and better nutritional status and health in elderly individuals (Hollister *et al.* 2014; Claesson *et al.* 2012). The basic concepts of intestinal microbial physiology and function in energy metabolism and host immunity will be briefly reviewed to facilitate the discussion of FMT in various disease states.

The intestinal microbiota is located adjacent to the intestinal epithelium and secretes factors that modulate intestinal permeability, epithelial cell function, innate and adaptive immunity, motility, and neurotransmission (Hollister *et al.* 2014). There are some bacterial taxa that are known to confer enhanced metabolism, immunomodulation, anti-inflammatory, and antineoplastic effects, notably *Bacteroides, Bifidobacterium, Clostridium clusters XIVa and IVa, Eubacterium, Faecalibacterium, Lactobacillus*, and *Roseburia* (Hollister *et al.* 2014). The microbial communities are distinct in composition and function based on their location, both along the gastrointestinal tract and axially from the lumen to the mucosa, as well as on the age, sex, ethnicity, and dietary intake of the host (Harrell *et al.* 2012). It is hypothesized that the bacteria in proximity to the mucosal surfaces influence the immune system, while the luminal microbes are involved with digestion, nutrient assimilation, signaling, and metabolic activities (Nieuwdorp *et al.* 2014). In healthy adults, the gut microbiome is relatively stable with similar composition and phenotypes among those living together (Gilbert *et al.* 2014). It is, however, influenced by environmental factors such as antibiotics, travel, and acute illness (Hollister *et al.* 2014; Jalanka-Tuovinen *et al.* 2011). Diet has been recognized as a major influential factor. For example, *Prevotella* and *Xylanibacter* genera are found amongst children on fiber- and complex carbohydrate-enriched diets compared to those on a typical Western diet (Wu *et al.* 2011; Hollister *et al.* 2014).

10.2.1. Role of the Microbiome In Immune Function

The gastrointestinal (GI) microbiota plays a pivotal role in the regulation of the host immune system and in turn the composition and function of the gut ecosystem is regulated by the immune system. A conceptual understanding of the role of the gut microbiota in immunity development and regulation might elucidate the mechanism whereby breaches in this interaction lead to the pathogenesis of diseases such as inflammatory bowel disease, gastrointestinal malignancies, *Clostridium difficile* infections, and other related diseases.

Intestinal bacteria influence the development and function of the immune system. For instance, GF mice have shown impaired development of gut-associated lymphoid tissues (GALTs) such as Peyer's patches and isolated lymphoid follicles (Gordon *et al.* 1966; Kamada and Núñez 2013; Hamada *et al.* 2002; Bouskra *et al.* 2008) as mediated by NOD1 signaling (Bouskra *et al.* 2008). Intestinal microbes influence the intestinal T-helper (Th) cell-mediated immunity that is essential for autoimmunity and defense against extracellular microorganisms. For instance, in GF mice, the number of Th1 and Th17 cells are reduced (Mazmanian *et al.* 2005; Chung *et al.* 2012; Littman and Rudensky 2010). However, these numbers increase upon colonization by commensal Clostridiales-related bacteria, called segmented filamentous bacteria, that have Th17 cell-inducing capabilities (Luzza *et al.* 2000; Ivanov *et al.* 2008, 2009; Gaboriau-Routhiau *et al.* 2009; Littman and Rudensky 2010). Colonic T regulatory cells (T_{reg}) and IL-10 production in GF mouse models are induced by intestinal commensals such as *Clostridium* species belonging to the clusters IV, XIVa, and XVIII, mouse intestinal *Bacteroides* species, Altered Schaedler Flora (containing a mixture of eight defined murine commensal species), and enteric pathogens such as *Citrobacter rodentium* (Atarashi *et al.* 2011, 2013; Round and Mazmanian 2010; Round *et al.* 2011; Geuking *et al.* 2011; Kamada *et al.* 2012). For instance, microbiota induced T_{reg} cells play a protective role as shown in mice studies whereby *Clostridium*-induced T_{reg} cells suppress colitis and allergic diarrhea in mice, and *B. fragilis* induced IL10-producing T_{reg} cells reduce intestinal inflammation caused by *Helicobacter hepaticus* (Atarashi *et al.* 2013, 2011; Round *et al.* 2011; Mazmanian *et al.* 2008). The intestinal ecology also regulates the development of gut-specific B cell receptors, the immunoglobulin response, and intestinal mononuclear phagocytes within the mucosa (Wesemann *et al.* 2013; Franchi *et al.* 2012). Specifically, the commensal bacteria regulate the resident macrophage function by controlling the production of the anti-inflammatory cytokine IL-10, and modulating neutrophil function and systemic neutrophil responses via expression of pro-IL1β (Kamada *et al.* 2005; Denning *et al.* 2007; Rivollier *et al.* 2012; Clarke *et al.* 2010).

While there might be several mechanisms whereby intestinal bacteria induce the development of T_{reg} cells, there is mounting evidence for the role of bacterial metabolites such as short chain fatty acids (SCFAs). In particular, butyrate seems to be pivotal for the regulation of host immunity. SCFAs are hypothesized to directly regulate the development and function of colonic regulatory T cells via epigenetic regulation of the *Foxp3* gene (Smith *et al.* 2013; Furusawa *et al.* 2013; Arpaia *et al.*

2013), and via subsets of dendritic cells and macrophages in the lamina propria by secreting TGF-β (Coombes *et al.* 2007; Sun *et al.* 2007; Mucida *et al.* 2007). SCFAs also directly activate neutrophils and increase their bactericidal activity (Maslowski *et al.* 2009).

On the other hand, the host immune system regulates the gut microbiota and its composition to avoid excessive stimulation that may lead to deleterious intestinal inflammation. First, the intestinal immune system is physically separated from the luminal microbes by a mucus layer. Second, there are factors that limit the penetration of luminal bacteria into the intestinal mucosa via RegIIIγ, secreted IgA, and neutrophils. Third, macrophage hypo-responsiveness to microbial ligands and suppression of commensal effector T cells by T_{reg} cells circumvents excessive inflammatory responses preventing immune reactivity against commensal bacteria (Franchi *et al.* 2012; Kamada *et al.* 2005; Denning *et al.* 2007; Johansson *et al.* 2008; Bergstrom *et al.* 2010). The influence of the immune system on the gut microbial composition is illustrated in IL-22 deficient mice that were found to have an altered intestinal microbiota that was transmissible and colitogenic (Fagarasan *et al.* 2002; Wei *et al.* 2011; Qiu *et al.* 2013; Zenewicz *et al.* 2013).

10.3. INTRODUCTION TO FECAL MICROBIOTA TRANSPLANTATION

FMT is the administration of a fecal suspension from a healthy donor into the gastrointestinal tract of a recipient to re-establish the intestinal microbiota, increase microbial diversity, and normalize fecal composition. While the exact constituents of FMT are somewhat ill-defined, it can be considered a biologically active complex mixture of living organisms and their metabolites, with therapeutic potential.

FMT is distinct from a probiotic, which is defined as "live microorganisms which when administered in adequate amounts confer a health benefit on the host," (Marteau *et al.* 2002; Reid 2001). Probiotics, in contrast to FMT, are often lab-adapted organisms employed in the dairy industry, or in the production of fermented foods. In 1907, Elie Metchnikoff first conceptualized the concept of probiotics when regular consumption of lactic acid bacteria (LAB) in fermented dairy products was found to be associated with enhanced health and longevity in the rural population of Bulgaria (Metchnikoff 1908). Besides LAB, other common probiotic organisms are the *Bifidobacterium* species, *Escherichia coli* Nissle, and the yeast *Saccharomyces boulardii* (Gupta and Garg continued antibiotics and this can range as high as 40–65% in patients previously treated for rCDI (McFarland 2005). Recurrent CDI reflects the inability of the intestinal microbiota to recover and re-establish itself to homeostasis with the host after the initial insult (Allen-Vercoe *et al.* 2012; Chang *et al.* 2008). Spore recrudescence or persistence from initial infection is a proposed mechanism for recurrent CDI. A recognized therapeutic challenge, existing treatment options for rCDI are somewhat limited (Allen-Vercoe *et al.* 2012). The failure rate of metronidazole for uncomplicated CDI is more than 18%, and the risk of recurrence after two recurrent episodes is more than 60% with antibiotic therapy (Kelly and LaMont 2008; Pepin *et al.* 2005; Cohen *et al.* 2010; Petrella *et al.* 2012). Alternative therapies are being explored, none of which have thus far proved to be as highly effective as FMT. By virtue of restoring the phylogenetic diversity of the intestinal microbiota composition, FMT has been shown to be generally safe, inexpensive, and highly effective with reported cure rates of greater than 90% with negligible adverse effects (Bakken *et al.* 2011).

While the ability of FMT to cure recurrent CDI is supportive for the proof of concept of restoring the intestinal diversity as a means to ameliorating CDI, the precise mechanism is not entirely clear. Indigenous intestinal microbiota species are essential for colonization resistance and this homeostasis is altered by antibiotic administration facilitating a change in the community size, composition, and function that correlates with reduced resistance to CDI. In fact, the intestinal microbiota composition differs among patients with primary or recurrent symptomatic CDI compared to those with asymptomatic colonization by *C. difficile* (Rousseau *et al.* 2012). Specifically, fecal samples from patients with CDI after antibiotic therapy show decreased overall microbial diversity compared to those from patients who did not develop CDI (Dethlefsen and Relman 2011) and a similar observation of lower microbial diversity was seen in those with recurrent disease (Chang *et al.* 2008). Reeves *et al.* (2011) have shown a correlated decrease in the numbers of indigenous cecal community of Lachnospiraceae and a relative abundance of Enterobacteriaceae in mouse CDI colitis leading the authors to suggest that the severity of disease is related to these bacterial community dynamics. Patients who developed CDI were also found to have reduced proportions of the family Clostridiales Incertae Sedis XI compared with those who did not develop CDI (Merrigan *et al.* 2003; Sambol *et al.* 2002; Vincent *et al.* 2013).

One proposed mechanism of FMT effectiveness relates to the stimulation of *C. difficile* sporulation by bile salt metabolism, whereby primary bile acids stimulate germination of spores (Sorg and Sonenshein 2008, 2009). Taurocholic acid, for example, is a primary bile acid and is such a potent germinant that it is routinely used in *C. difficile* growth media, to promote germination and growth of the organism (Wilson 1983). Because a high germination efficiency has been positively linked with recurrent CDI (Moore *et al.* 2013) and spores are notoriously resistant to antibiotics such as vancomycin, spore germination efficacy has long been suspected to contribute to rCDI. Of interest, secondary bile acids such as lithocholate serve as potent inhibitors of germination and many native species of the healthy intestinal microbiota (e.g., Lachnospiraceae and *Ruminococcaceae* of *Clostridium* clusters XIVa and IV) possess 7α-dehydroxylation activity, allowing them to convert primary bile acids to secondary bile acids (Doerner *et al.* 1997; Hirano *et al.* 1981; Kitahara *et al.* 2000; Ridlon *et al.* 2006; Stellwag and Hylemon 1978; Takamine and Imamura 1995). This accumulation of secondary bile acids, in turn, likely has an inhibitory effect on both germination and growth of *C. difficile*. A study of bile salt acids in rCDI patients pre- and post-FMT seems to corroborate the bile salt hypothesis: pre-FMT fecal samples from patients with recurrent CDI contained nearly undetectable levels of secondary bile acids and high concentrations of primary bile acids, whereas post-FMT fecal samples showed the reverse pattern with mostly secondary bile acids, as did non-CDI stool (Weingarden *et al.* 2014). These changes in bile salts correlated with the changes in the intestinal microbiota, suggesting that bile salt metabolism likely plays a major role in rCDI and is directly linked to the composition of the gut microbiota.

Besides spore germination, the pathogenesis of CDI also involves production of large clostridial toxins, namely TcdA and TcdB (Voth and Ballard 2005; Pruitt and Lacy 2012; Shen 2012) that have local virulence effects as well as systemic effects (Hamm *et al.* 2006). *C. difficile* utilizes proteases, flagella, and adhesins to penetrate the mucous layer and adhere to the colonic epithelial cells. The cytotoxins inactivate the host Rho and Ras family guanosine triphosphatases (GTPases) via glucosylation which induces an inflammatory cellular response and collapse of the cellular cytoskeleton (Pruitt and Lacy 2012; Paredes-Sabja *et al.* 2008; Denève *et al.* 2009). In addition, some strains such as the NAP1/027/BI strain produce a binary toxin (McDonald *et al.* 2005). The binary toxin is an ADP-ribosylating toxin consisting of two separate components: CDTa possesses the enzymatic ADP-ribosyltransferase activity which modifies actin, whereas CDTb binds to the LSR receptor (lipolysis-stimulated lipoprotein receptor) of host cells and allows its entry into the cell (Carter *et al.* 2012; Papatheodorou *et al.* 2011). Presence of binary toxin has been associated with worse outcome overall and with a higher likelihood of developing recurrent CDI (Stewart *et al.* 2013, 2014). Future studies are needed to determine whether FMT has any direct inhibitory effect on TcdA, TcdB, or binary toxin.

Another important contributing factor in rCDI that is corrected by FMT is the concept of niche, or resource competition. This is well illustrated by the fact that colonization of hamsters by a nontoxigenic strain of *C. difficile* can prevent CDI (Merrigan *et al.* 2003; Sambol *et al.* 2002; Vincent *et al.* 2013). The nontoxigenic strain likely competes for the same nutrients and same environmental niche as the pathogenic strain, and thus "outcompetes" its pathogenic counterpart.

Antibiotics can also disrupt the composition of the microbiota and change the balance of available nutrient sources in a manner that can promote CDI. Antibiotic use decreases the overall diversity of the intestinal microbiota, leading to reduced competition for nutrient resources utilized by *C. difficile* and conditions that allow the opportunistic pathogen to thrive. For example, using animal models Ng and colleagues have shown that antibiotic treatment results in an increase in sialic acid, which in turn leads to an increase in *C. difficile* growth in the gut. The authors of the study propose that antibiotics spare enough bacteria with sialidase capability to liberate sialic acid from mucosal carbohydrates (an energy source used by many gut commensals as well as by *C. difficile*), but the antibiotics destroy important commensal bacteria that would normally compete with *C. difficile* and consume the sialic acids. This overabundance of free sialic acid, caused by disruption of the gut microbiota by antibiotics, thus provides ideal conditions and access to an abundant energy source to allow the expansion of *C. difficile* in the gut. It is therefore quite likely that one of the primary mechanisms behind success of FMT may involve increased utilization of sialic acid by commensal bacteria that results in inhibition of *C. difficile* growth and expansion.

FMT is usually employed for rCDI patients who have failed conventional antibiotic therapy with metronidazole and/or vancomycin therapy. Systematic reviews including 317 patients and 124 patients respectively had shown a resolution rate of 92% and 83% respectively with few adverse events (Gough *et al.* 2011; Guo *et al.* 2012; Palmer 2011). A robust systematic review by Kassam and colleagues (2013) included 11 studies, none of which were RCTs, found that 245 out of 273 patients had achieved clinical resolution with an unweighted and weighted pool resolution rate of 89.7% and 89.1%, respectively (Kassam et al. 2013). Support for FMT largely came from small, observational studies until van Nood et al. (2013) reported the first small, open-label randomized control trial of infusion of donor feces against rCDI in patients pretreated with vancomycin and bowel lavage. The trial compared an initial high dose oral vancomycin regimen (500 mg four times daily for 4 days) followed by bowel lavage and subsequent FMT through a nasoduodenal tube to a standard vancomycin regimen or a standard vancomycin regimen with bowel lavage. This study was stopped after the interim analysis given that there was an 81% clinical resolution rate after the first infusion of donor feces. In comparison, only 31% of patients in the vancomycin arm and 23% of patients in the vancomycin and bowel lavage arm had resolution of their CDI (van Nood et al. 2013). Furthermore, the fecal microbiota in rCDI patients had reduced microbial diversity, but this profile showed a persistent improvement after FMT with a notable increase in *Bacteroidetes species* and *Clostridium clusters IV and XIVa* (*Firmicutes*) while there was a decrease in *Proteobacteria* species.

The Fecal Microbiota Transplantation Workgroup has recommended indications for FMT, which include recurrent or relapsing CDI currently defined as at least three episodes of mild to moderate CDI and failure of 6- to 8-weeks vancomycin with or without an alternative antibiotic (e.g., rifaximin and nitazoxanide) (Bakken 2009). As yet, there is no consensus on route of FMT delivery, that is, nasogastric/nasojejunal, gastroscopy, colonscopy, or retention enema, but limited evidence suggests that lower gastrointestinal FMT might be preferred (Lund-Tønnesen et al. 1998; Garborg et al. 2010; Kassam et al. 2012; Rohlke et al. 2010; Aas et al. 2003). Beyond its relative ease of administration, feasiblity, and economical nature, FMT via retention enemas had a 93% clinical response rate in recurrent or refractory CDI with none of the inherent procedural risk factors of endoscopic administration (Kassam et al. 2012). In 2011, Brandt et al. proposed that endoscopic FMT in fact become the first line treatment for patients with deteriorating and severe CDI at the earliest possible time given its potential to eradicate CDI and restore the underlying microbiota (Brandt etl al. 2011). In critically ill patients, however, they propose FMT as an adjunctive therapy to restore the microbiota to facilitate the therapeutic effect of antibiotics (Brandt et al. 2011).

Some have expressed concerns about the potential of systemic pathogenic transmission associated with FMT. Synthetic stool may be an effective and feasible alternative to defecated donor fecal matter and can potentially offset some of these concerns. The exact bacterial species composition in synthetic stool substitutes is known because the material undergoes thorough screening for viruses, and pathogens, is reproducible and controlled, and is amenable for selection based on the antimicrobial sensitivity profiling of the bacterial strains, thus enhancing the overall safety profile of the mixture. Tvede and Rask-Madsen (1989) found that a cocktail of 10 facultative aerobes and anaerobes administered to five patients was effective against recurrent CDI with detectable levels of *Bacteroides* species post treatment. In 2013, Petrof *et al.* demonstrated that a microbial community or synthetic stool substitute mixture ("RePOOPulate") consisting of 33 different intestinal bacterial species isolated in pure culture from a single donor was able to revert patients with hyper virulent ribotype 078 CDI infected patients who had failed standard therapy back to normal bowel pattern. After treatment, the two patients were also found to have stool microbial profiles reflecting the synthetic mixture isolates, and despite having received multiple courses of antibiotics for unrelated infections, they maintained clinical remission at 6 months (Petrof *et al.* 2013). In other studies, FMT success in NAP-1/ribotype 027 strain has been found to 89% in comparison to 100% resolution in non-027 ribotype CDI (Mattila *et al.* 2012). Lawley *et al.* (2012) administered a six-species community of microbes to a mouse model of CDI and increased the intestinal microbial diversity but none of the species were able to colonize the gut over long time periods.

Nevertheless, there is increasing acceptance of FMT as a therapeutic option amongst physicians and patients. Bakken *et al.* (2013) surveyed infectious disease physicians about their treatment approaches toward patients with CDI and found that 80% of the respondents would consider FMT for patients with recurrent CDI and 72% had treated more than one patient with FMT in the past year. Given the overwhelming burden of evidence and acceptance for the role of FMT in *C. difficile* infections as a biologic therapeutic agent, it may soon become mainstream therapy for this disease.

10.4.2. Role of FMT in Inflammatory Bowel Disease

Inflammatory Bowel Disease (IBD) is a chronic relapsing inflammatory intestinal disorder that includes both ulcerative colitis (UC) and Crohn's disease (CD). IBD is a multifaceted disorder known to affect 3.6 million people that can be viewed as the product of an imbalance in the intestinal microbiota, the epithelium, and the immune system in genetically susceptible individuals. While the exact etiology is unknown, it is hypothesized to be the result of continuous antigenic stimulation by nonpathogenic commensals, leading to an exaggerated sustained immune response in a genetically predisposed host. Environmental factors such as diet, smoking, and drug treatments are also thought to play an important role in pathogenesis (Loftus 2004).

The role of the gut microbiota in the pathogenesis of IBD was suggested by the fact that antibiotics such as rifaximin have shown potential to reduce intestinal inflammation, induce remission, or prevent relapse (Sartor 2004, 2008; Casellas *et al.* 1998; Khan *et al.* 2011). However, repeated perturbation by antibiotics also results in cumulative and persistent changes in the microbial ecology of the gut, and thinning of the protective mucus layer (Dethlefsen and Relman 2011; Relman 2012; Wlodarska *et al.* 2011). For instance, antibiotics have been shown to cause a rapid increase in *E. coli* followed by a gradual decline (Looft and Allen 2012). An understanding of the mechanisms for the pathogenesis of IBD as associated with intestinal microbiota dysfunction has been obtained from 16S-gene sequencing, metagenomics, and metabolomic methods.

The composition of the GI microbiota in patients with inflammatory bowel disease is distinct from healthy controls such as sharp reductions in members of the phylum *Firmicutes* and *Clostridial clusters IV and XIV* (Ott *et al.* 2004). While some microbes may have a protective effect in IBD, other species that are found to be enriched in IBD may play a role in the pathogenesis of the disease. Several studies have reported a decrease in the gut bacterial biodiversity among IBD patients. Notably, these patients have shown a reduction in the representation of *Firmicutes* phylum and an increase in *Gammaproteobacteria* (Frank *et al.* 2011; Morgan *et al.* 2012). Other gut microbes that have been implicated in the pathogenesis of IBD include *Enterobacteriaceae*. For instance, adherent-invasive *E. coli* strains have been isolated from ileal CD (iCD) mucosal biopsy samples, CD granulomas, and UC patients (Darfeuille-Michaud *et al.* 2004; Sokol *et al.* 2006; Chassaing and Darfeuille-Michaud 2011). Furthermore, *Fusobacterium* species have been found to be increased in the colonic mucosa samples of IBD patients and have been found to induce colonic mucosal erosion in UC mouse models (Ohkusa *et al.* 2002, 2003, 2009; Strauss *et al.* 2011). Mesalamine treatment for IBD has been associated with a decrease in *Escherichia/Shigella* (Benjamin *et al.* 2012). Also, butyrate-producing bacteria such as *Faecalibacterium, Phascolarctobacterium,* and *Roseburia* species are decreased in IBD, especially the former being in reduced quantities in ileal CD, whilst *Odoribacter* and *Leuconostocaceae* are reduced in UC (Sokol *et al.* 2008; Duncan *et al.* 2002; Frank *et al.* 2007; Morgan *et al.* 2012; Kang *et al.* 2010; Willing *et al.* 2009). Therefore, Gevers and colleagues (2014) in the microbiome analysis of their 1,000 treatment-naïve new onset Crohn's disease pediatric cohort have suggested that assessing the rectal mucosa-associated microbiota has diagnostic potential in CD.

From a functional perspective, the IBD microbiome shows an overall preponderance of an inflammation-promoting microbiome. This is seen with increased glutathione transport and riboflavin metabolism genes corresponding to an increased propensity for oxidative stress, an increase in type II toxin secretion systems, and increased virulence-related bacterial genes (Erickson *et al.* 2012; Morgan *et al.* 2012). Metagenomic studies have shown a decrease in butanoate and propanoate metabolism genes in ileal CD, corresponding with lower levels of butyrate and other SCFAs, that similarly corresponds with a decreased proportion of SCFA-producing *Firmicutes*. Also, the IBD microbiome has been shown to have a decrease in pathways involved in biosynthesis of amino acids, but has increased amino acid transporter genes alluding to the ability of the microbiota to source nutrients from other sites. The IBD microbiome also has an increase in sulfate-reducing bacteria such as *Desulfovibrio* and associated genes involved in the metabolism of sulfur-containing amino acid cysteine and sulfate transport (Morgan *et al.* 2012; Erickson *et al.* 2012; Rowan *et al.* 2010).

It should be noted that the intestinal microbiota may also have a protective effect against IBD. There is considerable spotlight on the underrepresentation of *Faecalibacterium prausnitzii* in IBD (Sokol *et al.* 2009) which is found to be depleted in CD biopsy samples. A reduction in *F. prausnitzii* is associated with higher risk of recurrent CD post-surgery. Among UC patients in remission, Varela *et al.* (2013) have shown defective gut colonization by *F. prausnitzii* and that recovery of this species after relapse is associated with maintenance of clinical remission of UC (Sokol *et al.* 2008, 2009; Varela *et al.* 2013; Willing *et al.* 2009). The intestinal microbiota may confer direct beneficial functional effects by modulating host cytokine gene expression. For instance, bacterial species such as *Bifidobacterium, Lactobacillus,* and *Faecalibacterium* genera protect the host from mucosal inflammation by down-regulation of inflammatory cytokines or up-regulation of the anti-inflammatory cytokine IL-10 (Sokol *et al.* 2008; Llopis *et al.* 2009). Ileal specimens obtained during surgery from CD patients incubated with *Lactobacillus casei* DN-114 001 showed significantly decreased secretion of TNF-alpha, IFN-gamma, and IL-2. Also, the expression of IL-8, IL-6 and CXCL 1 was down-regulated by *L. casei* while *E. coli* significantly up-regulated the expression of theses cytokines (Llopis *et al.* 2009). Other mechanisms include induction of T_{reg} cells as illustrated by *Bacteroides* and *Clostridium* species, whilst other microbes inhibit NF-κB activation to reduce intestinal inflammation (Kitajima *et al.* 2001; Callaway *et al.* 2008; Kamada *et al.* 2012; Kane *et al.* 2011; Medellin-Peña *et al.* 2007; Atarashi *et al.* 2013; Kelly *et al.* 2004).

The intestinal microbial composition in IBD is also influenced by long-term dietary intake such that strict vegan or vegetarian diets have been associated with significant reductions in *Bacteroides* species, *Bifidobacterium* species, and the *Enterobacteriaceae* (Wu *et al.* 2011; Zimmer *et al.* 2012). Furthermore, the onset of IBD has a bimodal age related onset paralleling the age related changes in the composition witnessed by the gut microbiota. For instance, early onset of IBD corresponds to the low complexity and low stability microbiota composition found in younger patients (Dominguez-Bello *et al.* 2011). NOD2, expressed in Paneth cells in the terminal ileum producing antimicrobial defensins, was the first susceptibility gene linked to Crohn's disease (Ogura *et al.* 2001; Stappenbeck *et al.* 2002). IBD patients with NOD2 mutations have increased numbers of mucosa-adherent bacteria and decreased transcription of IL 10 (Philpott and Girardin 2009; Swidsinski *et al.* 2002) whilst IBD patients with both NOD2 and ATG16L1 risk alleles have decreased levels of *Faecalibacterium* and increased levels of *Escherichia coli* (Frank *et al.* 2011).

Currently available treatment options are often ineffective for certain IBD patients and this portends a poor quality of life and increasing morbidity. Alternative therapeutic options are therefore gaining recognition. Thus far, in terms of microbial therapeutics, probiotics have shown mixed results with some limited success to facilitate clinical remission in UC (Quattropani *et al.* 2003; Markowitz *et al.* 2004; Wong *et al.* 2009; Bennet and Brinkman 1989; Whelan and Quigley 2013). However, its durability to modify gut dysbiosis, given that probiotics have a limited number and diversity of species, is unknown. The potential of antibiotics to reduce intestinal inflammation and induce remission or prevent relapse elucidates the role of altering gut microbial demographics as a treatment option in IBD. To circumvent the undesirable effect of repeated antibiotic exposure on microbial diversity, fecal microbiota transplantation was explored to repopulate the gut with complex microbiota.

The literature available for FMT in IBD has thus far been limited. Bennet and Brinkman published the first case report of FMT for UC in 1989 with Bennet having treated himself via an FMT retention enema and becoming symptom free by 6 months post-treatment. Borody *et al.* (1989) initially described two cases of FMT in UC patients, both of whom were found to be refractory to conventional ASA therapy and were symptom free at the 3–4 month mark after FMT retention enemas. Thereafter, Borody *et al.* (2003) published a retrospective case series of six adult patients with UC who were able to discontinue conventional therapy and were in long-term clinical remission with documented mucosal healing after having undergone FMT. Borody and Campbell (2012) have further suggested that serial FMT infusions are required to achieve prolonged remission or cure in UC. Angelberger *et al.* (2013), in their small case series with moderate to severely active UC patients refractory to standard therapy, found that donor fecal microbiota is difficult to maintain in FMT recipients. The authors used time-resolved fecal microbiota analysis and found that relative abundance in the families *Enterobacteriaceae* and *Lachnospiraceae* can perhaps be a useful diagnostic indicator of clinical response post-FMT in UC patients. More recently, Kump and colleagues (2013) have shown that *Proteobacteria* was reduced, and *Bacteroides* was increased after FMT. However, these patients failed to reach or maintain remission (Kump *et al.* 2013) for which Kunde and colleagues (2013) have suggested multiple consecutive infusions as a means to increase chances of clinical remission.

Overall, FMT in IBD management is rare and several questions still remain; as such, there is a lack of proven efficacy and safety of FMT in UC patients (Anderson *et al.* 2012; Kelly *et al.* 2014). In their multicenter retrospective series, Kelly *et al.* investigated adverse events in addition to the rates of CDI cure after FMT in immuncompromised patients. The authors found that 14% of patients with IBD experienced disease exacerbation post FMT, however, it remains uncertain if this was attributable to FMT, CDI, or progression of the underlying disease. IBD exacerbations after FMT to treat CDI have been reported previously as seen in a patient with quiescent UC; in addition to the occurrence of fevers with temporary elevation of C-reactive protein levels in a small cohort of UC patients treated with FMT for IBD (De Leon *et al.* 2013; Angelberger *et al.* 2013; Kump *et al.* 2013). That said, FMT as an alternative therapeutic option for IBD is increasingly gaining recognition (Khoruts *et al.* 2010; Borody and Campbell 2012). Focus group studies have found that UC patients were eager and receptive to the concept of FMT as a therapeutic strategy (Kahn *et al.* 2012). Hence, well-designed RCTs to expand on the potential of microbial therapeutics in IBD need to be imminently pursued (Smits *et al.* 2013; Damman *et al.* 2012; Anderson *et al.* 2012).

10.4.3. Role of FMT in Obesity, Diabetes and Metabolic Syndrome

The development of obesity and metabolic syndrome is a complex process involving both genetic and environmental factors (Tilg 2010). The mechanistic association between obesity and metabolic syndrome is unclear but approximately 50% of all obese individuals develop metabolic syndrome (Eckel *et al.* 2010). The integration of metagenomics and metabolomics has unraveled the metabolic capabilities of the gut microbial genetic material or metagenome to elucidate the link between intestinal microbial ecology and metabolic disorders.

Recent studies have suggested that the intestinal microbiota plays an important role in obesity and related diseases, as well as in the development of insulin resistance (Kong *et al.* 2013). Evidence for this comes from antibiotic administration that alters the commensal gut composition and the associated detrimental effects on host metabolism. For instance, administration of subtherapeutic antibiotic doses to young mice promotes metabolic dysfunction and increases adiposity (Cho *et al.* 2012). The intestinal microbiota mediates host metabolism through its interaction with host nutrient intake, bile acids, and the luminal mucus layer (Nieuwdorp *et al.* 2014). Profound changes have been found in the composition and metabolic function of the gut microbiota in obesity that facilitate the "obese microbiota" to extract more energy from the diet (Tilg 2010; Turnbaugh *et al.* 2006). Hence, GF mice are found to be leaner when compared to conventional mice and the introduction of gut microbiota in GF mice causes an increase in their fat mass and insulin resistance (Bäckhed *et al.* 2004). Also, the gut microbiota interacts with host epithelial cells to indirectly control energy expenditure and storage (Tilg 2010).

The gut microbiota is a reservoir of inflammatory ligands such as lipopolysaccharide (LPS) and peptidoglycans that can contribute to the etiology of metabolic disorders (Cani *et al.* 2007; Moreno-Navarrete *et al.* 2012). One of the proposed mechanisms for the role of gut microbiota in the onset of insulin resistance and type 2 diabetes is through low-grade inflammation. For instance, the intestinal microbiota produces proinflammatory bacterial inflammatory ligands and endotoxins such as peptidoglycans, flagellins, and plasma lipopolysaccharide—binding protein that activates inflammatory pathways leading to obesity and insulin resistance (Clarke *et al.* 2010; Vijay-Kumar *et al.* 2010). LPS is derived from the cell wall of Gram-negative bacteria and is a potent proinflammatory molecule (Cani and Delzenne 2007; Cani *et al.* 2007) that can activate immunogenic pathways via Toll-like receptor 4 (Cani *et al.* 2007; Moreno-Navarrete *et al.* 2012; Karlsson *et al.* 2013; Wentworth *et al.* 2010; Cani and Delzenne 2007; Clarke *et al.* 2010; Vijay-Kumar *et al.* 2010). Metabolic endotoxemia has since been conceptualized as a systemic inflammatory state associated with metabolic syndrome that is mediated by bacterial translocation through the gastrointestinal tract resulting in increased levels of lipopolysaccharide (Cani *et al.* 2007; Amar *et al.* 2011; Amar *et al.* 2008). For instance, a high-fat diet was found to increase plasma levels of LPS (Cani *et al.* 2007). Furthermore, dietary derivatives metabolized by the commensal bacteria such as SCFAs play a protective effect on immunomodulation by reducing intestinal permeability (Maslowski *et al.* 2009).

The intestinal metagenome as an environmental factor is implicated in obesity by its effect on host lipid metabolism. This was strongly demonstrated by showing that significantly greater adiposity was conferred to GF recipients who received intestinal microbiota from obese mice rather than from lean donors (Bäckhed *et al.* 2007; Ridaura *et al.* 2013). The greater adiposity was perhaps derived by increased energy extraction from indigestible carbohydrates metabolized into SCFAs such as butyrate, acetate, and propionate (Tilg 2010). Furthermore, insulin resistance in obese mice fed on high-calorie diets was prevented and reversed on a diet supplemented with butyrate (Gao *et al.* 2009). However, higher levels of SCFAs have also been found by the small intestinal microbiota of lean individuals suggesting that altered SCFA composition may be responsible for some of the metabolic and immunologic effects of the intestinal microbiota (Vrieze *et al.* 2012).

Obesity has not only been associated with alterations in the gut metagenome but also with reduced bacterial diversity at the phylogenetic and the metagenomic level (de Vos and de Vos 2012; Turnbaugh *et al.* 2009; Le Chatelier *et al.* 2013). Obese mice have a significant increase in the *Firmicutes* to *Bacteroidetes* ratio (Ley *et al.* 2005; Turnbaugh *et al.* 2006), a trend also seen in the fecal microbiota of obese human individuals, and this composition was found to change with weight loss (Ley *et al.* 2006). Separate studies also found an increased proportion of *Lactobacillus gasseri, Streptococcus mutans*, and *Escherichia coli* in cohorts with type 2 diabetes mellitus (Karlsson *et al.* 2013; Qin *et al.* 2012). A decrease in *Bifidobacterium* species was found in obesity, type 2 diabetes mellitus, and in children who become overweight, suggesting that this species may confer a protective effect (Schwiertz *et al.* 2010; Kalliomäki *et al.* 2008). Reduced numbers of butyrate-producing *Clostridiales* (*Roseburia* and *F. prausnitzii*) were also seen in individuals with type 2 diabetes mellitus whilst nonbutyrate producing *Clostridiales* and pathogens such as *Clostridium clostridioforme* were present in greater proportions (Larsen *et al.* 2010; Furet *et al.* 2010). These data further reinforce the role of reduced butyrate production by the microbiota as a contributory factor to the development of insulin resistance.

Categorizing individuals as high or low gene count based on metagenomic DNA sequencing of microbial genes has corresponded with lower prevalence of obesity and metabolic disorders among the high gene cohort. High gene cohort individuals were also found to have an increased proportion of *Anaerotruncus coliohominis, Butyrovibrio crossotus, Akkermansia* species, and *Faecalibacterium* species (Le Chatelier *et al.* 2013). Also, they have a relative abundance of genes involved in metabolic pathways for hydrogen and methane production, and lower proportion of pathways in hydrogen sulfide production (Hollister *et al.* 2014; Le Chatelier *et al.* 2013). Gene function studies in type 2 diabetes showed an increase in genes involved in membrane transport, sulfate reduction, and oxidative stress resistance functions and a decrease in functions involving cofactor and vitamin metabolism and butyrate production (Qin *et al.* 2012).

Given the physiological role of the proximal intestine in carbohydrate and fat uptake, Vrieze and colleagues (2012) hypothesized that restoring the small intestinal microbiota in obese individuals by infusions from lean donors would affect host metabolism and insulin sensitivity in patients with metabolic syndrome. The authors found that peripheral insulin sensitivity in patients with metabolic syndrome improved 6 weeks after having received feces from healthy, lean donors compared to those who had autologous fecal transplants (Vrieze *et al.* 2012). Obese individuals were found to have decreased fecal microbial diversity, particularly higher amounts of *Bacteroides* and decreased amounts of *Clostridium cluster XIVa* compared with lean donor subjects. This microbial diversity increased significantly after allogenic gut microbiota transfer, but the same trend was not seen in the autogenic treatment group. There was a two-and-a-half-fold increase in the number of bacteria related to the butyrate-producing *R. intestinalis* after an allogenic microbiota infusion, and the bacteria related to the similarly butyrate-producing *E. hallii* were increased in the small bowel mucosa after the allogenic transfer. As mentioned earlier, butyrate is an essential intestinal substrate required for energy and signaling purposes (Maslowski *et al.* 2009; Donohoe *et al.* 2011). Overall, these data suggest that that increased bacterial diversity is associated with reduced insulin resistance, and that butyrate derived from gut microbial metabolism might play a regulatory role in improved insulin sensitivity (Vrieze *et al.* 2012).

Landmark studies showing the ability to transfer metabolic phenotypes among individuals elucidate the potential for FMT as a therapeutic strategy for metabolic disorders. For instance, increasing murine gut colonization of *A. muciniphila* reversed gain in fat mass, metabolic endotoxemia, adipose tissue inflammation, and insulin resistance (Everard *et al.* 2013) and provides proof of concept for treatment strategies involving defined intestinal bacterial species. While the metabolic effects of FMT in humans are still being studied, the demonstrated changes in glycemic control and bile acid metabolism as a consequence of microbiota alterations presents FMT as a viable option in the future to prevent and treat metabolic dysregulation in obesity and insulin resistance (Weingarden *et al.* 2014; Vrieze *et al.* 2012).

10.4.4. Other Potential Applications of FMT

10.4.4.1. Necrotizing Enterocolitis

There has been abundant literature on the importance of the microbiome in pediatric disease in recent years. In particular, there has been resurgence in the data suggesting that intestinal bacterial colonization plays a fundamental role in the multifactorial pathogenesis of necrotizing enterocolitis (NEC). NEC is a common disease of premature infants, such that 7% of infants weighing less than 1,500 grams will develop NEC. It is a lethal disease that is characterized by intestinal inflammation with the potential for rapid progression to necrosis and multisystem organ failure. NEC has an associated mortality rate of 20–30% and significant morbidity among the 20–40% who require surgical intervention, including neurodevelopmental delay and short gut syndrome (Neu and Walker 2011; Pike et al. 2012; Lin et al. 2008) The main risk factor for NEC is prematurity while other contributory factors such as genetic predisposition, enteral feeds, intestinal ischemia, and colonization by pathogenic bacteria result in an intestinal dysbiosis that likely contributes to the pathogenesis (Neu and Walker 2011; Henry and Moss 2009; Petrosyan et al. 2009; Hintz et al. 2005; Martin and Walker 2006).

Intestinal colonization by bacterial species begins at birth and it is known that the microbiota of premature infants differs from that of term infants. There is an increase in microbial diversity over the first year with a microbial composition that is similar to the adult microbiota. Multiple factors such as mode of delivery, gestational age, antibiotic exposure, immature immune system, feeding regimens, and the environment influence the early microbiota composition (Westerbeek et al. 2006). As such, the decreased diversity in preterm infant gut microbiota is influenced by exposure to the nosocomial microbes in the neonatal intensive care unit and by the early introduction of antibiotics (Gewolb et al. 1999; Hall et al. 1990). For instance, early antibiotic exposure in particular reduces gut microbial colonization (Cotton 2010; Kuppala et al. 2011) and increases the risk for NEC. Paradoxically, parenteral antibiotics remain the cornerstone for management of NEC.

The microbiota composition is an important risk factor for the development of NEC in premature infants (Hoy et al. 2000; Gewolb et al. 1999; Stark and Lee 1982; Claud and Walker 2001; Musemeche et al. 1986; Jilling et al. 2006; Bjornvad et al. 2008; Cilieborg et al. 2011; Carlisle et al. 2011). For instance, Jilling et al. (2006) have shown that the use of sterilized feeding catheters in murine models prevented NEC and that the receptor for bacterial cell component recognition, Toll-like receptor 4 (TLR4), was integral for development of NEC. However, no specific bacterial species have been linked to the etiology of NEC and the exact mechanism of this association remains unclear (Jilling et al. 2006).

Studies evaluating the specific microbiota differences in infants with NEC have not been consistent. Independent studies by Millar *et al.* (1996) and recently by Normann *et al.* (2013) demonstrated no significant microbiota differences (Mshvildadze *et al.* 2008, 20102010). Other studies, however, have shown a decreased microbial diversity, an abundance of the phylum *Proteobacteria*, and reduced numbers of *Enterococcus* in the fecal microbiota of infants with NEC (Millar *et al.* 1996; Wang *et al.* 2009; Mai *et al.* 2011; Stewart *et al.* 2012; LaTuga *et al.* 2011; Spor *et al.* 2011). Intestinal samples of infants with NEC demonstrated a similar abundance of *Proteobacteria* and the genus *Ralstonia*. Torrazza *et al.* (2013) in their study including 18 NEC cases and 35 controls found a different pattern of microbial colonization among the infants with NEC. As with prior studies, there was a higher proportion of *Proteobacteria* two weeks prior and *Actinobacteria* one week prior to the onset of NEC, as well as lower numbers of *Bifidobacteria* and *Bacteroidetes*. Furthermore, they found a novel signature sequence distinct from but closely related to *Klebsiella pneumoniae* that was associated with the development of NEC (Torrazza *et al.* 2013).

Bjorkstrom *et al.* (2009) characterized the intestinal microbiota in the first four weeks of life of 48 very low-birth weight (VLBW) infants and in select cases prior to the onset of NEC. They found an early and predominant colonization with lactic acid bacteria (LAB) likely due to the early introduction of nonpasteurized breast milk (Björkström *et al.* 2009). The authors showed that bifidobacteria and lactobacilli were present in 55% and 71% of the cases within the first week of life respectively and this increased to 82% and 98% respectively in the second week of life. Furthermore, all cases of NEC had an early colonization of LAB, and this suggests that the presence of a nonpathogenic species alone does not clearly prevent NEC. Of note, prior to the onset of NEC, all VLBW infants had a high colonization of non-*E. coli* Gram-negative species such as *Klebsiella* spp., *Pseudomonas* spp., and *Proteus* spp. (Björkström *et al.* 2009). This supports an earlier finding by Hoy *et al.* (1990) that there are pertinent changes in the gut microbiota prior to the diagnosis of NEC. For instance, several studies have isolated *Clostridium perfringens*, *Klebsiella* spp., and *E. coli* (de la Cochetiere *et al.* 2004; Bell *et al.* 1979). Stewart *et al.* (2013), in their study including preterm twins at risk of NEC, demonstrated shared similar microbiome development between the twins suggesting shared genetic and immunomodulatory factors. Furthermore, the gut microbiota of the affected twin showed a reduction in diversity and increasing dominance of *Escherichia* spp. prior to the onset of NEC compared to the healthy twin. Furthermore, antibiotics had a substantial reduction in the proportion of *Escherichia* spp. and an increase in other *Enterobacteriaceae* (Stewart *et al.* 2013).

Gut microbiota manipulation as a therapeutic option in NEC has been explored in recent years. Recent evidence suggested that probiotics play a role in the prevention of NEC with oral administration of probiotic microorganisms, likely by altering the composition of the microbiota (Indrio and Neu 2011; Lin *et al.* 2005, 2008; Deshpande *et al.* 2007, 2010; Wang *et al.* 2012). Underwood and colleagues (2013) conducted an open dose-escalation and crossover trial where premature infants were randomly assigned to receive two strains of bifidobacteria, *Bifidobacterium longum* ssp. *infantis* (Binf), and *Bifidobacterium animalis* ssp. *lactis* (Blac). There was a greater increase in fecal *bifidobacteria* among those receiving Binf than those receiving Blac among both formula-fed and human milk-fed premature infants with increased microbial diversity over time per dose. The combination of human milk and Binf was the most effective at normalizing the fecal microbiota and resulted in the greatest fecal levels of bifidobacteria (Underwood *et al.* 2013). Based on available evidence, the American Pediatric Surgical Association has now supported the use of probiotics to prevent NEC in low birth weight infants (Downard *et al.* 2012). Fecal microbiota transplantation has not yet been trialed in premature infants to prevent NEC. However, Claud *et al.* (2013) suggest that there is a difference in microbial community structure taxonomy and function, at 3 weeks prior to the onset of NEC. Therefore, further characterization of the key microbial components and functions at critical time points such as this could present as a window of opportunity. As such, FMT has the potential to improve clinical outcomes for premature infants in the context of specific diseases such as NEC and further studies involving the manipulation of the intestinal microbiota in the pediatric population are warranted.

10.4.4.2. Colon Cancer

Gastrointestinal tract malignancies are a leading cause of mortality and morbidity; among which colorectal cancer is the second leading cause of cancer related mortality in the United States (Siegel *et al.* 2013). In addition, the incidence of colon cancer is on the rise and is higher than other intestinal malignancies (Jemal *et al.* 2009). Besides environmental factors, advances in sequencing technology have now identified microbial factors that play a role in the pathogenesis of colorectal carcinoma (CRC). As discussed earlier, the colon microbiota composed primarily of Gram-negative *Bacteroidetes* and Gram-positive *Firmicutes*, along with *Proteobacteria, Actinobacteria,* and *Fusobacteria,* is essential for metabolic homeostasis and immune regulation. Historically, pathogens associated with gastrointestinal malignancies were primarily *Streptococcus bovis,* now known as *Streptococcus gallolyticus, Helicobacter pylori,* and adherent invasive *E. coli* (Ma *et al.* 2012; Arthur *et al.* 2012; Boleij *et al.* 2011). However, recent studies with fecal and biopsy specimens from individuals with colon adenomas and colorectal cancer have demonstrated that multiple bacterial species contribute to the development of CRC. It is likely that colonic microbiota changes predispose the microenvironment towards inflammation, dysplasia, and ultimately, carcinoma. Overall, there is an increase in *Bacteroides* (Qin *et al.* 2010; Sobhani *et al.* 2011; Wu *et al.* 2013) and *Fusobacterium* (Wu *et al.* 2013 Castellarin *et al.* 2012; Kostic *et al.* 2012, 2013 2013) species and a reduction in butyrate producing bacteria (Weir *et al.* 2013; Qin *et al.* 2010; Wu *et al.* 2013) among fecal and biopsy samples from CRC (Grivennikov *et al.* 2012; Sobhani *et al.* 2011; Castellarin *et al.* 2012; Kostic *et al.* 2012; Marchesi *et al.* 2011) with an overall reduced bacterial diversity in cancer tissues (Chen *et al.* 2012). Mucosa-associated bacteria studies found that CRC tumor tissue had more numbers of *Coriobacteridae* (Chen *et al.* 2013; Marchesi *et al.* 2011). While the mechanism is unknown, there are a few postulated models for the mechanisms by which bacterial pathogens contribute to the pathogenesis of CRC. Of them, the "driver-passenger" model describes certain populations of bacteria as drivers of intestinal epithelium damage that initiate tumorigenesis thereby promoting the growth of opportunistic bacteria known as passengers. In contrast, the α "bug model" proposes virulent bacteria as α bugs initiating tumorigenesis and alteration of the gut microbiota, thereby creating helper bugs to dominate the colonic ecosystem.

Diet is an important factor associated with the development of CRC. Recent studies have therefore elucidated the association between diet, intestinal microbiota, and the risk for CRC (Chen *et al.* 2013; Ou *et al.* 2013). For instance, Western diets that are low in fiber generate lower levels of antitumorigenic butyrate producing bacteria and other SCFAs that are associated with an incidence of CRC. This trend is seen among cohorts at high risk for CRC, namely those on low-fiber diets, African

Americans consuming Western diets, and those with advanced colorectal adenoma (Ou *et al.* 2013; Chen *et al.* 2013). Western diets rich in meat, protein, and fat are metabolized into inflammatory and carcinogenic metabolites such as phenols, ammonia, branched-chain SCFAs, hydrogen sulfide, lithocholic acid, and deoxycholic acid (Wells and Hylemon 2000; Abreu and Peek 2014).

Alterations in the intestinal microbiota leading to chronic inflammation most likely promotes colon carcinogenesis given that individuals with inflammatory bowel disease are at an increased risk for cancers as a result of chronic inflammation induced dysplasia and carcinogenesis. It has also been demonstrated that mice given antibiotics have a reduced incidence and severity of colitis-associated cancer (Couturier-Maillard *et al.* 2013; Gianotti *et al.* 2010). The innate immune system has pattern recognition receptors (PRRs) that include Nod-like receptors and TLRs that recognize microbial products such as LPS, flagellin, and peptidoglycans. Several studies have shown that colon cancer tissues have increased expression of TLR2, 4, 7, 8, and 9 messenger RNAs (Moossavi and Rezaei 2013) and TLR4 deletion in mice has been shown to protect against CRC while overexpression results in increased colitis-associated carcinogenesis after administration of mutagenic and colitis inducing agents such as azocymethane and dextran sulfate sodium (Santaolalla *et al.* 2013; Fukata *et al.* 2007, 2011). Hence there is evidence that bacterial activation of PRRs result in inflammation, and that immune activation predisposes to colonic tumor growth (Grivennikov *et al.* 2012).

Enterotoxigenic *Bacteroides fragilis* and *Fusobacterium nucleatum* have been associated with CRC and have been found to be highly prevalent in this cohort. Animal studies have identified several mechanisms by which these species promote tumorigenesis by generating genotoxic compounds such as the metalloprotease *B. fragilis* toxin, and the *Fusobacterium* adhesin, FadA, which both have effects on E-cadherin signaling to activate β-catenin (Rubinstein *et al.* 2013; Wu *et al.* 2009). Infection of $APC^{min/+}$ mice with these species generates colorectal tumors. Other bacterial agents produce reactive oxygen and nitric oxide species that damage DNA thereby increasing the risk of CRC. For instance, *Enterococcus faecalis* produces extracellular superoxide and hydrogen peroxide, damaging colonic epithelial cell DNA (Huycke *et al.* 2002; Huycke and Moore 2002; Wang and Huycke 2007). Carroll *et al.* (2007) showed that *Lactobacillus gasseri* producing manganese superoxide dismutase had significant anti-inflammatory activity that reduced the severity of colitis in IL-10-deficient mice by suppressing reactive oxygen damage that may be beneficial for patients predisposed to CRC.

$APC^{min/+}$ mice are genetically engineered mice that carry a heterozygous mutation in the APC gene and develop spontaneous tumors. GF $APC^{min/+}$ mice, developed to elucidate the link between intestinal microbiota and CRC were found to develop fewer small intestinal colon polyps than the controls (Li *et al.* 2012). Further support of the role of microbes in intestinal neoplasia comes from infection of GF $Rag2^{-/-}$ or $IL10^{-/-}$ mice infected with *H. hepaticus* that are found to have colitis-associated cancer (Erdman *et al.* 2003; Nagamine *et al.* 2008) and this risk was reduced by mono-association with *Bacteroides vulgatus* (Uronis *et al.* 2009). The latter risk reduction role of *B. vulgatus* suggests that different microbial species have unique effects on carcinogenesis.

The evidence supporting the role of beneficial commensal bacteria in the prevention of CRC comes largely from laboratory-based studies and limited clinical trials using probiotics. For example, a small clinical trial including patients treated with *Lactobacillus casei* showed suppressed colorectal tumor growth after 2–4 years of treatment (Ishikawa *et al.* 2005). *Lactobacillus johnsonii* was found to have an effect in terms of decreasing the proportion of *Enterobacteriaceae* and modulating the intestinal immune response in patients with CRC, whereas *Bifidobacterium longum* was found to have no response (Gianotti *et al.* 2010). Some of the proposed mechanisms of the benefits of probiotic administration include reduction of oxidative stress, and reduction of apoptosis, as well as direct modulation of the intestinal composition that influences the homeostasis of gut metabolism and innate immune system (Yan and Polk 2002; Gill and Prasad 2008).

Although no clinical trials using FMT to treat GI malignancies have been performed in humans, some rationale for this approach can be gleaned from animal studies. Couturier-Maillard *et al.* (2013) and Riley *et al.* (2013) have shown that $Nod2^{-/-}$ mice have reduced susceptibility to tumors when treated with antibiotics or stool transplants from wild-type mice. In contrast, the tumorigenic potential was increased when the wild-type mice were colonized by dysbiotic $Nod2^{-/-}$ microbiota (Couturier-Maillard *et al.* 2013). These data lend support to the need for further studies identifying the specific agents potentiating the risk for CRC thereby circumventing the development of CRC by future applications of fecal microbiota transplantation.

10.4.4.3. Other GI Malignancies (Esophageal and Gastric Adenocarcinoma)

Up to 15% of the global malignancy cases can be attributed to infectious pathogens (de Martel et al. 2012). This lends support to the potential role of the intestinal microbiota in the pathogenesis of certain cancers. For instance, chronic infection with *H. pylori*, a Gram-negative bacterial species, is the main risk factor for gastric adenocarcinoma and is designated as a class I carcinogen by the World Health Organization. On the contrary, *H. pylori* carriage is associated with a significant risk reduction in the development of esophageal adenocarcinoma (Lagergren and Lagergren 2013).

The risk of gastric cancer as a result of *H. pylori* infection is influenced by the host genetic predisposition, and the interaction between the host and other environmental factors. In comparison to the colon, the gastric microbiota density is less and approximates to 10^1–10^3 colony forming units/gram (Sheh and Fox 2013). Infection with *H. pylori* reduces the diversity of the gastric microbiota such that the predominant phyla are found to be *Streptococcus*, *Prevotella*, and *Gemella* (Andersson et al. 2008). Whereas *H. pylori* negative individuals have richer gastric microbiota with the common phyla being *Firmicutes*, *Bacteroidetes*, and *Actinobacteria* (Cho and Blaser 2012; Bik et al. 2006). Maldonado-Contreras et al. (2011) found that *H. pylori* infection increased the proportion of *Proteobacteria*, *Spirochetes*, and *Acidobacteria*, while reducing the abundance of *Actinobacteria*, *Bacteriodetes*, and *Firmicutes*.

H. pylori features that have been found to influence carcinogenesis are identified as the cytotoxin associated gene (*cag*) pathogenicity island (PAI) that encodes antigenic effector proteins that have been associated with an increased risk of distal gastric cancer; as well as the secreted vacuolating cytotoxin A (VacA) toxin that results in changes to the gastric epithelial cells (Odenbreit et al. 2000; Fischer et al. 2001; Kwok et al. 2007; Shaffer et al. 2011; Blaser et al. 1995; Cover and Blanke 2005). VacA is a multifunctional bacterial toxin that contributes to *H. pylori* colonization and persistence *in vivo* leading to development of gastroduodenal disease. The toxin mediates its effects via induction of cellular vacuolation, reduction of the mitochondrial transmembrane potential, and release of cytochrome *c*, and its effects on cellular signal-transduction pathways and epithelial monolayer permeability. VacA is also known to modulate immune cell function by its effects on phagocytosis, antigen presentation, *T. lymphocytes*, and production of proinflammatory cytokines (TNF alpha, IL6) by mast cells (Cover and Blanke 2005). Furthermore, dietary factors such as iron depletion may accelerate gastric dysplasia by potentiating the *H. pylori* cytotoxin associated gene (*cag*) (Noto *et al.* 2013). Virulent *H. pylori* strains and host genetic factors such as IL1β gene polymorphisms that result in increased IL1β production are also associated with an increased risk of gastric cancer. Increased IL1β is associated with hypochlorhydria and gastric atrophy; the latter are classical features of histological progression in gastric cancer due to loss of parietal cells (Polk and Peek 2010; El-Omar *et al.* 2000; Figueiredo *et al.* 2002; Garza-González *et al.* 2005). Research by Ma *et al.* (2012) has suggested that the development of gastric cancer may not be limited to *H. pylori*. In the *H. pylori* seropositive cohort, there was a significant reduction in the incidence of gastric cancer 15 years after being treated with antimicrobial therapy; however only 47% of this cohort was found to have eradicated *H. pylori* (Ma *et al.* 2012). Nevertheless, certain strains of *H. pylori* can lead to distal gastric adenocarcinoma in Mongolian gerbil models even in the absence of carcinogens (Franco *et al.* 2005; Watanabe *et al.* 1998; Honda *et al.* 1998; Ogura *et al.* 2000). *Lactobacillus* is the predominant genus in the gastric microbiota of uninfected gerbils and mice and *H. pylori* colonization has been suggested to change the microbiota composition such that in *H. pylori* infected gerbils, the proportion of *Clostridium coccoides* becomes abundant (Sun *et al.* 2003; Aebischer *et al.* 2006). Of interest, Rolig *et al.* (2013) found that administration of antibiotics to *H. pylori* infected mice altered the gastric microbiota and decreased the severity of gastric inflammation, and these changes were then reversed when the gastric microbiota from antibiotic naïve mice was transferred to this cohort of mice. Lofgren *et al.* (2011), in their study with GF genetically susceptible mice known as INS-GAS mice, have shown that infection with *H. pylori* increased the abundance of *Firmicutes* and decreased the proportion of *Bacteroidetes*. However, there were no associated changes in the colonic microbiota. It is essential to recognize that the gastric microbiota composition and the features of *H. pylori* colonization among mouse models and humans are different. For instance, studies among rhesus monkeys have shown that infection with *H. pylori* does not alter the relative abundance of other taxa (Martin *et al.* 2013). As such, there is paucity of detailed molecular characterization of the differences in gastric microbiota among individuals with gastric cancer. Dicksved *et al.* (2009) characterized the microbiota of patients with gastric cancer via terminal restriction fragment length polymorphism and found no significant differences with the control group. However, their findings supported a preponderance of *Streptococcus*, *Lactobacillus*, *Veillonella*, and *Prevotella* among the gastric cancer microbiota (Dicksved *et al.* 2009).

Examination of the esophageal bacterial composition by Yang *et al.* (2009) identified that there are two distinct microbiome ecosystems designated as type I and type II. Type I is characterized by Gram-positive bacteria, predominantly *Firmicutes*, and was associated with a normal esophagus on endoscopy. In contrast, the type II microbiome mainly consists of Gram-negative bacteria with higher proportions of the phyla *Bacteroidetes*, *Proteobacteria*, *Fusobacteria*, and *Spirochaetes* and is associated with esophagitis and Barrett's esophagus, known to increase the risk for esophageal adenocarcinoma (Yang *et al.* 2009). Putative mechanisms for increased risk of esophageal carcinoma are proposed to be production of LPS by Gram-negative bacteria that activates the innate immune system Toll-like receptor 4, activating NF-κB, that is in turn found to be increased in esophageal samples from individuals with Barrett's esophagus, and esophageal adenocarcinoma (Yang *et al.* 2012). Furthermore, LPS is associated with delayed gastric emptying, increased levels of inflammatory cytokines such as IL-1b, IL-6, IL-8, and TNF-α, and downstream activation of inducible nitric oxide synthase which in turn lowers the lower esophageal sphincter tone potentiating the risk for gastroesophageal reflux disease (Yang *et al.* 2012; Fan *et al.* 2001; Calatayud *et al.* 2002).

Although there is abundant data suggesting that the microbiome composition can influence the pathogenesis of gastric and esophageal adenocarcinoma, the exact mechanism of these interactions remains unclear. Further characterization of the relationship between these bacterial communities can facilitate the administration of microbial therapeutics to select locations to ameliorate the risk of adenocarcinoma at these sites.

10.4.4.4. Liver Disease

Commensal microbial ecology is essential for metabolic homeostasis. Several studies have found associations between alterations in the intestinal microbiome composition and liver disease.

Nonalcoholic fatty liver disease (NAFLD) encompasses the entire spectrum of fatty liver disease in individuals without significant alcoholic consumption, and ranges from fatty liver to steatohepatitis and cirrhosis. NAFLD is associated with metabolic risk factors such as obesity, diabetes mellitus, and dyslipidemia. Nonalcoholic fatty liver (NAFL) and nonalcoholic steatohepatitis (NASH) are histologic categories of NAFLD, of which NASH requires evidence of hepatic steatosis and inflammation with hepatocyte injury with or without fibrosis. The worldwide prevalence of NAFLD ranges from 6.3–33% with prevalence of NASH ranging from 3–5% while the prevalence of NASH cirrhosis is unknown (Vernon et al. 2011; O'Shea et al. 2010). NASH portends an increased liver-related mortality rate given that it may histologically progress to cirrhotic-stage disease (Vernon et al. 2011, Musso et al. 2011). Individuals with NAFLD or obesity have a higher prevalence of small intestinal bacterial overgrowth and this seems to correlate with severity of steatosis (Miele et al. 2009; Sabaté et al. 2008). In particular, studies have found a lower proportion of the *Ruminococcaceae* family among patients with NAFLD or NASH compared to healthy individuals (Zhu et al. 2013; Raman et al. 2013). Furthermore, Zhu et al. (2013) found *Escherichia* as the abundant genus in the intestinal microbiota of obese children with NASH while the species *Clostridium coccoides* was higher among adults with NAFLD (Mouzaki et al. 2013).

The spectrum of alcoholic liver disease ranges from simple steatosis to cirrhosis. In 2003, 44% of all deaths from liver disease were attributed to alcohol (O'Shea et al. 2010). Continued alcohol use increased the risk of progression of alcoholic liver disease to cirrhosis to 30% (Teli et al. 1995; Yoon and Yi 2006). While data on the importance of the microbiome in liver disease is limited, there is some evidence of its role from mice placed on the Lieber DeCarli diet model or the Tsukamoto-French model, whereby mice are given intragastric ethanol infusions. These studies have shown that excessive alcohol intake is associated with intestinal bacterial overgrowth and dysbiosis, as characterized by reduced numbers of *Lactobacillus*, *Pediococcus*, *Leuconostoc*, and *Lactococcus*. In animal models and in small clinical trials, there is some evidence to support reduced features of alcoholic liver disease by the administration of probiotic *Lactobacillus* (Nanji et al. 1994, Forsyth et al. 2009; Wang et al. 2012; Kirpich et al. 2008). Mechanistically, both bacterial overgrowth and dysbiosis are integral factors in the pathogenesis of alcoholic liver disease in the context of increased intestinal permeability, with the dysbiosis perpetuating colonic inflammation and bacterial translocation (Henao-Mejia et al. 2012). Alcohol, synergistically with microbiota-derived products, promotes liver disease. On the other hand, the mechanism for the change in microbiota as a consequence of alcoholic liver disease is not entirely understood. It is possibly related to intestinal dysmotility, alterations in gastric acid secretion, and innate immune response (Wegener et al. 1991; Bode and Bode 1997). Llopis et al. (2014) have provided evidence for a causal role of intestinal microbiota in alcohol-induced inflammation. The authors have shown that transplantation of human fecal microbiota from a patient with severe alcoholic hepatitis into GF mice generated a higher inflammatory process in the liver, adipose tissue, mesenteric lymph nodes, and greater intestinal mucosa disruption compared to GF mice that received microbiota from the patient without alcoholic hepatitis. These authors also identified two *Clostridium* species that produced ethanol *in vitro* and that were systematically associated with liver inflammation (Llopis et al. 2014).

Cirrhosis, often referred to as end stage liver disease, is the eighth leading cause of death in the United States (Asrani et al. 2012). The intestinal microbiota of patients with cirrhosis depends on the etiology. For instance, alcoholic cirrhosis is associated with a significant increase in the *Prevotella*ceae family when compared to patients with hepatitis B cirrhosis or controls (Chen et al. 2011). Nevertheless, most patients with cirrhosis have intestinal bacterial overgrowth contributed by impaired small intestinal motility, alterations in bile flow and immunoglobulin A secretion, and compromised antimicrobial defense mediated by deficiency of Paneth cell defensins (Chen et al. 2011; Bauer et al. 2001, 2002; Chang et al. 1998; Teltschik et al. 2012; Schnabl and Brenner 2014). Bacterial translocation plays a pivotal role in the induction of localized and systemic inflammation that can lead to lethal infections such as spontaneous bacterial peritonitis and bacteremia in the context of leaky intestinal barriers and immune deficits (Arvaniti et al. 2010; Wiest et al. 2014). Small intestinal overgrowth is a major contributor to translocation and consequent systemic endotoxemia. For instance, high serum concentrations of microbial endotoxins have been observed in patients with NASH and children with NAFLD (Alisi et al. 2010) perpetuating hepatic inflammation. This has clinical relevance as seen by the role of a nonabsorbable antibiotic such as rifaximin in the reduction of endotoxemia and reduced severity of liver disease (Kalambokis and Tsianos 2012). Hepatic encephalopathy is a neurocognitive impairment seen among patients with decompensated cirrhosis. Among the several suggested mechanisms, one of them relates to the differences in the gut microbiota. Bajaj *et al.* (2012) found that cirrhotic patients with hepatic encephalopathy had mucosal microbiota with significantly lower genera of beneficial autochthonous such as *Dorea, Subdoligranulum, Incertae Sedis XIV, Blautia, Roseburia, Faecalibacterium*, and an overgrowth of pathogeneic genera such as *Enterococcus, Burkholderia*, and *Proteus* (Bajaj *et al.* 2012). Based on prior studies, nonabsorbable disaccharides such as lacitol and lactulose, and broad-spectrum antibiotics such as rifaximin, have been the mainstay of therapy for hepatic encephalopathy by virtue of their role in the reduction of neurotoxic ammonia levels. Despite mixed results for lactulose, it is still the first-line treatment; with clinical evidence supportive for the adjunctive use of rifaximin in severe overt hepatic encephalopathy (Rahimi and Rockey 2014).

The pathways of microbiota metabolism contributing to the development of NAFLD are being investigated with animal and human studies. Host and diet-induced changes in the intestinal microbiota contribute to the development of NAFLD and NASH. The predominant mechanism derived from studies thus far is that the intestinal microbiota regulates hepatic production of triglycerides by suppression of lipoprotein lipase inhibitors such as fasting-induced adipocyte factor (Fiaf) and its secretion from the small intestine, thereby resulting in continued expression of lipoprotein lipase and accumulation of triglycerides in the liver (Koliwad *et al.* 2009; Bäckhed *et al.* 2004, 2007). Backhed and colleagues (2004) analyzed circulating GF and conventionalized, normal, and Fiaf knockout mice and showed that suppression of Fiaf, a circulating lipoprotein lipase inhibitor is essential for the microbiota-induced deposition of triglycerides in adipocytes through effects mediated by transcription factors such as carbohydrate response element binding protein (ChREBP) and through transcriptional suppression of an intestinal epithelial gene encoding a circulating lipoprotein lipase (LPL) inhibitor. GF Fiaf$^{-/-}$ animals have reduced expression of genes encoding the peroxisomal proliferator activated receptor coactivator (Pgc-1α) and enzymes involved in fatty acid oxidation (Bäckhed *et al.* 2007). The gut microbiota also influences the production of inflammatory cytokines such as TNF alpha that might facilitate the pathogenesis of NAFLD (Henao-Mejia *et al.* 2012). Inflammasomes are cytoplasmic complexes that sense endogenous or exogenous pathogen-associated molecular patterns or damage-associated molecular patterns that facilitate cleavage of effector proinflammatory cytokines, and are composed of one of several NLR and PYHIN proteins such as NLRP1, NLRP3, NLRC4, and AIM2. Mouse models have shown that inflammasome-deficiency-associated changes in the gut microbiota are associated with accumulation of TLR4 and TLR9 agonists into the portal circulation leading to enhanced hepatic TNF alpha expression that leads to progression of NASH (Henao-Mejia et al. 2012). Metabolized dietary substrates influence the gut microbiota composition such that microbial conversion of dietary choline to trimethylamine contributes to NAFLD in susceptible mice (Dumas et al. 2006). The proposed mechanism for this effect is likely via triglyceride accumulation in hepatocytes, secondary to lower hepatic secretion of very low density lipoprotein (Cope et al. 2000; Hartmann et al. 2012; Jiang et al. 2005; Wang et al. 2011). Besides being associated with hepatic steatosis in rodents and humans, microbiota induced choline deficiency has also been associated with low levels of gamma-*Proteobacteria* and high levels of *Erysipelotrichi* (Lührs et al. 2002). Ethanol, a product of the intestinal microbiome, is found in higher concentrations in obese animals than lean controls, possibly causing triglyceride accumulation in hepatocytes and initiating inflammation via production of reactive oxygen species.

As such, the viability of FMT to treat liver disease is far from reality yet. Given the abundant literature support, however, the role of FMT holds significant promise warranting further research.

10.5. SUMMARY

With unparalleled cure rates and ability to circumvent the undesirable effects of repeated antibiotic exposure on microbial diversity, there is insurmountable evidence for a role of fecal microbiota transplantation in *C. difficile* infections. It is also apparent that disturbances in the delicate balance of the composition of the intestinal microbiota likely play a role in the pathogenesis of a variety of other gastrointestinal diseases. Within the realm of gastrointestinal diseases, the role of FMT as a therapeutic option has generated interest in inflammatory bowel disease, obesity, metabolic syndrome, liver disease, gastrointestinal malignancies, and necrotizing enterocolitis. As knowledge in the field progresses and the interactions between the host, intestinal microbiota, and various disease states become better understood, further development of FMT and "next generation" FMT-derived products will likely be utilized for an increasingly broad range of intestinal disorders where gut ecosystem dysfunction plays a major role.

10.6. REFERENCES

Aas, J., C.E. Gessert, and J.S. Bakken. 2003. Recurrent *Clostridium difficile* colitis: Case series involving 18 patients treated with donor stool administered via a nasogastric tube. *Clin Infect Dis 36* (5):580-5. doi: 10.1086/367657.

Abreu, M.T. and R.M. Peek. 2014. Gastrointestinal malignancy and the microbiome. *Gastroenterology 146* (6):1534-1546.e3. doi: 10.1053/j.gastro.2014.01.001.

Aebischer, T., A. Fischer, A. Walduck, C. Schlötelburg, M. Lindig, S. Schreiber, *et al.* 2006. Vaccination prevents *Helicobacter pylori*-induced alterations of the gastric flora in mice. *FEMS Immunol Med Microbiol 46* (2):221-9. doi: 10.1111/j.1574-695X.2005.00024.x.

Alisi, A., M. Manco, R. Devito, F. Piemonte, and V. Nobili. 2010. Endotoxin and plasminogen activator inhibitor-1 serum levels associated with nonalcoholic steatohepatitis in children. *J Pediatr Gastroenterol Nutr 50* (6):645-9. doi: 10.1097/MPG.0b013e3181c7bdf1.

Allen-Vercoe, E., G. Reid, N. Viner, G.B. Gloor, S. Hota, P. Kim, *et al.* 2012. A Canadian Working Group report on fecal microbial therapy: Microbial ecosystems therapeutics. *Can J Gastroenterol 26* (7):457-62.

Alvarez-Olmos, M.I. and R.A. Oberhelman. 2001. Probiotic agents and infectious diseases: A modern perspective on a traditional therapy. *Clin Infect Dis 32* (11):1567-76. doi: 10.1086/320518.

Amar, J., R. Burcelin, J.B. Ruidavets, P.D. Cani, J. Fauvel, M.C. Alessi, *et al.* 2008. Energy intake is associated with endotoxemia in apparently healthy men. *Am J Clin Nutr 87* (5):1219-23.

Amar, J., M. Serino, C. Lange, C. Chabo, J. Iacovoni, S. Mondot, *et al.* 2011. Involvement of tissue bacteria in the onset of diabetes in humans: Evidence for a concept. *Diabetologia 54* (12):3055-61. doi: 10.1007/s00125-011-2329-8.

Anderson, J.L., R.J. Edney, and K. Whelan. 2012. Systematic review: Faecal microbiota transplantation in the management of inflammatory bowel disease. *Aliment Pharmacol Ther 36* (6):503-16. doi: 10.1111/j.1365-2036.2012.05220.x.

Andersson, A.F., M. Lindberg, H. Jakobsson, F. Bäckhed, P. Nyrén, and L. Engstrand. 2008. Comparative analysis of human gut microbiota by barcoded pyrosequencing. *PLoS One 3* (7):e2836. doi: 10.1371/journal.pone.0002836.

Angelberger, S., W. Reinisch, A. Makristathis, C. Lichtenberger, C. Dejaco, P. Papay, *et al.* 2013. Temporal bacterial community dynamics vary among ulcerative colitis patients after fecal microbiota transplantation. *Am J Gastroenterol 108* (10):1620-30. doi: 10.1038/ajg.2013.257.

Appell, Rainer G. 2008. Christian Franz Paullini und die homöopathische Dreckapotheke. Allgemeine Homöopathische Zeitung.

Aroniadis, O.C. and L.J. Brandt. 2014. Intestinal microbiota and the efficacy of fecal microbiota transplantation in gastrointestinal disease. Gastroenterol Hepatol (N Y) 10 (4):230-7.

Arpaia, N., C. Campbell, X. Fan, S. Dikiy, J. van der Veeken, P. deRoos, *et al.* 2013. Metabolites produced by commensal bacteria promote peripheral regulatory T-cell generation. *Nature 504* (7480):451-5. doi: 10.1038/nature12726.

Arthur, J.C., E. Perez-Chanona, M. Mühlbauer, S. Tomkovich, J.M. Uronis, T.J. Fan, et al. 2012. Intestinal inflammation targets cancer-inducing activity of the microbiota. *Science 338* (6103):120-3. doi: 10.1126/science.1224820.

Arvaniti, V., G. D'Amico, G. Fede, P. Manousou, E. Tsochatzis, M. Pleguezuelo, and A.K. Burroughs. 2010. Infections in patients with cirrhosis increase mortality fourfold and should be used in determining prognosis. *Gastroenterology 139* (4):1246-56, 1256.e1-5. doi: 10.1053/j.gastro.2010.06.019.

Asrani, S.K., N.S. Asrani, D.K. Freese, S.D. Phillips, C.A. Warnes, J. Heimbach, and P.S. Kamath. 2012. Congenital heart disease and the liver. *Hepatology 56* (3):1160-9. doi: 10.1002/hep.25692.

Atarashi, K., T. Tanoue, K. Oshima, W. Suda, Y. Nagano, H. Nishikawa, et al. 2013. T_{reg} induction by a rationally selected mixture of Clostridia strains from the human microbiota. *Nature 500* (7461):232-6. doi: 10.1038/nature12331.

Atarashi, K., T. Tanoue, T. Shima, A. Imaoka, T. Kuwahara, Y. Momose, et al. 2011. Induction of colonic regulatory T cells by indigenous Clostridium species. *Science 331* (6015):337-41. doi: 10.1126/science.1198469.

Bajaj, J.S., P.B. Hylemon, J.M. Ridlon, D.M. Heuman, K. Daita, M.B. White, et al. 2012. Colonic mucosal microbiome differs from stool microbiome in cirrhosis and hepatic encephalopathy and is linked to cognition and inflammation. *Am J Physiol Gastrointest Liver Physiol 303* (6):G675-85. doi: 10.1152/ajpgi.00152.2012.

Bakken, J.S. 2009. Fecal bacteriotherapy for recurrent *Clostridium difficile* infection. *Anaerobe 15* (6):285-9. doi: 10.1016/j.anaerobe.2009.09.007.

Bakken, J.S., T. Borody, L.J. Brandt, J.V. Brill, D.C. Demarco, M.A. Franzos, et al. 2011. Treating *Clostridium difficile* infection with fecal microbiota transplantation. *Clin Gastroenterol Hepatol 9* (12):1044-9. doi: 10.1016/j.cgh.2011.08.014.

Bakken, J.S., P.M. Polgreen, S.E. Beekmann, F.X. Riedo, and J.A. Streit. 2013. Treatment approaches including fecal microbiota transplantation for recurrent *Clostridium difficile* infection (RCDI) among infectious disease physicians. *Anaerobe 24*:20-4. doi: 10.1016/j.anaerobe.2013.08.007.

Bartlett, J.G. and D.N. Gerding. 2008. Clinical recognition and diagnosis of *Clostridium difficile* infection. *Clin Infect Dis 46 Suppl 1*:S12-8. doi: 10.1086/521863.

Bauer, M.P., D.W. Notermans, B.H. van Benthem, J.S. Brazier, M.H. Wilcox, M. Rupnik, et al. 2011. *Clostridium difficile* infection in Europe: A hospital-based survey. *Lancet 377* (9759):63–73. doi: 10.1016/S0140-6736(10)61266-4.

Bauer, T.M., H. Schwacha, B. Steinbrückner, F.E. Brinkmann, A.K. Ditzen, J.J. Aponte, et al. 2002. Small intestinal bacterial overgrowth in human cirrhosis is associated with systemic endotoxemia. *Am J Gastroenterol 97* (9):2364–70. doi: 10.1111/j.1572-0241.2002.05791.x.

Bauer, T.M., B. Steinbrückner, F.E. Brinkmann, A.K. Ditzen, H. Schwacha, J.J. Aponte, et al. 2001. Small intestinal bacterial overgrowth in patients with cirrhosis: Prevalence and relation with spontaneous bacterial peritonitis. *Am J Gastroenterol 96* (10):2962–7. doi: 10.1111/j.1572-0241.2001.04668.x.

Bell, M.J., R.D. Feigin, and J.L. Ternberg. 1979. Changes in the incidence of necrotizing enterocolitis associated with variation of the gastrointestinal microflora in neonates. *Am J Surg 138* (5):629–31.

Benjamin, J.L., C.R. Hedin, A. Koutsoumpas, S.C. Ng, N.E. McCarthy, N.J. Prescott, et al. 2012. Smokers with active Crohn's disease have a clinically relevant dysbiosis of the gastrointestinal microbiota. *Inflamm Bowel Dis 18* (6):1092-100. doi: 10.1002/ibd.21864.

Bennet, J.D. and M. Brinkman. 1989. Treatment of ulcerative colitis by implantation of normal colonic flora. *Lancet 1* (8630):164.

Bergstrom, K.S., V. Kissoon-Singh, D.L. Gibson, C. Ma, M. Montero, H.P. Sham, et al. 2010. Muc2 protects against lethal infectious colitis by disassociating pathogenic and commensal bacteria from the colonic mucosa. *PLoS Pathog 6* (5):e1000902. doi: 10.1371/journal.ppat.1000902.

Bik, E.M., P.B. Eckburg, S.R. Gill, K.E. Nelson, E.A. Purdom, F. Francois, et al. 2006. Molecular analysis of the bacterial microbiota in the human stomach. *Proc Natl Acad Sci USA 103* (3):732-7. doi: 10.1073/pnas.0506655103.

Bjornvad, C.R., T. Thymann, N.E. Deutz, D.G. Burrin, S.K. Jensen, B.B. Jensen, et al. 2008. Enteral feeding induces diet-dependent mucosal dysfunction, bacterial proliferation, and necrotizing enterocolitis in preterm pigs on parenteral nutrition. *Am J Physiol Gastrointest Liver Physiol 295* (5):G1092-103. doi: 10.1152/ajpgi.00414.2007.

Björkström, M.V., L. Hall, S. Söderlund, E.G. Håkansson, S. Håkansson, and M. Domellöf. 2009. Intestinal flora in very low-birth weight infants. *Acta Paediatr 98* (11):1762-7. doi: 10.1111/j.1651-2227.2009.01471.x.

Blaser, M.J., G.I. Perez-Perez, H. Kleanthous, T.L. Cover, R.M. Peek, P.H. Chyou, et al. 1995. Infection with *Helicobacter pylori* strains possessing cagA is associated with an increased risk of developing adenocarcinoma of the stomach. *Cancer Res 55* (10):2111–5.

Bode, C. and J.C. Bode. 1997. Alcohol's role in gastrointestinal tract disorders. *Alcohol Health Res World 21* (1):76–83.

Boleij, A., M.M. van Gelder, D.W. Swinkels, and H. Tjalsma. 2011. Clinical Importance of Streptococcus gallolyticus infection among colorectal cancer patients: Systematic review and meta-analysis. *Clin Infect Dis 53* (9):870–8. doi: 10.1093/cid/cir609.

Borody, T.J. and J. Campbell. 2012. Fecal microbiota transplantation: Techniques, applications, and issues. *Gastroenterol Clin North Am 41* (4):781–803. doi: 10.1016/j.gtc.2012.08.008.

Borody, T.J., L. George, P. Andrews, S. Brandl, S. Noonan, P. Cole, et al. 1989. Bowel-flora alteration: A potential cure for inflammatory bowel disease and irritable bowel syndrome? *Med J Aust 150* (10):604.

Borody, T.J., E.F. Warren, S.M. Leis, R. Surace, O. Ashman, and S. Siarakas. 2004. Bacteriotherapy using fecal flora: Toying with human motions. *J Clin Gastroenterol 38* (6):475–83.

Borody, T.J., E.F. Warren, S. Leis, R. Surace, and O. Ashman. 2003. Treatment of ulcerative colitis using fecal bacteriotherapy. *J Clin Gastroenterol 37* (1):42–7.

Bouskra, D., C. Brézillon, M. Bérard, C. Werts, R. Varona, I.G. Boneca, and G. Eberl. 2008. Lymphoid tissue genesis induced by commensals through NOD1 regulates intestinal homeostasis. *Nature 456* (7221):507–10. doi: 10.1038/nature07450.

Brandt, L.J., T.J. Borody, and J. Campbell. 2011. Endoscopic fecal microbiota transplantation: "first-line" treatment for severe *Clostridium difficile* infection? *J Clin Gastroenterol* 45 (8):655–7. doi: 10.1097/MCG.0b013e3182257d4f.

Bäckhed, F., H. Ding, T. Wang, L.V. Hooper, G.Y. Koh, A. Nagy, et al. 2004. The gut microbiota as an environmental factor that regulates fat storage. *Proc Natl Acad Sci USA* 101 (44):15718–23. doi: 10.1073/pnas.0407076101.

Bäckhed, F., J.K. Manchester, C.F. Semenkovich, and J.I. Gordon. 2007. Mechanisms underlying the resistance to diet-induced obesity in germ-free mice. *Proc Natl Acad Sci USA* 104 (3):979–84. doi: 10.1073/pnas.0605374104.

Calatayud, S., E. García-Zaragozá, C. Hernández, E. Quintana, V. Felipo, J.V. Esplugues, and M.D. Barrachina. 2002. Downregulation of nNOS and synthesis of PGs associated with endotoxin-induced delay in gastric emptying. *Am J Physiol Gastrointest Liver Physiol* 283 (6):G1360-7. doi: 10.1152/ajpgi.00168.2002.

Callaway, T.R., T.S. Edrington, R.C. Anderson, R.B. Harvey, K.J. Genovese, C.N. Kennedy, et al. 2008. Probiotics, prebiotics and competitive exclusion for prophylaxis against bacterial disease. *Anim Health Res Rev* 9 (2):217–25. doi: 10.1017/S1466252308001540.

Cammarota, G., G. Ianiro, and A. Gasbarrini. 2014. Fecal Microbiota Transplantation for the Treatment of *Clostridium difficile* Infection: A Systematic Review. *J Clin Gastroenterol* 48 (8):693–702. doi: 10.1097/MCG.0000000000000046.

Cani, P.D., J. Amar, M.A. Iglesias, M. Poggi, C. Knauf, D. Bastelica, et all. 2007. Metabolic endotoxemia initiates obesity and insulin resistance. *Diabetes* 56 (7):1761–72. doi: 10.2337/db06-1491.

Cani, P.D. and N.M. Delzenne. 2007. Gut microflora as a target for energy and metabolic homeostasis. *Curr Opin Clin Nutr Metab Care* 10 (6):729–34. doi: 10.1097/MCO.0b013e3282efdebb.

Carlisle, E.M., V. Poroyko, M.S. Caplan, J.A. Alverdy, and D. Liu. 2011. Gram negative bacteria are associated with the early stages of necrotizing enterocolitis. *PLoS One* 6 (3):e18084. doi: 10.1371/journal.pone.0018084.

Carroll, I.M., J.M. Andrus, J.M. Bruno-Bárcena, T.R. Klaenhammer, H.M. Hassan, and D.S. Threadgill. 2007. Anti-inflammatory properties of Lactobacillus gasseri expressing manganese superoxide dismutase using the interleukin 10-deficient mouse model of colitis. *Am J Physiol Gastrointest Liver Physiol* 293 (4):G729-38. doi: 10.1152/ajpgi.00132.2007.

Carter, G.P., J.I. Rood, and D. Lyras. 2012. The role of toxin A and toxin B in the virulence of *Clostridium difficile*. *Trends Microbiol* 20 (1):21-9. doi: 10.1016/j.tim.2011.11.003.

Casellas, F., N. Borruel, M. Papo, F. Guarner, M. Antolín, S. Videla, and J.R. Malagelada. 1998. Antiinflammatory effects of enterically coated amoxicillin-clavulanic acid in active ulcerative colitis. *Inflamm Bowel Dis* 4 (1):1–5.

Castellarin, M., R.L. Warren, J.D. Freeman, L. Dreolini, M. Krzywinski, J. Strauss, et al. 2012. *Fusobacterium nucleatum* infection is prevalent in human colorectal carcinoma. *Genome Res* 22 (2):299–306. doi: 10.1101/gr.126516.111.

Chang, C.S., G.H. Chen, H.C. Lien, and H.Z. Yeh. 1998. Small intestine dysmotility and bacterial overgrowth in cirrhotic patients with spontaneous bacterial peritonitis. *Hepatology* 28 (5):1187–90. doi: 10.1002/hep.510280504.

Chang, J.Y., D.A. Antonopoulos, A. Kalra, A. Tonelli, W.T. Khalife, T.M. Schmidt, and V.B. Young. 2008. Decreased diversity of the fecal Microbiome in recurrent *Clostridium difficile*-associated diarrhea. *J Infect Dis 197* (3):435–8. doi: 10.1086/525047.

Chassaing, B. and A. Darfeuille-Michaud. 2011. The commensal microbiota and enteropathogens in the pathogenesis of inflammatory bowel diseases. *Gastroenterology 140* (6):1720–28. doi: 10.1053/j.gastro.2011.01.054.

Chen, H.M., Y.N. Yu, J.L. Wang, Y.W. Lin, X. Kong, C.Q. Yang, *et al.* 2013. Decreased dietary fiber intake and structural alteration of gut microbiota in patients with advanced colorectal adenoma. *Am J Clin Nutr 97* (5):1044–52. doi: 10.3945/ajcn.112.046607.

Chen, W., F. Liu, Z. Ling, X. Tong, and C. Xiang. 2012. Human intestinal lumen and mucosa-associated microbiota in patients with colorectal cancer. *PLoS One 7* (6):e39743. doi: 10.1371/journal.pone.0039743.

Chen, Y., F. Yang, H. Lu, B. Wang, D. Lei, Y. Wang, *et al.* Characterization of fecal microbial communities in patients with liver cirrhosis. *Hepatology 54* (2):562–72. doi: 10.1002/hep.24423.

Cho, I. and M.J. Blaser. 2012. The human microbiome: At the interface of health and disease. *Nat Rev Genet 13* (4):260–70. doi: 10.1038/nrg3182.

Cho, I., S. Yamanishi, L. Cox, B.A. Methé, J. Zavadil, K. Li, *et al.* 2012. Antibiotics in early life alter the murine colonic microbiome and adiposity. *Nature 488* (7413):621–6. doi: 10.1038/nature11400.

Chung, H., S.J. Pamp, J.A. Hill, N.K. Surana, S.M. Edelman, E.B. Troy, *et al.* 2012. Gut immune maturation depends on colonization with a host-specific microbiota. *Cell 149* (7):1578–93. doi: 10.1016/j.cell.2012.04.037.

Cilieborg, M.S., M. Boye, L. Mølbak, T. Thymann, and P.T. Sangild. 2011. Preterm birth and necrotizing enterocolitis alter gut colonization in pigs. *Pediatr Res 69* (1):10-6. doi: 10.1203/PDR.0b013e3181ff2a89.

Claesson, M.J., I.B. Jeffery, S. Conde, S.E. Power, E.M. O'Connor, S. Cusack, *et al.* 2012. Gut microbiota composition correlates with diet and health in the elderly. *Nature 488* (7410):178–84. doi: 10.1038/nature11319.

Clarke, T.B., K.M. Davis, E.S. Lysenko, A.Y. Zhou, Y. Yu, and J.N. Weiser. 2010. Recognition of peptidoglycan from the microbiota by Nod1 enhances systemic innate immunity. *Nat Med 16* (2):228–31. doi: 10.1038/nm.2087.

Claud, E.C., K.P. Keegan, J.M. Brulc, L. Lu, D. Bartels, E. Glass, *et al.* 2013. "Bacterial community structure and functional contributions to emergence of health or necrotizing enterocolitis in preterm infants." *Microbiome 1* (1):20. doi: 10.1186/2049-2618-1-20.

Claud, E.C. and W.A. Walker. 2001. Hypothesis: Inappropriate colonization of the premature intestine can cause neonatal necrotizing enterocolitis. *FASEB J 15* (8):1398-403.

Cohen, M.B. 2009. *Clostridium difficile* infections: Emerging epidemiology and new treatments." *J Pediatr Gastroenterol Nutr 48 Suppl 2*:S63-5. doi: 10.1097/MPG.0b013e3181a118c6.

Cohen, S.H., D.N. Gerding, S. Johnson, C.P. Kelly, V.G. Loo, L.C. McDonald, *et al.* 2010. Clinical practice guidelines for *Clostridium difficile* infection in adults: 2010 update by the society for healthcare epidemiology of America (SHEA) and the infectious diseases society of America (IDSA). *Infect Control Hosp Epidemiol 31* (5):431–55. doi: 10.1086/651706.

Coombes, J.L., K.R. Siddiqui, C.V. Arancibia-Cárcamo, J. Hall, C.M. Sun, Y. Belkaid, and F. Powrie. 2007. A functionally specialized population of mucosal CD103+ DCs induces Foxp3+ regulatory T cells via a TGF-beta and retinoic acid-dependent mechanism. *J Exp Med 204* (8):1757–64. doi: 10.1084/jem.20070590.

Cope, K., T. Risby, and A.M. Diehl. 2000. Increased gastrointestinal ethanol production in obese mice: Implications for fatty liver disease pathogenesis. *Gastroenterology 119* (5):1340–7.

Cotton, C.M. 2010. Early, prolonged use of postnatal antibiotics increased the risk of necrotising enterocolitis. *Arch Dis Child Educ Pract Ed 95* (3):94. doi: 10.1136/adc.2010.187732.

Couturier-Maillard, A., T. Secher, A. Rehman, S. Normand, A. De Arcangelis, R. Haesler, et al. 2013. NOD2-mediated dysbiosis predisposes mice to transmissible colitis and colorectal cancer. *J Clin Invest 123* (2):700–11. doi: 10.1172/JCI62236.

Cover, T.L. and S.R. Blanke. 2005. *Helicobacter pylori* VacA, a paradigm for toxin multifunctionality. *Nat Rev Microbiol 3* (4):320–32. doi: 10.1038/nrmicro1095.

Damman, C.J., S.I. Miller, C.M. Surawicz, and T.L. Zisman. 2012. The microbiome and inflammatory bowel disease: Is there a therapeutic role for fecal microbiota transplantation? *Am J Gastroenterol 107* (10):1452–9. doi: 10.1038/ajg.2012.93.

Darfeuille-Michaud, A., J. Boudeau, P. Bulois, C. Neut, A.L. Glasser, N. Barnich, et al. 2004. High prevalence of adherent-invasive *Escherichia coli* associated with ileal mucosa in Crohn's disease. *Gastroenterology 127* (2):412–21.

de la Cochetiere, M.F., H. Piloquet, C. des Robert, D. Darmaun, J.P. Galmiche, and J.C. Roze. 2004. Early intestinal bacterial colonization and necrotizing enterocolitis in premature infants: The putative role of Clostridium. *Pediatr Res 56* (3):366–70. doi: 10.1203/01.PDR.0000134251.45878.D5.

De Leon, L.M., J.B. Watson, and C.R. Kelly. 2013. Transient flare of ulcerative colitis after fecal microbiota transplantation for recurrent *Clostridium difficile* infection. *Clin Gastroenterol Hepatol 11* (8):1036–8. doi: 10.1016/j.cgh.2013.04.045.

de Martel, C., J. Ferlay, S. Franceschi, J. Vignat, F. Bray, D. Forman, and M. Plummer. 2012. Global burden of cancers attributable to infections in 2008: A review and synthetic analysis. *Lancet Oncol 13* (6):607–15. doi: 10.1016/S1470-2045(12)70137-7.

de Vos, W.M. and E.A. de Vos. 2012. Role of the intestinal microbiome in health and disease: From correlation to causation. *Nutr Rev 70 Suppl 1*:S45–56. doi: 10.1111/j.1753-4887.2012.00505.x.

de Vrese, M. and P.R. Marteau. 2007. Probiotics and prebiotics: Effects on diarrhea. *J Nutr 137* (3 Suppl 2):803S–11S.

Denning, T.L., Y.C. Wang, S.R. Patel, I.R. Williams, and B. Pulendran. 2007. Lamina propria macrophages and dendritic cells differentially induce regulatory and interleukin 17-producing T cell responses. *Nat Immunol 8* (10):1086–94. doi: 10.1038/ni1511.

Denève, C., C. Janoir, I. Poilane, C. Fantinato, and A. Collignon. 2009. New trends in

Clostridium difficile virulence and pathogenesis. *Int J Antimicrob Agents* 33 Suppl 1:S24-8. doi: 10.1016/S0924-8579(09)70012-3.

Deshpande, G., S. Rao, and S. Patole. 2007. Probiotics for prevention of necrotising enterocolitis in preterm neonates with very low birthweight: A systematic review of randomised controlled trials. *Lancet* 369 (9573):1614–20. doi: 10.1016/S0140-6736(07)60748-X.

Deshpande, G., S. Rao, S. Patole, and M. Bulsara. 2010. Updated meta-analysis of probiotics for preventing necrotizing enterocolitis in preterm neonates. *Pediatrics* 125 (5):921–30. doi: 10.1542/peds.2009-1301.

Dethlefsen, L. and D.A. Relman. 2011. Incomplete recovery and individualized responses of the human distal gut microbiota to repeated antibiotic perturbation. *Proc Natl Acad Sci USA* 108 Suppl 1:4554–61. doi: 10.1073/pnas.1000087107.

Dicksved, J., M. Lindberg, M. Rosenquist, H. Enroth, J. K. Jansson, and L. Engstrand. 2009. Molecular characterization of the stomach microbiota in patients with gastric cancer and in controls. *J Med Microbiol* 58 (Pt 4):509–16. doi: 10.1099/jmm.0.007302-0.

Doerner, K.C., F. Takamine, C.P. LaVoie, D.H. Mallonee, and P.B. Hylemon. 1997. Assessment of fecal bacteria with bile acid 7 alpha-dehydroxylating activity for the presence of bai-like genes. *Appl Environ Microbiol* 63 (3):1185–8.

Dominguez-Bello, M.G., M.J. Blaser, R.E. Ley, and R. Knight. 2011. Development of the human gastrointestinal microbiota and insights from high-throughput sequencing. *Gastroenterology* 140 (6):1713–9. doi: 10.1053/j.gastro.2011.02.011.

Donohoe, D.R., N. Garge, X. Zhang, W. Sun, T.M. O'Connell, M.K. Bunger, and S.J. Bultman. 2011. The microbiome and butyrate regulate energy metabolism and autophagy in the mammalian colon. *Cell Metab* 13 (5):517–26. doi: 10.1016/j.cmet.2011.02.018.

Downard, C.D., E. Renaud, S.D. St Peter, F. Abdullah, S. Islam, J.M. Saito, *et al.* 2012. Treatment of necrotizing enterocolitis: An American Pediatric Surgical Association Outcomes and Clinical Trials Committee systematic review. *J Pediatr Surg* 47 (11):2111–22. doi: 10.1016/j.jpedsurg.2012.08.011.

Dubberke, E. 2012. *Clostridium difficile* infection: The scope of the problem. *J Hosp Med* 7 Suppl 3:S1–4. doi: 10.1002/jhm.1916.

Dubberke, E.R. and M.A. Olsen. 2012. Burden of *Clostridium difficile* on the healthcare system. *Clin Infect Dis* 55 Suppl 2:S88–92. doi: 10.1093/cid/cis335.

Dumas, M.E., R.H. Barton, A. Toye, O. Cloarec, C. Blancher, A. Rothwell, *et al.* 2006. Metabolic profiling reveals a contribution of gut microbiota to fatty liver phenotype in insulin-resistant mice. *Proc Natl Acad Sci USA* 103 (33):12511-6. doi: 10.1073/pnas.0601056103.

Duncan, S.H., G.L. Hold, A. Barcenilla, C.S. Stewart, and H.J. Flint. 2002. Roseburia intestinalis sp. nov., a novel saccharolytic, butyrate-producing bacterium from human faeces. *Int J Syst Evol Microbiol* 52 (Pt 5):1615–20.

Eckel, R.H., K.G. Alberti, S.M. Grundy, and P.Z. Zimmet. 2010. The metabolic syndrome. *Lancet* 375 (9710):181–3. doi: 10.1016/S0140-6736(09)61794-3.

Eiseman, B., W. Silen, G.S. Bascom, and A.J. Kauvar. 1958. Fecal enema as an adjunct in the treatment of *pseudomembranous enterocolitis*. *Surgery* 44 (5):854–9.

El-Omar, E.M., M. Carrington, W.H. Chow, K.E. McColl, J.H. Bream, H.A. Young, et al. 2000. Interleukin-1 polymorphisms associated with increased risk of gastric cancer. *Nature* 404 (6776):398–402. doi: 10.1038/35006081.

Erdman, S.E., T. Poutahidis, M. Tomczak, A.B. Rogers, K. Cormier, B. Plank, et al. 2003. CD4+ CD25+ regulatory T lymphocytes inhibit microbially induced colon cancer in Rag2-deficient mice. *Am J Pathol* 162 (2):691–702. doi: 10.1016/S0002-9440(10)63863-1.

Erickson, A.R., B.L. Cantarel, R. Lamendella, Y. Darzi, E.F. Mongodin, C. Pan, et al. 2012. Integrated metagenomics/metaproteomics reveals human host-microbiota signatures of Crohn's disease. *PLoS One* 7 (11):e49138. doi: 10.1371/journal.pone.0049138.

Everard, A., C. Belzer, L. Geurts, J.P. Ouwerkerk, C. Druart, L.B. Bindels, et al. 2013. Cross-talk between *Akkermansia muciniphila* and intestinal epithelium controls diet-induced obesity. *Proc Natl Acad Sci USA* 110 (22):9066–71. doi: 10.1073/pnas.1219451110.

Eyre, D.W., A.S. Walker, D. Wyllie, K.E. Dingle, D. Griffiths, J. Finney, et al. 2012. Predictors of first recurrence of *Clostridium difficile* infection: Implications for initial management. *Clin Infect Dis* 55 Suppl 2:S77–87. doi: 10.1093/cid/cis356.

Fagarasan, S., M. Muramatsu, K. Suzuki, H. Nagaoka, H. Hiai, and T. Honjo. 2002. Critical roles of activation-induced cytidine deaminase in the homeostasis of gut flora. *Science* 298 (5597):1424–7. doi: 10.1126/science.1077336.

Fan, Y.P., S. Chakder, F. Gao, and S. Rattan. 2001. Inducible and neuronal nitric oxide synthase involvement in lipopolysaccharide-induced sphincteric dysfunction. *Am J Physiol Gastrointest Liver Physiol* 280 (1):G32–42.

Figueiredo, C., J.C. Machado, P. Pharoah, R. Seruca, S. Sousa, R. Carvalho, et al. 2002. *Helicobacter pylori* and interleukin 1 genotyping: An opportunity to identify high-risk individuals for gastric carcinoma. *J Natl Cancer Inst* 94 (22):1680–7.

Fischer, W., J. Püls, R. Buhrdorf, B. Gebert, S. Odenbreit, and R. Haas. 2001. Systematic mutagenesis of the *Helicobacter pylori* cag pathogenicity island: Essential genes for CagA translocation in host cells and induction of interleukin-8. *Mol Microbiol* 42 (5):1337–48.

Forsyth, C.B., A. Farhadi, S.M. Jakate, Y. Tang, M. Shaikh, and A. Keshavarzian. 2009. Lactobacillus GG treatment ameliorates alcohol-induced intestinal oxidative stress, gut leakiness, and liver injury in a rat model of alcoholic steatohepatitis. *Alcohol* 43 (2):163–72. doi: 10.1016/j.alcohol.2008.12.009.

Franchi, L., N. Kamada, Y. Nakamura, A. Burberry, P. Kuffa, S. Suzuki, et al. 2012. NLRC4-driven production of IL-1β discriminates between pathogenic and commensal bacteria and promotes host intestinal defense. *Nat Immunol* 13 (5):449–56. doi: 10.1038/ni.2263.

Franco, A.T., D.A. Israel, M.K. Washington, U. Krishna, J.G. Fox, A.B. Rogers, et al. 2005. Activation of beta-catenin by carcinogenic *Helicobacter pylori*. *Proc Natl Acad Sci USA* 102 (30):10646–51. doi: 10.1073/pnas.0504927102.

Frank, D.N., C.E. Robertson, C.M. Hamm, Z. Kpadeh, T. Zhang, H. Chen, et al. 2011. Disease phenotype and genotype are associated with shifts in intestinal-associated microbiota in inflammatory bowel diseases. *Inflamm Bowel Dis* 17 (1):179–84. doi: 10.1002/ibd.21339.

Frank, D.N., A.L. St Amand, R.A. Feldman, E.C. Boedeker, N. Harpaz, and N.R. Pace. 2007. Molecular-phylogenetic characterization of microbial community imbalances in human inflammatory bowel diseases. *Proc Natl Acad Sci USA 104* (34):13780–5. doi: 10.1073/pnas.0706625104.

Fukata, M., A. Chen, A.S. Vamadevan, J. Cohen, K. Breglio, S. Krishnareddy, *et al.* 2007. Toll-like receptor-4 promotes the development of colitis-associated colorectal tumors. *Gastroenterology 133* (6):1869–81. doi: 10.1053/j.gastro.2007.09.008.

Fukata, M., L. Shang, R. Santaolalla, J. Sotolongo, C. Pastorini, C. España, *et al.* 2011. Constitutive activation of epithelial TLR4 augments inflammatory responses to mucosal injury and drives colitis-associated tumorigenesis. *Inflamm Bowel Dis 17* (7):1464–73. doi: 10.1002/ibd.21527.

Furet, J.P., L.C. Kong, J. Tap, C. Poitou, A. Basdevant, J.L. Bouillot, *et al.* 2010. Differential adaptation of human gut microbiota to bariatric surgery-induced weight loss: Links with metabolic and low-grade inflammation markers. *Diabetes 59* (12):3049–57. doi: 10.2337/db10-0253.

Furusawa, Y., Y. Obata, S. Fukuda, T.A. Endo, G. Nakato, D. Takahashi, *et al.* 2013. Commensal microbe-derived butyrate induces the differentiation of colonic regulatory T cells. *Nature 504* (7480):446–50. doi: 10.1038/nature12721.

Gaboriau-Routhiau, V., S. Rakotobe, E. Lécuyer, I. Mulder, A. Lan, C. Bridonneau, *et al.* 2009. The key role of segmented filamentous bacteria in the coordinated maturation of gut helper T cell responses. *Immunity 31* (4):677–89. doi: 10.1016/j.immuni.2009.08.020.

Gao, Z., J. Yin, J. Zhang, R.E. Ward, R.J. Martin, M. Lefevre, *et al.* 2009. Butyrate improves insulin sensitivity and increases energy expenditure in mice. *Diabetes 58* (7):1509–17. doi: 10.2337/db08-1637.

Garborg, K., B. Waagsbø, A. Stallemo, J. Matre, and A. Sundøy. 2010. Results of faecal donor instillation therapy for recurrent *Clostridium difficile*-associated diarrhoea. *Scand J Infect Dis 42* (11–12):857–61. doi: 10.3109/00365548.2010.499541.

Garey, K.W., T.K. Dao-Tran, Z.D. Jiang, M.P. Price, L.O. Gentry, and H.L. Dupont. 2008. A clinical risk index for *Clostridium difficile* infection in hospitalised patients receiving broad-spectrum antibiotics. *J Hosp Infect 70* (2):142–7. doi: 10.1016/j.jhin.2008.06.026.

Garza-González, E., F.J. Bosques-Padilla, E. El-Omar, G. Hold, R. Tijerina-Menchaca, H.J. Maldonado-Garza, and G.I. Pérez-Pérez. 2005. Role of the polymorphic IL-1B, IL-1RN and TNF-A genes in distal gastric cancer in Mexico. *Int J Cancer 114* (2):237–41. doi: 10.1002/ijc.20718.

Geuking, M.B., J. Cahenzli, M.A. Lawson, D.C. Ng, E. Slack, S. Hapfelmeier, *et al.* 2011. Intestinal bacterial colonization induces mutualistic regulatory T cell responses. *Immunity 34* (5):794–806. doi: 10.1016/j.immuni.2011.03.021.

Gevers, D., S. Kugathasan, L.A. Denson, Y. Vázquez-Baeza, W. Van Treuren, B. Ren, *et al.* 2014. The treatment-naive microbiome in new-onset Crohn's disease. *Cell Host Microbe 15* (3):382–92. doi: 10.1016/j.chom.2014.02.005.

Gewolb, I.H., R.S. Schwalbe, V.L. Taciak, T.S. Harrison, and P. Panigrahi. 1999. Stool microflora in extremely low birthweight infants. *Arch Dis Child Fetal Neonatal Ed 80* (3):F167–73.

Gianotti, L., L. Morelli, F. Galbiati, S. Rocchetti, S. Coppola, A. Beneduce, et al. 2010. A randomized double-blind trial on perioperative administration of probiotics in colorectal cancer patients. *World J Gastroenterol 16* (2):167–75.

Gibson, G.R., H.M. Probert, J.V. Loo, R.A. Rastall, and M.B. Roberfroid. 2004. Dietary modulation of the human colonic microbiota: Updating the concept of prebiotics. *Nutr Res Rev 17* (2):259–75. doi: 10.1079/NRR200479.

Gilbert, J.A., J.K. Jansson, and R. Knight. 2014. The Earth Microbiome project: Successes and aspirations. *BMC Biol 12* (1):69. doi: 10.1186/s12915-014-0069-1.

Gill, H. and J. Prasad. 2008. Probiotics, immunomodulation, and health benefits. *Adv Exp Med Biol 606*:423–54. doi: 10.1007/978-0-387-74087-4_17.

Gordon, H.A., E. Bruckner-Kardoss, and B.S. Wostmann. 1966. Aging in germ-free mice: Life tables and lesions observed at natural death. *J Gerontol 21* (3):380–7.

Gough, E., H. Shaikh, and A.R. Manges. 2011. Systematic review of intestinal microbiota transplantation (fecal bacteriotherapy) for recurrent *Clostridium difficile* infection. *Clin Infect Dis 53* (10):994–1002. doi: 10.1093/cid/cir632.

Grivennikov, S.I., K. Wang, D. Mucida, C.A. Stewart, B. Schnabl, D. Jauch, et al. 2012. Adenoma-linked barrier defects and microbial products drive IL-23/IL-17-mediated tumour growth. *Nature 491* (7423):254–8. doi: 10.1038/nature11465.

Guo, B., C. Harstall, T. Louie, S. Veldhuyzen van Zanten, and L. A. Dieleman. 2012. Systematic review: Faecal transplantation for the treatment of *Clostridium difficile*-associated disease. *Aliment Pharmacol Ther 35* (8):865–75. doi: 10.1111/j.1365-2036.2012.05033.x.

Gupta, V. and R. Garg. 2009. Probiotics. *Indian J Med Microbiol 27* (3):202–9. doi: 10.4103/0255-0857.53201.

Guy, PR. Coprophagy in the African elephant (Loxadonta africana Blumenbach). *African Journal of Ecology*.

Hall, M.A., C.B. Cole, S.L. Smith, R. Fuller, and C.J. Rolles. 1990. Factors influencing the presence of faecal lactobacilli in early infancy. *Arch Dis Child 65* (2):185–8.

Hamada, H., T. Hiroi, Y. Nishiyama, H. Takahashi, Y. Masunaga, S. Hachimura, et al. 2002. Identification of multiple isolated lymphoid follicles on the antimesenteric wall of the mouse small intestine. *J Immunol 168* (1):57–64.

Hamm, E.E., D.E. Voth, and J.D. Ballard. 2006. Identification of *Clostridium difficile* toxin B cardiotoxicity using a zebrafish embryo model of intoxication. *Proc Natl Acad Sci USA 103* (38):14176–81. doi: 10.1073/pnas.0604725103.

Harrell, L., Y. Wang, D. Antonopoulos, V. Young, L. Lichtenstein, Y. Huang, et al. 2012. Standard colonic lavage alters the natural state of mucosal-associated microbiota in the human colon. *PLoS One 7* (2):e32545. doi: 10.1371/journal.pone.0032545.

Hartmann, P., W.C. Chen, and B. Schnabl. 2012. The intestinal microbiome and the leaky gut as therapeutic targets in alcoholic liver disease. *Front Physiol 3*:402. doi: 10.3389/fphys.2012.00402.

Henao-Mejia, J., E. Elinav, C. Jin, L. Hao, W.Z. Mehal, T. Strowig, et al. 2012. Inflammasome-mediated dysbiosis regulates progression of NAFLD and obesity. *Nature 482* (7384):179–85. doi: 10.1038/nature10809.

Henry, M.C. and R.L. Moss. 2009. Necrotizing enterocolitis. *Annu Rev Med 60*:111–24. doi: 10.1146/annurev.med.60.050207.092824.

Hintz, S.R., D.E. Kendrick, B.J. Stoll, B.R. Vohr, A.A. Fanaroff, E.F. Donovan, et al. 2005. Neurodevelopmental and growth outcomes of extremely low birth weight infants after necrotizing enterocolitis. *Pediatrics 115* (3):696–703. doi: 10.1542/peds.2004-0569.

Hirano, S., R. Nakama, M. Tamaki, N. Masuda, and H. Oda. 1981. Isolation and characterization of thirteen intestinal microorganisms capable of 7 alpha-dehydroxylating bile acids. *Appl Environ Microbiol 41* (3):737–45.

Hollister, E.B., C. Gao, and J. Versalovic. 2014. Compositional and functional features of the gastrointestinal microbiome and their effects on human health. *Gastroenterology 146* (6):1449–58. doi: 10.1053/j.gastro.2014.01.052.

Honda, S., T. Fujioka, M. Tokieda, R. Satoh, A. Nishizono, and M. Nasu. 1998. Development of *Helicobacter pylori*-induced gastric carcinoma in Mongolian gerbils. *Cancer Res 58* (19):4255-9.

Hoy, C.M., C.M. Wood, P.M. Hawkey, and J.W. Puntis. 2000. Duodenal microflora in very-low-birth-weight neonates and relation to necrotizing enterocolitis. *J Clin Microbiol 38* (12):4539–47.

Hoy, C., M.R. Millar, P. MacKay, P.G. Godwin, V. Langdale, and M.I. Levene. 1990. Quantitative changes in faecal microflora preceding necrotising enterocolitis in premature neonates. *Arch Dis Child 65* (10 Spec No):1057-9.

Hu, M.Y., K. Katchar, L. Kyne, S. Maroo, S. Tummala, V. Dreisbach, et al. 2009. Prospective derivation and validation of a clinical prediction rule for recurrent *Clostridium difficile* infection. *Gastroenterology 136* (4):1206–14. doi: 10.1053/j.gastro.2008.12.038.

Huycke, M.M., V. Abrams, and D.R. Moore. 2002. Enterococcus faecalis produces extracellular superoxide and hydrogen peroxide that damages colonic epithelial cell DNA. *Carcinogenesis 23* (3):529–36.

Huycke, M.M. and D.R. Moore. 2002. In vivo production of hydroxyl radical by Enterococcus faecalis colonizing the intestinal tract using aromatic hydroxylation. *Free Radic Biol Med 33* (6):818–26.

Indrio F. and J. Neu. 2011. The intestinal microbiome of infants and the use of probiotics. *Curr Opin Pediatr 23* (2):145–50. doi: 10.1097/MOP.0b013e3283444ccb.

Ishikawa, H., I. Akedo, T. Otani, T. Suzuki, T. Nakamura, I. Takeyama, et al. 2005. Randomized trial of dietary fiber and Lactobacillus casei administration for prevention of colorectal tumors. *Int J Cancer 116* (5):762–7. doi: 10.1002/ijc.21115.

Ivanov, I.I., K. Atarashi, N. Manel, E.L. Brodie, T. Shima, U. Karaoz, et al. 2009. Induction of intestinal Th17 cells by segmented filamentous bacteria. *Cell 139* (3):485–98. doi: 10.1016/j.cell.2009.09.033.

Ivanov, I.I., R. el Frutos, N. Manel, K. Yoshinaga, D.B. Rifkin, R.B. Sartor, et al. 2008. Specific microbiota direct the differentiation of IL-17-producing T-helper cells in the mucosa of the small intestine. *Cell Host Microbe 4* (4):337–49. doi: 10.1016/j.chom.2008.09.009.

Jalanka-Tuovinen, J., A. Salonen, J. Nikkilä, O. Immonen, R. Kekkonen, L. Lahti, et al. 2011. Intestinal microbiota in healthy adults: Temporal analysis reveals individual and common core and relation to intestinal symptoms. *PLoS One 6* (7):e23035. doi: 10.1371/journal.pone.0023035.

Jemal, A., R. Siegel, E. Ward, Y. Hao, J. Xu, and M.J. Thun. 2009. Cancer statistics, 2009. *CA Cancer J Clin* 59 (4):225–49. doi: 10.3322/caac.20006.

Jiang, X.C., Z. Li, R. Liu, X.P. Yang, M. Pan, L. Lagrost, *et al.* 2005. Phospholipid transfer protein deficiency impairs apolipoprotein-B secretion from hepatocytes by stimulating a proteolytic pathway through a relative deficiency of vitamin E and an increase in intracellular oxidants. *J Biol Chem* 280 (18):18336–40. doi: 10.1074/jbc.M500007200.

Jilling, T., D. Simon, J. Lu, F.J. Meng, D. Li, R. Schy, *et al.* 2006. The roles of bacteria and TLR4 in rat and murine models of necrotizing enterocolitis. *J Immunol* 177 (5):3273–82.

Jin, L.Z., R.R. Marquardt, and X. Zhao. 2000. A strain of Enterococcus faecium (18C23) inhibits adhesion of enterotoxigenic *Escherichia coli* K88 to porcine small intestine mucus. *Appl Environ Microbiol* 66 (10):4200–4.

Johansson, M.E., M. Phillipson, J. Petersson, A. Velcich, L. Holm, and G.C. Hansson. 2008. The inner of the two Muc2 mucin-dependent mucus layers in colon is devoid of bacteria. *Proc Natl Acad Sci USA* 105 (39):15064–9. doi: 10.1073/pnas.0803124105.

Johnson, S. 2009. Recurrent *Clostridium difficile* infection: Causality and therapeutic approaches. Int J Antimicrob Agents 33 Suppl 1:S33-6. doi: 10.1016/S0924-8579(09)70014-7.

Kahn, S.A., R. Gorawara-Bhat, and D.T. Rubin. 2012. Fecal bacteriotherapy for ulcerative colitis: Patients are ready, are we? *Inflamm Bowel Dis* 18 (4):676–84. doi: 10.1002/ibd.21775.

Kalambokis, G.N. and E.V. Tsianos. 2012. Rifaximin reduces endotoxemia and improves liver function and disease severity in patients with decompensated cirrhosis. *Hepatology* 55 (2):655–6. doi: 10.1002/hep.24751.

Kalliomäki, M., M.C. Collado, S. Salminen, and E. Isolauri. 2008. Early differences in fecal microbiota composition in children may predict overweight. *Am J Clin Nutr* 87 (3):534–8.

Kamada, N., G. Chen, and G. Núñez. 2012. A complex microworld in the gut: Harnessing pathogen-commensal relations. Nat Med 18 (8):1190–1. doi: 10.1038/nm.2900.

Kamada, N., T. Hisamatsu, S. Okamoto, T. Sato, K. Matsuoka, K. Arai, *et al.* 2005. Abnormally differentiated subsets of intestinal macrophage play a key role in Th1-dominant chronic colitis through excess production of IL-12 and IL-23 in response to bacteria. *J Immunol* 175 (10):6900–8.

Kamada, N., Y.G. Kim, H.P. Sham, B.A. Vallance, J.L. Puente, E.C. Martens, and G. Núñez. 2012. Regulated virulence controls the ability of a pathogen to compete with the gut microbiota. *Science* 336 (6086):1325–9. doi: 10.1126/science.1222195.

Kamada, N. and G. Núñez. 2013. Role of the gut microbiota in the development and function of lymphoid cells. *J Immunol* 190 (4):1389–95. doi: 10.4049/jimmunol.1203100.

Kane, M., L.K. Case, K. Kopaskie, A. Kozlova, C. MacDearmid, A.V. Chervonsky, and T.V. Golovkina. 2011. Successful transmission of a retrovirus depends on the commensal microbiota. *Science* 334 (6053):245–9. doi: 10.1126/science.1210718.

Kang, S., S.E. Denman, M. Morrison, Z. Yu, J. Dore, M. Leclerc, and C.S. McSweeney.

2010. Dysbiosis of fecal microbiota in Crohn's disease patients as revealed by a custom phylogenetic microarray. *Inflamm Bowel Dis 16* (12):2034–42. doi: 10.1002/ibd.21319.

Karlsson, F.H., V. Tremaroli, I. Nookaew, G. Bergström, C.J. Behre, B. Fagerberg, *et al.* 2013. Gut metagenome in European women with normal, impaired and diabetic glucose control. *Nature 498* (7452):99–103. doi: 10.1038/nature12198.

Kassam, Z., R. Hundal, J.K. Marshall, and C.H. Lee. 2012. Fecal transplant via retention enema for refractory or recurrent *Clostridium difficile* infection. *Arch Intern Med 172* (2):191–3. doi: 10.1001/archinte.172.2.191.

Kassam, Z., C.H. Lee, Y. Yuan, and R.H. Hunt. 2013. Fecal microbiota transplantation for *Clostridium difficile* infection: Systematic review and meta-analysis. *Am J Gastroenterol 108* (4):500–8. doi: 10.1038/ajg.2013.59.

Kelly, C.P. and J.T. LaMont. 2008. *Clostridium difficile*--more difficult than ever. *N Engl J Med 359* (18):1932–40. doi: 10.1056/NEJMra0707500.

Kelly, C.R., C. Ihunnah, M. Fischer, A. Khoruts, C. Surawicz, A. Afzali, *et al.* 2014. Fecal microbiota transplant for treatment of *Clostridium difficile* infection in immunocompromised patients. *Am J Gastroenterol 109* (7):1065–71. doi: 10.1038/ajg.2014.133.

Kelly, D., J.I. Campbell, T.P. King, G. Grant, E.A. Jansson, A.G. Coutts, *et al.* 2004. Commensal anaerobic gut bacteria attenuate inflammation by regulating nuclear-cytoplasmic shuttling of PPAR-gamma and RelA. *Nat Immunol 5* (1):104–12. doi: 10.1038/ni1018.

Kelsen, J.R., J. Kim, D. Latta, S. Smathers, K.L. McGowan, T. Zaoutis, *et al.* 2011. Recurrence rate of *Clostridium difficile* infection in hospitalized pediatric patients with inflammatory bowel disease. *Inflamm Bowel Dis 17* (1):50–5. doi: 10.1002/ibd.21421.

Khan, K.J., T.A. Ullman, A.C. Ford, M.T. Abreu, A. Abadir, J.K. Marshall, *et al.* 2011. Antibiotic therapy in inflammatory bowel disease: A systematic review and meta-analysis. *Am J Gastroenterol 106* (4):661–73. doi: 10.1038/ajg.2011.72.

Khoruts, A., J. Dicksved, J.K. Jansson, and M.J. Sadowsky. 2010. Changes in the composition of the human fecal microbiome after bacteriotherapy for recurrent *Clostridium difficile*-associated diarrhea. *J Clin Gastroenterol 44* (5):354–60. doi: 10.1097/MCG.0b013e3181c87e02.

Kirpich, I.A., N.V. Solovieva, S.N. Leikhter, N.A. Shidakova, O.V. Lebedeva, P.I. Sidorov, *et al.* 2008. Probiotics restore bowel flora and improve liver enzymes in human alcohol-induced liver injury: A pilot study. *Alcohol 42* (8):675–82. doi: 10.1016/j.alcohol.2008.08.006.

Kitahara, M., F. Takamine, T. Imamura, and Y. Benno. 2000. Assignment of Eubacterium sp. VPI 12708 and related strains with high bile acid 7alpha-dehydroxylating activity to Clostridium scindens and proposal of Clostridium hylemonae sp. nov., isolated from human faeces. *Int J Syst Evol Microbiol 50 Pt 3*:971–8.

Kitajima, S., M. Morimoto, E. Sagara, C. Shimizu, and Y. Ikeda. 2001. Dextran sodium sulfate-induced colitis in germ-free IQI/Jic mice. *Exp Anim 50* (5):387–95.

Koliwad, S.K., T. Kuo, L.E. Shipp, N.E. Gray, F. Backhed, A.Y. So, *et al.* 2009. Angiopoietin-like 4 (ANGPTL4, fasting-induced adipose factor) is a direct glucocorticoid receptor target and participates in glucocorticoid-regulated triglyceride metabolism. *J Biol Chem* 284 (38):25593–601. doi: 10.1074/jbc.M109.025452.

Kong, L.C., J. Tap, J. Aron-Wisnewsky, V. Pelloux, A. Basdevant, J.L. Bouillot, *et al.* 2013. Gut microbiota after gastric bypass in human obesity: Increased richness and associations of bacterial genera with adipose tissue genes. *Am J Clin Nutr* 98 (1):16–24. doi: 10.3945/ajcn.113.058743.

Kostic, A.D., E. Chun, L. Robertson, J.N. Glickman, C.A. Gallini, M. Michaud, *et al.* 2013. *Fusobacterium nucleatum* potentiates intestinal tumorigenesis and modulates the tumor-immune microenvironment. *Cell Host Microbe* 14 (2):207–15. doi: 10.1016/j.chom.2013.07.007.

Kostic, A.D., D. Gevers, C.S. Pedamallu, M. Michaud, F. Duke, A.M. Earl, *et al.* 2012. Genomic analysis identifies association of Fusobacterium with colorectal carcinoma. *Genome Res* 22 (2):292–8. doi: 10.1101/gr.126573.111.

Kump, P.K., H.P. Gröchenig, S. Lackner, S. Trajanoski, G. Reicht, K.M. Hoffmann, *et al.* 2013. Alteration of intestinal dysbiosis by fecal microbiota transplantation does not induce remission in patients with chronic active ulcerative colitis. *Inflamm Bowel Dis* 19 (10):2155–65. doi: 10.1097/MIB.0b013e31829ea325.

Kunde, S., A. Pham, S. Bonczyk, T. Crumb, M. Duba, H. Conrad, *et al.* 2013. Safety, tolerability, and clinical response after fecal transplantation in children and young adults with ulcerative colitis. *J Pediatr Gastroenterol Nutr* 56 (6):597–601. doi: 10.1097/MPG.0b013e318292fa0d.

Kuppala, V.S., J. Meinzen-Derr, A.L. Morrow, and K.R. Schibler. 2011. Prolonged initial empirical antibiotic treatment is associated with adverse outcomes in premature infants. *J Pediatr* 159 (5):720–5. doi: 10.1016/j.jpeds.2011.05.033.

Kwok, T., D. Zabler, S. Urman, M. Rohde, R. Hartig, S. Wessler, *et al.* 2007. Helicobacter exploits integrin for type IV secretion and kinase activation. *Nature* 449 (7164):862–6. doi: 10.1038/nature06187.

Lagergren, J. and P. Lagergren. 2013. Recent developments in esophageal adenocarcinoma. *CA Cancer J Clin* 63 (4):232–48. doi: 10.3322/caac.21185.

Larsen, N., F.K. Vogensen, F.W. van den Berg, D.S. Nielsen, A.S. Andreasen, B.K. *et al.* 2010. Gut microbiota in human adults with type 2 diabetes differs from non-diabetic adults. *PLoS One* 5 (2):e9085. doi: 10.1371/journal.pone.0009085.

LaTuga, M.S., J.C. Ellis, C.M. Cotton, R.N. Goldberg, J.L. Wynn, R.B. Jackson, and P.C. Seed. 2011. Beyond bacteria: A study of the enteric microbial consortium in extremely low birth weight infants. *PLoS One* 6 (12):e27858. doi: 10.1371/journal.pone.0027858.

Lawley, T.D., S. Clare, A.W. Walker, M.D. Stares, T.R. Connor, C. Raisen, *et al.* 2012. Targeted restoration of the intestinal microbiota with a simple, defined bacteriotherapy resolves relapsing *Clostridium difficile* disease in mice. *PLoS Pathog* 8 (10):e1002995. doi: 10.1371/journal.ppat.1002995.

Le Chatelier, E., T. Nielsen, J. Qin, E. Prifti, F. Hildebrand, G. Falony, *et al.* 2013. Richness of human gut microbiome correlates with metabolic markers. *Nature* 500 (7464):541–6. doi: 10.1038/nature12506.

Lehrer, S. 2013. Duodenal infusion of feces for recurrent *Clostridium difficile*. *N Engl J Med* 368 (22):2144. doi: 10.1056/NEJMc1303919#SA4.

Ley, R.E., F. Bäckhed, P. Turnbaugh, C.A. Lozupone, R.D. Knight, and J.I. Gordon. 2005. Obesity alters gut microbial ecology. *Proc Natl Acad Sci USA 102* (31):11070–5. doi: 10.1073/pnas.0504978102.

Ley, R.E., C.A. Lozupone, M. Hamady, R. Knight, and J.I. Gordon. 2008. Worlds within worlds: Evolution of the vertebrate gut microbiota. *Nat Rev Microbiol 6* (10):776–88. doi: 10.1038/nrmicro1978.

Ley, R.E., P.J. Turnbaugh, S. Klein, and J.I. Gordon. 2006. Microbial ecology: Human gut microbes associated with obesity. *Nature 444* (7122):1022–3. doi: 10.1038/4441022a.

Li, Y., P. Kundu, S.W. Seow, C.T. de Matos, L. Aronsson, K.C. Chin, *et al.* 2012. Gut microbiota accelerate tumor growth via c-jun and STAT3 phosphorylation in APC-Min/+ mice. *Carcinogenesis 33* (6):1231–8. doi: 10.1093/carcin/bgs137.

Lin, H.C., C.H. Hsu, H.L. Chen, M.Y. Chung, J.F. Hsu, R.I. Lien, *et al.* 2008. Oral probiotics prevent necrotizing enterocolitis in very low birth weight preterm infants: A multicenter, randomized, controlled trial. *Pediatrics 122* (4):693–700. doi: 10.1542/peds.2007-3007.

Lin, H.C., B.H. Su, A.C. Chen, T.W. Lin, C.H. Tsai, T.F. Yeh, and W. Oh. 2005. Oral probiotics reduce the incidence and severity of necrotizing enterocolitis in very low birth weight infants. *Pediatrics 115* (1):1–4. doi: 10.1542/peds.2004-1463.

Lin, P.W., T.R. Nasr, and B.J. Stoll. 2008. Necrotizing enterocolitis: Recent scientific advances in pathophysiology and prevention. *Semin Perinatol 32* (2):70–82. doi: 10.1053/j.semperi.2008.01.004.

Littman, D.R. and A.Y. Rudensky. 2010. Th17 and regulatory T cells in mediating and restraining inflammation. *Cell 140* (6):845–58. doi: 10.1016/j.cell.2010.02.021.

Llopis, M., A.M. Cassard-Doulcier, L. Boschat, A. Bruneau, F. Cailleux, S. Rabot, *et al.* 2014. Intestinal Dysbiosis explains inter-individual differences in susceptibility to alcoholic liver disease. The International Liver Congress, London.

Llopis, M., M. Antolin, M. Carol, N. Borruel, F. Casellas, C. Martinez, *et al.* 2009. Lactobacillus casei downregulates commensals' inflammatory signals in Crohn's disease mucosa. *Inflamm Bowel Dis 15* (2):275–83. doi: 10.1002/ibd.20736.

Lofgren, J.L., M.T. Whary, Z. Ge, S. Muthupalani, N.S. Taylor, M. Mobley, *et al.* 2011. Lack of commensal flora in *Helicobacter pylori*-infected INS-GAS mice reduces gastritis and delays intraepithelial neoplasia. *Gastroenterology 140* (1):210–20. doi: 10.1053/j.gastro.2010.09.048.

Loftus, E.V. 2004. Clinical epidemiology of inflammatory bowel disease: Incidence, prevalence, and environmental influences. *Gastroenterology 126* (6):1504–17.

Loo, V.G., L. Poirier, M.A. Miller, M. Oughton, M.D. Libman, S. Michaud, *et al.* 2005. A predominantly clonal multi-institutional outbreak of *Clostridium difficile*-associated diarrhea with high morbidity and mortality. *N Engl J Med 353* (23):2442–9. doi: 10.1056/NEJMoa051639.

Looft, T. and H.K. Allen. 2012. Collateral effects of antibiotics on mammalian gut microbiomes. *Gut Microbes 3* (5):463–7. doi: 10.4161/gmic.21288.

Lund-Tønnesen, S., A. Berstad, A. Schreiner, and T. Midtvedt. 1998. *Clostridium difficile*-associated diarrhea treated with homologous faeces. *Tidsskr Nor Laegeforen 118* (7):1027–30.

Luzza, F., T. Parrello, G. Monteleone, L. Sebkova, M. Romano, R. Zarrilli, et al. 2000. Up-regulation of IL-17 is associated with bioactive IL-8 expression in *Helicobacter pylori*-infected human gastric mucosa. *J Immunol 165* (9):5332–7.

Lührs, H., T. Gerke, J.G. Müller, R. Melcher, J. Schauber, F. Boxberge, et al. 2002. Butyrate inhibits NF-κB activation in lamina propria macrophages of patients with ulcerative colitis. *Scand J Gastroenterol 37* (4):458–66.

Ma, J.L., L. Zhang, L.M. Brown, J.Y. Li, L. Shen, K.F. Pan, et al. 2012. Fifteen-year effects of *Helicobacter pylori*, garlic, and vitamin treatments on gastric cancer incidence and mortality. *J Natl Cancer Inst 104* (6):488–92. doi: 10.1093/jnci/djs003.

Mai, V., C.M. Young, M. Ukhanova, X. Wang, Y. Sun, G. Casella, et al. 2011. Fecal microbiota in premature infants prior to necrotizing enterocolitis. *PLoS One 6* (6):e20647. doi: 10.1371/journal.pone.0020647.

Maldonado-Contreras, A., K.C. Goldfarb, F. Godoy-Vitorino, U. Karaoz, M. Contreras, M.J. Blaser, et al. 2011. Structure of the human gastric bacterial community in relation to *Helicobacter pylori* status. *ISME J 5* (4):574–9. doi: 10.1038/ismej.2010.149.

Marchesi, J.R., B.E. Dutilh, N. Hall, W.H. Peters, R. Roelofs, A. Boleij, and H. Tjalsma. 2011. Towards the human colorectal cancer microbiome. *PLoS One 6* (5):e20447. doi: 10.1371/journal.pone.0020447.

Markowitz, J.E., P. Mamula, J.F. delRosario, R.N. Baldassano, J.D. Lewis, A.F. Jawad, et al. 2004. Patterns of complementary and alternative medicine use in a population of pediatric patients with inflammatory bowel disease. *Inflamm Bowel Dis 10* (5):599–605.

Marteau, P., E. Cuillerier, S. Meance, M.F. Gerhardt, A. Myara, M. Bouvier, et al. 2002. Bifidobacterium animalis strain DN-173 010 shortens the colonic transit time in healthy women: A double-blind, randomized, controlled study. *Aliment Pharmacol Ther 16* (3):587–93.

Martin, C.R. and W.A. Walker. 2006. Intestinal immune defences and the inflammatory response in necrotising enterocolitis. *Semin Fetal Neonatal Med 11* (5):369–77. doi: 10.1016/j.siny.2006.03.002.

Martin, M.E., S. Bhatnagar, M.D. George, B.J. Paster, D.R. Canfield, J.A. Eisen, and J.V. Solnick. 2013. The impact of *Helicobacter pylori* infection on the gastric microbiota of the rhesus macaque. *PLoS One 8* (10):e76375. doi: 10.1371/journal.pone.0076375.

Maslowski, K.M., A.T. Vieira, A. Ng, J. Kranich, F. Sierro, D. Yu, et al. 2009. Regulation of inflammatory responses by gut microbiota and chemoattractant receptor GPR43. *Nature 461* (7268):1282–6. doi: 10.1038/nature08530.

Mattila, E., R. Uusitalo-Seppälä, M. Wuorela, L. Lehtola, H. Nurmi, M. Ristikankare, et al. 2012. Fecal transplantation, through colonoscopy, is effective therapy for recurrent *Clostridium difficile* infection. *Gastroenterology 142* (3):490–6. doi: 10.1053/j.gastro.2011.11.037.

Mazmanian, S.K., C.H. Liu, A.O. Tzianabos, and D.L. Kasper. 2005. An immunomodulatory molecule of symbiotic bacteria directs maturation of the host immune system. *Cell 122* (1):107–18. doi: 10.1016/j.cell.2005.05.007.

Mazmanian, S.K., J.L. Round, and D.L. Kasper. 2008. A microbial symbiosis factor prevents intestinal inflammatory disease. *Nature 453* (7195):620–5. doi: 10.1038/nature07008.

McDonald, L.C., G.E. Killgore, A. Thompson, R.C. Owens, S.V. Kazakova, S.P. Sambol, *et al.* 2005. An epidemic, toxin gene-variant strain of *Clostridium difficile*. *N Engl J Med 353* (23):2433–41. doi: 10.1056/NEJMoa051590.

McFarland, L.V. 2005. Alternative treatments for *Clostridium difficile* disease: What really works? *J Med Microbiol 54* (Pt 2):101–11.

McGovern, K. 2013. *Approach to the adult horse with chronic diarrhea.* Livestock.

Medellin-Peña, M.J., H. Wang, R. Johnson, S. Anand, and M.W. Griffiths. 2007. Probiotics affect virulence-related gene expression in *Escherichia coli* O157:H7. *Appl Environ Microbiol 73* (13):4259–67. doi: 10.1128/AEM.00159-07.

Merrigan, M., S. Sambol, S. Johnson, and D.N. Gerding. 2003. Susceptibility of hamsters to human pathogenic *Clostridium difficile* strain B1 following clindamycin, ampicillin or ceftriaxone administration. *Anaerobe 9* (2):91–5. doi: 10.1016/S1075-9964(03)00063-5.

Metchnikoff, E. 1908. *The prolongation of life: Optimistic Studies.* New York: Putman's Sons.

Miele, L., V. Valenza, G. La Torre, M. Montalto, G. Cammarota, R. Ricci, *et al.* 2009. Increased intestinal permeability and tight junction alterations in nonalcoholic fatty liver disease. *Hepatology 49* (6):1877–87. doi: 10.1002/hep.22848.

Millar, M.R., C.J. Linton, A. Cade, D. Glancy, M. Hall, and H. Jalal. 1996. Application of 16S rRNA gene PCR to study bowel flora of preterm infants with and without necrotizing enterocolitis. *J Clin Microbiol 34* (10):2506–10.

Moore, P., L. Kyne, A. Martin, and K. Solomon. 2013. Germination efficiency of clinical *Clostridium difficile* spores and correlation with ribotype, disease severity and therapy failure. *J Med Microbiol 62* (Pt 9):1405–13. doi: 10.1099/jmm.0.056614-0.

Moossavi, S. and N. Rezaei. 2013. Toll-like receptor signalling and their therapeutic targeting in colorectal cancer. *Int Immunopharmacol 16* (2):199–209. doi: 10.1016/j.intimp.2013.03.017.

Moreno-Navarrete, J.M., F. Ortega, M. Serino, E. Luche, A. Waget, G. Pardo, *et al.* 2012. Circulating lipopolysaccharide-binding protein (LBP) as a marker of obesity-related insulin resistance. *Int J Obes (Lond) 36* (11):1442–9. doi: 10.1038/ijo.2011.256.

Morgan, X.C., T.L. Tickle, H. Sokol, D. Gevers, K.L. Devaney, D.V. Ward, *et al.* 2012. Dysfunction of the intestinal microbiome in inflammatory bowel disease and treatment. *Genome Biol 13* (9):R79. doi: 10.1186/gb-2012-13-9-r79.

Mouzaki, M., E.M. Comelli, B.M. Arendt, J. Bonengel, S.K. Fung, S.E. Fischer, *et al.* 2013. Intestinal microbiota in patients with nonalcoholic fatty liver disease. *Hepatology 58* (1):120–7. doi: 10.1002/hep.26319.

Mshvildadze, M., J. Neu, and V. Mai. 2008. Intestinal microbiota development in the premature neonate: Establishment of a lasting commensal relationship? *Nutr Rev 66* (11):658–63. doi: 10.1111/j.1753-4887.2008.00119.x.

Mshvildadze, M., J. Neu, J. Shuster, D. Theriaque, N. Li, and V. Mai. 2010. Intestinal microbial ecology in premature infants assessed with non-culture-based techniques. *J Pediatr 156* (1):20–5. doi: 10.1016/j.jpeds.2009.06.063.

Mucida, D., Y. Park, G. Kim, O. Turovskaya, I. Scott, M. Kronenberg, and H. Cheroutre. 2007. Reciprocal TH17 and regulatory T cell differentiation mediated by retinoic acid. *Science 317* (5835):256–60. doi: 10.1126/science.1145697.

Musemeche, C.A., A.M. Kosloske, S.A. Bartow, and E.T. Umland. 1986. Comparative effects of ischemia, bacteria, and substrate on the pathogenesis of intestinal necrosis. *J Pediatr Surg 21* (6):536–8.

Musso, G., R. Gambino, M. Cassader, and G. Pagano. 2011. Meta-analysis: Natural history of non-alcoholic fatty liver disease (NAFLD) and diagnostic accuracy of non-invasive tests for liver disease severity. *Ann Med 43* (8):617–49. doi: 10.3109/07853890.2010.518623.

Nagamine, C.M., A.B. Rogers, J.G. Fox, and D.B. Schauer. 2008. Helicobacter hepaticus promotes azoxymethane-initiated colon tumorigenesis in BALB/c-IL10-deficient mice. *Int J Cancer 122* (4):832–8. doi: 10.1002/ijc.23175.

Nanji, A.A., U. Khettry, and S.M. Sadrzadeh. 1994. Lactobacillus feeding reduces endotoxemia and severity of experimental alcoholic liver (disease). *Proc Soc Exp Biol Med 205* (3):243–7.

Neu, J. and W.A. Walker. 2011. Necrotizing enterocolitis. *N Engl J Med 364* (3):255–64. doi: 10.1056/NEJMra1005408.

Nieuwdorp, M., P.W. Gilijamse, N. Pai, and L.M. Kaplan. 2014. Role of the microbiome in energy regulation and metabolism. *Gastroenterology 146* (6):1525–33. doi: 10.1053/j.gastro.2014.02.008.

Normann, E., A. Fahlén, L. Engstrand, and H.E. Lilja. 2013. Intestinal microbial profiles in extremely preterm infants with and without necrotizing enterocolitis. *Acta Paediatr 102* (2):129–36. doi: 10.1111/apa.12059.

Noto, J.M., J.A. Gaddy, J.Y. Lee, M.B. Piazuelo, D.B. Friedman, D.C. Colvin, et al. 2013. Iron deficiency accelerates *Helicobacter pylori*-induced carcinogenesis in rodents and humans. *J Clin Invest 123* (1):479–92. doi: 10.1172/JCI64373.

O'Shea, R.S., S. Dasarathy, and A.J. McCullough. 2010. Alcoholic liver disease. *Am J Gastroenterol 105* (1):14–32; quiz 33. doi: 10.1038/ajg.2009.593.

Odenbreit, S., J. Püls, B. Sedlmaier, E. Gerland, W. Fischer, and R. Haas. 2000. Translocation of *Helicobacter pylori* CagA into gastric epithelial cells by type IV secretion. *Science 287* (5457):1497–500.

Ogura, K., S. Maeda, M. Nakao, T. Watanabe, M. Tada, T. Kyutoku, et al. 2000. Virulence factors of *Helicobacter pylori* responsible for gastric diseases in Mongolian gerbil. *J Exp Med 192* (11):1601–10.

Ogura, Y., D.K. Bonen, N. Inohara, D.L. Nicolae, F.F. Chen, R. Ramos, et al. 2001. A frameshift mutation in NOD2 associated with susceptibility to Crohn's disease. *Nature 411* (6837):603–6. doi: 10.1038/35079114.

Ohkusa, T., I. Okayasu, T. Ogihara, K. Morita, M. Ogawa, and N. Sato. 2003. Induction of experimental ulcerative colitis by Fusobacterium varium isolated from colonic mucosa of patients with ulcerative colitis. *Gut 52* (1):79–83.

Ohkusa, T., N. Sato, T. Ogihara, K. Morita, M. Ogawa, and I. Okayasu. 2002. Fusobacterium varium localized in the colonic mucosa of patients with ulcerative colitis stimulates species-specific antibody. *J Gastroenterol Hepatol 17* (8):849–53.

Ohkusa, T., T. Yoshida, N. Sato, S. Watanabe, H. Tajiri, and I. Okayasu. 2009. Commensal bacteria can enter colonic epithelial cells and induce proinflammatory cytokine secretion: A possible pathogenic mechanism of ulcerative colitis. *J Med Microbiol* 58 (Pt 5):535–45. doi: 10.1099/jmm.0.005801-0.

Ott, S.J., M. Musfeldt, D.F. Wenderoth, J. Hampe, O. Brant, U.R. Fölsch, *et al.* 2004. Reduction in diversity of the colonic mucosa associated bacterial microflora in patients with active inflammatory bowel disease. *Gut* 53 (5):685–93.

Ou, J., F. Carbonero, E.G. Zoetendal, J.P. DeLany, M. Wang, K. Newton, *et al.* 2013. Diet, microbiota, and microbial metabolites in colon cancer risk in rural Africans and African Americans. *Am J Clin Nutr* 98 (1):111-20. doi: 10.3945/ajcn.112.056689.

Palmer, R. 2011. Fecal matters. *Nat Med* 17 (2):150–2. doi: 10.1038/nm0211-150.

Papatheodorou, P., J.E. Carette, G.W. Bell, C. Schwan, G. Guttenberg, T.R. Brummelkamp, and K. Aktories. 2011. Lipolysis-stimulated lipoprotein receptor (LSR) is the host receptor for the binary toxin *Clostridium difficile* transferase (CDT). *Proc Natl Acad Sci USA* 108 (39):16422–7. doi: 10.1073/pnas.1109772108.

Paredes-Sabja, D., C. Bond, R.J. Carman, P. Setlow, and M.R. Sarker. 2008. Germination of spores of *Clostridium difficile* strains, including isolates from a hospital outbreak of *Clostridium difficile*-associated disease (CDAD). *Microbiology* 154 (Pt 8):2241–50. doi: 10.1099/mic.0.2008/016592-0.

Pepin, J., M.E. Alary, L. Valiquette, E. Raiche, J. Ruel, K. Fulop, *et al.* 2005. Increasing risk of relapse after treatment of *Clostridium difficile* colitis in Quebec, Canada. *Clin Infect Dis* 40 (11):1591–7. doi: 10.1086/430315.

Petrella, L.A., S.P. Sambol, A. Cheknis, K. Nagaro, Y. Kean, P.S. Sears, *et al.* 2012. Decreased cure and increased recurrence rates for *Clostridium difficile* infection caused by the epidemic C. difficile BI strain. *Clin Infect Dis* 55 (3):351–7. doi: 10.1093/cid/cis430.

Petrof, E.O., G.B. Gloor, S.J. Vanner, S.J. Weese, D. Carter, M.C. Daigneault, *et al.* 2013. Stool substitute transplant therapy for the eradication of *Clostridium difficile* infection: 'RePOOPulating' the gut. *Microbiome* 1 (1):3. doi: 10.1186/2049-2618-1-3.

Petrosyan, M., Y.S. Guner, M. Williams, A. Grishin, and H.R. Ford. 2009. Current concepts regarding the pathogenesis of necrotizing enterocolitis. *Pediatr Surg Int* 25 (4):309–18. doi: 10.1007/s00383-009-2344-8.

Philpott, D.J. and S.E. Girardin. 2009. Crohn's disease-associated Nod2 mutants reduce IL10 transcription. *Nat Immunol* 10 (5):455–7. doi: 10.1038/ni0509-455.

Pike, K., P. Brocklehurst, D. Jones, S. Kenyon, A. Salt, D. Taylor, and N. Marlow. 2012. Outcomes at 7 years for babies who developed neonatal necrotising enterocolitis: The ORACLE Children Study. *Arch Dis Child Fetal Neonatal Ed* 97 (5):F318–22. doi: 10.1136/fetalneonatal-2011-300244.

Polk, D.B. and R.M. Peek. 2010. *Helicobacter pylori*: Gastric cancer and beyond. *Nat Rev Cancer* 10 (6):403–14. doi: 10.1038/nrc2857.

Pruitt, R.N. and D.B. Lacy. 2012. Toward a structural understanding of *Clostridium difficile* toxins A and B. *Front Cell Infect Microbiol* 2:28. doi: 10.3389/fcimb.2012.00028.

Pépin, J., S. Routhier, S. Gagnon, and I. Brazeau. 2006. Management and outcomes of a first recurrence of *Clostridium difficile*-associated disease in Quebec, Canada. *Clin Infect Dis 42* (6):758–64. doi: 10.1086/501126.

Qin, J., R. Li, J. Raes, M. Arumugam, K.S. Burgdorf, C. Manichanh, *et al.* 2010. A human gut microbial gene catalogue established by metagenomic sequencing. *Nature 464* (7285):59–65. doi: 10.1038/nature08821.

Qin, J., Y. Li, Z. Cai, S. Li, J. Zhu, F. Zhang, *et al.* 2012. A metagenome-wide association study of gut microbiota in type 2 diabetes. *Nature 490* (7418):55–60. doi: 10.1038/nature11450.

Qiu, J., X. Guo, Z.M. Chen, L. He, G.F. Sonnenberg, D. Artis, *et al.* 2013. Group 3 innate lymphoid cells inhibit T-cell-mediated intestinal inflammation through aryl hydrocarbon receptor signaling and regulation of microflora. *Immunity 39* (2):386–99. doi: 10.1016/j.immuni.2013.08.002.

Quattropani, C., B. Ausfeld, A. Straumann, P. Heer, and F. Seibold. 2003. Complementary alternative medicine in patients with inflammatory bowel disease: Use and attitudes. *Scand J Gastroenterol 38* (3):277–82.

Rahimi, R.S. and D.C. Rockey. 2014. Hepatic encephalopathy: How to test and treat. *Curr Opin Gastroenterol 30* (3):265–71. doi: 10.1097/MOG.0000000000000066.

Raman, M., I. Ahmed, P.M. Gillevet, C.S. Probert, N.M. Ratcliffe, S. Smith, *et al.* 2013. Fecal microbiome and volatile organic compound metabolome in obese humans with nonalcoholic fatty liver disease. *Clin Gastroenterol Hepatol 11* (7):868–75.e1-3. doi: 10.1016/j.cgh.2013.02.015.

Reeves, A.E., C.M. Theriot, I.L. Bergin, G.B. Huffnagle, P.D. Schloss, and V.B. Young. 2011. The interplay between microbiome dynamics and pathogen dynamics in a murine model of *Clostridium difficile* Infection. *Gut Microbes 2* (3):145–58.

Reid, G., J. Jass, M.T. Sebulsky, and J.K. McCormick. 2003. Potential uses of probiotics in clinical practice. *Clin Microbiol Rev 16* (4):658–72.

Reid, G. and Food and Agricultural Organization of the United Nation and the WHO. 2005. The importance of guidelines in the development and application of probiotics. *Curr Pharm Des 11* (1):11–6.

Reid, Gregor. 2001. Regulatory and clinical aspects of dairy probiotics [online]. Background paper for the Joint FAO/WHOExpert Consultation on Evaluation of Health and Nutritional Properties of Probiotics in Food Including Powder Milk with Live Lactic Acid Bacteria. Rome, Italy. Food and Agriculture Organization of the United Nations (FAO) Accessed September 1, 2014. ftp://ftp.fao.org/es/esn/food/reid.pdf

Relman, D.A. 2012. The human microbiome: Ecosystem resilience and health. *Nutr Rev 70* Suppl 1:S2-9. doi: 10.1111/j.1753-4887.2012.00489.x.

Ridaura, V.K., J.J. Faith, F.E. Rey, J. Cheng, A.E. Duncan, A.L. Kau, *et al.* 2013. Gut microbiota from twins discordant for obesity modulate metabolism in mice. *Science 341* (6150):1241214. doi: 10.1126/science.1241214.

Ridlon, J.M., D.J. Kang, and P.B. Hylemon. 2006. Bile salt biotransformations by human intestinal bacteria. *J Lipid Res 47* (2):241–59. doi: 10.1194/jlr.R500013-JLR200.

Riley, D.R., K.B. Sieber, K.M. Robinson, J.R. White, A. Ganesan, S. Nourbakhsh, and J.C. Dunning Hotopp. 2013. Bacteria-human somatic cell lateral gene transfer is enriched in cancer samples. *PLoS Comput Biol 9* (6):e1003107. doi: 10.1371/journal. pcbi.1003107.

Rivollier, A., J. He, A. Kole, V. Valatas, and B.L. Kelsall. 2012. Inflammation switches the differentiation program of Ly6Chi monocytes from antiinflammatory macrophages to inflammatory dendritic cells in the colon. *J Exp Med 209* (1):139–55. doi: 10.1084/jem.20101387.

Rohlke, F., C.M. Surawicz, and N. Stollman. 2010. Fecal flora reconstitution for recurrent *Clostridium difficile* infection: Results and methodology. *J Clin Gastroenterol 44* (8):567–70. doi: 10.1097/MCG.0b013e3181dadb10.

Rolig, A.S., C. Cech, E. Ahler, J.E. Carter, and K.M. Ottemann. 2013. The degree of *Helicobacter pylori*-triggered inflammation is manipulated by preinfection host microbiota. *Infect Immun 81* (5):1382–9. doi: 10.1128/IAI.00044-13.

Round, J.L., S.M. Lee, J. Li, G. Tran, B. Jabri, T.A. Chatila, and S.K. Mazmanian. 2011. The Toll-like receptor 2 pathway establishes colonization by a commensal of the human microbiota. *Science 332* (6032):974–7. doi: 10.1126/science.1206095.

Round, J.L. and S.K. Mazmanian. 2010. Inducible Foxp3+ regulatory T-cell development by a commensal bacterium of the intestinal microbiota. *Proc Natl Acad Sci USA 107* (27):12204–9. doi: 10.1073/pnas.0909122107.

Rousseau, C., I. Poilane, L. De Pontual, A.C. Maherault, A. Le Monnier, and A. Collignon. 2012. *Clostridium difficile* carriage in healthy infants in the community: A potential reservoir for pathogenic strains. *Clin Infect Dis 55* (9):1209–15. doi: 10.1093/cid/cis637.

Rowan, F., N.G. Docherty, M. Murphy, B. Murphy, J. Calvin Coffey, and P.R. O'Connell. 2010. Desulfovibrio bacterial species are increased in ulcerative colitis. *Dis Colon Rectum 53* (11):1530–6. doi: 10.1007/DCR.0b013e3181f1e620.

Rubinstein, M.R., X. Wang, W. Liu, Y. Hao, G. Cai, and Y.W. Han. 2013. *Fusobacterium nucleatum* promotes colorectal carcinogenesis by modulating E-cadherin/β-catenin signaling via its FadA adhesin. *Cell Host Microbe 14* (2):195–206. doi: 10.1016/j.chom.2013.07.012.

Sabaté, J.M., P. Jouët, F. Harnois, C. Mechler, S. Msika, M. Grossin, and B. Coffin. 2008. High prevalence of small intestinal bacterial overgrowth in patients with morbid obesity: A contributor to severe hepatic steatosis. *Obes Surg 18* (4):371–7. doi: 10.1007/s11695-007-9398-2.

Sambol, S.P., M.M. Merrigan, J.K. Tang, S. Johnson, and D.N. Gerding. 2002. Colonization for the prevention of *Clostridium difficile* disease in hamsters. *J Infect Dis 186* (12):1781–9. doi: 10.1086/345676.

Santaolalla, R., D.A. Sussman, J.R. Ruiz, J.M. Davies, C. Pastorini, C.L. España, *et al.* 2013. TLR4 activates the β-catenin pathway to cause intestinal neoplasia. *PLoS One 8* (5):e63298. doi: 10.1371/journal.pone.0063298.

Sartor, R.B. 2004. Therapeutic manipulation of the enteric microflora in inflammatory bowel diseases: Antibiotics, probiotics, and prebiotics. *Gastroenterology 126* (6):1620–33.

Sartor, R.B. 2008. Microbial influences in inflammatory bowel diseases. *Gastroenterology 134* (2):577–94. doi: 10.1053/j.gastro.2007.11.059.

Schnabl, B. and D.A. Brenner. 2014. Interactions between the intestinal microbiome and liver diseases. *Gastroenterology* 146 (6):1513–24. doi: 10.1053/j.gastro.2014.01.020.

Schwiertz, A., D. Taras, K. Schäfer, S. Beijer, N.A. Bos, C. Donus, and P.D. Hardt. 2010. Microbiota and SCFA in lean and overweight healthy subjects. *Obesity (Silver Spring)* 18 (1):190–5. doi: 10.1038/oby.2009.167.

Shaffer, C.L., J.A. Gaddy, J.T. Loh, E.M. Johnson, S. Hill, E.E. Hennig, et al. 2011. *Helicobacter pylori* exploits a unique repertoire of type IV secretion system components for pilus assembly at the bacteria-host cell interface. *PLoS Pathog* 7 (9):e1002237. doi: 10.1371/journal.ppat.1002237.

Sheh, A. and J.G. Fox. 2013. The role of the gastrointestinal microbiome in *Helicobacter pylori* pathogenesis. *Gut Microbes* 4 (6):505–31. doi: 10.4161/gmic.26205.

Shen, A. 2012. *Clostridium difficile* toxins: Mediators of inflammation. *J Innate Immun* 4 (2):149–58. doi: 10.1159/000332946.

Siegel, R., D. Naishadham, and A. Jemal. 2013. Cancer statistics, 2013. *CA Cancer J Clin* 63 (1):11–30. doi: 10.3322/caac.21166.

Smith, P.M., M.R. Howitt, N. Panikov, M. Michaud, C.A. Gallini, M. Bohlooly-Y, et al. 2013. The microbial metabolites, short-chain fatty acids, regulate colonic T_{reg} cell homeostasis. *Science* 341 (6145):569–73. doi: 10.1126/science.1241165.

Smits, L.P., K.E. Bouter, W.M. de Vos, T.J. Borody, and M. Nieuwdorp. 2013. Therapeutic potential of fecal microbiota transplantation. *Gastroenterology* 145 (5):946–53. doi: 10.1053/j.gastro.2013.08.058.

Sobhani, I., J. Tap, F. Roudot-Thoraval, J.P. Roperch, S. Letulle, P. Langella, et al. G. Corthier, J. Tran Van Nhieu, and J.P. Furet. 2011. Microbial dysbiosis in colorectal cancer (CRC) patients. *PLoS One* 6 (1):e16393. doi: 10.1371/journal.pone.0016393.

Sokol, H., P. Lepage, P. Seksik, J. Doré, and P. Marteau. 2006. Temperature gradient gel electrophoresis of fecal 16S rRNA reveals active *Escherichia coli* in the microbiota of patients with ulcerative colitis. *J Clin Microbiol* 44 (9):3172–7. doi: 10.1128/JCM.02600-05.

Sokol, H., B. Pigneur, L. Watterlot, O. Lakhdari, L.G. Bermúdez-Humarán, J.J. Gratadoux, et al. 2008. *Faecalibacterium prausnitzii* is an anti-inflammatory commensal bacterium identified by gut microbiota analysis of Crohn disease patients. *Proc Natl Acad Sci USA* 105 (43):16731–6. doi: 10.1073/pnas.0804812105.

Sokol, H., P. Seksik, J.P. Furet, O. Firmesse, I. Nion-Larmurier, L. Beaugerie, et al. 2009. Low counts of *Faecalibacterium prausnitzii* in colitis microbiota. *Inflamm Bowel Dis* 15 (8):1183–9. doi: 10.1002/ibd.20903.

Sorg, J.A. and A.L. Sonenshein. 2008. Bile salts and glycine as cogerminants for *Clostridium difficile* spores. *J Bacteriol* 190 (7):2505–12. doi: 10.1128/JB.01765-07.

Sorg, J.A. and A.L. Sonenshein. 2009. Chenodeoxycholate is an inhibitor of *Clostridium difficile* spore germination. *J Bacteriol* 191 (3):1115–7. doi: 10.1128/JB.01260-08.

Spor, A., O. Koren, and R. Ley. 2011. Unravelling the effects of the environment and host genotype on the gut microbiome. *Nat Rev Microbiol* 9 (4):279–90. doi: 10.1038/nrmicro2540.

Stappenbeck, T.S., L.V. Hooper, and J.I. Gordon. 2002. Developmental regulation of intestinal angiogenesis by indigenous microbes via Paneth cells. *Proc Natl Acad Sci USA 99* (24):15451–5. doi: 10.1073/pnas.202604299.

Stark, P.L. and A. Lee. 1982. The bacterial colonization of the large bowel of pre-term low birth weight neonates. *J Hyg (Lond) 89* (1):59–67.

Stellwag, E.J. and P.B. Hylemon. 1978. Characterization of 7α-dehydroxylase in Clostridium leptum. *Am J Clin Nutr 31* (10 Suppl):S243–S247.

Stewart, C.J. E.C. Marrs, S. Magorrian, A. Nelson, C. Lanyon, J.D. Perry, *et al.* 2012. The preterm gut microbiota: Changes associated with necrotizing enterocolitis and infection. *Acta Paediatr 101* (11):1121–7. doi: 10.1111/j.1651-2227.2012.02801.x.

Stewart, C.J. E.C. Marrs, A. Nelson, C. Lanyon, J.D. Perry, N.D. Embleton, *et al.* 2013. Development of the preterm gut microbiome in twins at risk of necrotising enterocolitis and sepsis. *PLoS One 8* (8):e73465. doi: 10.1371/journal.pone.0073465.

Stewart, D.B., A. Berg, and J. Hegarty. 2013. Predicting recurrence of C. difficile colitis using bacterial virulence factors: Binary toxin is the key. *J Gastrointest Surg 17* (1):118–24; discussion p.124–5. doi: 10.1007/s11605-012-2056-6.

Stewart, D.B., A.S. Berg, and J.P. Hegarty. 2014. Single nucleotide polymorphisms of the tcdC gene and presence of the binary toxin gene predict recurrent episodes of *Clostridium difficile* infection. *Ann Surg 260* (2):299–304. doi: 10.1097/SLA.0000000000000469.

Strauss, J., G.G. Kaplan, P.L. Beck, K. Rioux, R. Panaccione, R. Devinney, *et al.* 2011. Invasive potential of gut mucosa-derived *Fusobacterium nucleatum* positively correlates with IBD status of the host. *Inflamm Bowel Dis 17* (9):1971–8. doi: 10.1002/ibd.21606.

Sun, C.M., J.A. Hall, R.B. Blank, N. Bouladoux, M. Oukka, J.R. Mora, and Y. Belkaid. 2007. Small intestine lamina propria dendritic cells promote de novo generation of Foxp3 T_{reg} cells via retinoic acid. *J Exp Med 204* (8):1775–85. doi: 10.1084/jem.20070602.

Sun, Y.Q., H.J. Monstein, L.E. Nilsson, F. Petersson, and K. Borch. 2003. Profiling and identification of eubacteria in the stomach of Mongolian gerbils with and without *Helicobacter pylori* infection. *Helicobacter 8* (2):149–57.

Swidsinski, A., A. Ladhoff, A. Pernthaler, S. Swidsinski, V. Loening-Baucke, M. Ortner, *et al.* 2002. Mucosal flora in inflammatory bowel disease. *Gastroenterology 122* (1):44–54.

Takamine, F. and T. Imamura. 1995. Isolation and characterization of bile acid 7α-dehydroxylating bacteria from human feces. *Microbiol Immunol 39* (1):11–8.

Teli, M.R., C.P. Day, A.D. Burt, M.K. Bennett, and O.F. James. 1995. Determinants of progression to cirrhosis or fibrosis in pure alcoholic fatty liver. *Lancet 346* (8981):987–90.

Teltschik, Z., R. Wiest, J. Beisner, S. Nuding, C. Hofmann, J. Schoelmerich, *et al.* 2012. Intestinal bacterial translocation in rats with cirrhosis is related to compromised Paneth cell antimicrobial host defense. *Hepatology 55* (4):1154–63. doi: 10.1002/hep.24789.

Tilg, H. 2010. Obesity, metabolic syndrome, and microbiota: Multiple interactions. *J Clin Gastroenterol 44* Suppl 1:S16–8. doi: 10.1097/MCG.0b013e3181dd8b64.

Torrazza, R.M., M. Ukhanova, X. Wang, R. Sharma, M.L. Hudak, J. Neu, and V. Mai.

2013. Intestinal microbial ecology and environmental factors affecting necrotizing enterocolitis. *PLoS One 8* (12):e83304. doi: 10.1371/journal.pone.0083304.

Turnbaugh, P.J., M. Hamady, T. Yatsunenko, B.L. Cantarel, A. Duncan, R.E. Ley, *et al.* 2009. A core gut microbiome in obese and lean twins. *Nature 457* (7228):480–4. doi: 10.1038/nature07540.

Turnbaugh, P.J., R.E. Ley, M.A. Mahowald, V. Magrini, E.R. Mardis, and J.I. Gordon. 2006. An obesity-associated gut microbiome with increased capacity for energy harvest. *Nature 444* (7122):1027–31. doi: 10.1038/nature05414.

Tvede, M. and J. Rask-Madsen. 1989. Bacteriotherapy for chronic relapsing *Clostridium difficile* diarrhoea in six patients. *Lancet 1* (8648):1156–60.

Underwood, M.A., K.M. Kalanetra, N.A. Bokulich, Z.T. Lewis, M. Mirmiran, D.J. Tancredi, and D. A. Mills. 2013. A comparison of two probiotic strains of bifidobacteria in premature infants. *J Pediatr 163* (6):1585–1591.e9. doi: 10.1016/j.jpeds.2013.07.017.

Uronis, J.M., M. Mühlbauer, H.H. Herfarth, T.C. Rubinas, G.S. Jones, and C. Jobin. 2009. Modulation of the intestinal microbiota alters colitis-associated colorectal cancer susceptibility. *PLoS One 4* (6):e6026. doi: 10.1371/journal.pone.0006026.

van Nood, E., A. Vrieze, M. Nieuwdorp, S. Fuentes, E.G. Zoetendal, W.M. de Vos, *et al.* 2013. Duodenal infusion of donor feces for recurrent *Clostridium difficile*. *N Engl J Med 368* (5):407–15. doi: 10.1056/NEJMoa1205037.

Varela, E., C. Manichanh, M. Gallart, A. Torrejón, N. Borruel, F. Casellas, *et al.* 2013. Colonisation by *Faecalibacterium prausnitzii* and maintenance of clinical remission in patients with ulcerative colitis. *Aliment Pharmacol Ther 38* (2):151–61. doi: 10.1111/apt.12365.

Vernon, G., A. Baranova, and Z.M. Younossi. 2011. Systematic review: The epidemiology and natural history of non-alcoholic fatty liver disease and non-alcoholic steatohepatitis in adults. *Aliment Pharmacol Ther 34* (3):274–85. doi: 10.1111/j.1365-2036.2011.04724.x.

Vijay-Kumar, M., J.D. Aitken, F.A. Carvalho, T.C. Cullender, S. Mwangi, S. Srinivasan, *et al.* 2010. Metabolic syndrome and altered gut microbiota in mice lacking Toll-like receptor 5. *Science 328* (5975):228–31. doi: 10.1126/science.1179721.

Vincent, C., D.A. Stephens, V.G. Loo, T.J. Edens, M.A. Behr, K. Dewar, and A.R. Manges. 2013. Reductions in intestinal Clostridiales precede the development of nosocomial *Clostridium difficile* infection. *Microbiome 1* (1):18. doi: 10.1186/2049-2618-1-18.

Vonberg, R.P., C. Reichardt, M. Behnke, F. Schwab, S. Zindler, and P. Gastmeier. 2008. Costs of nosocomial *Clostridium difficile*-associated diarrhoea. *J Hosp Infect 70* (1):15–20. doi: 10.1016/j.jhin.2008.05.004.

Voth, D.E. and J.D. Ballard. 2005. *Clostridium difficile* toxins: Mechanism of action and role in disease. *Clin Microbiol Rev 18* (2):247–63. doi: 10.1128/CMR.18.2.247-263.2005.

Vrieze, A., E. Van Nood, F. Holleman, J. Salojärvi, R.S. Kootte, J.F. Bartelsman, *et al.* 2012. Transfer of intestinal microbiota from lean donors increases insulin sensitivity in individuals with metabolic syndrome. *Gastroenterology 143* (4):913–6.e7. doi: 10.1053/j.gastro.2012.06.031.

Wang, Q., J. Dong, and Y. Zhu. 2012. Probiotic supplement reduces risk of necrotizing enterocolitis and mortality in preterm very low-birth-weight infants: An updated meta-analysis of 20 randomized, controlled trials. *J Pediatr Surg 47* (1):241–8. doi: 10.1016/j.jpedsurg.2011.09.064.

Wang, X. and M.M. Huycke. 2007. Extracellular superoxide production by Enterococcus faecalis promotes chromosomal instability in mammalian cells. *Gastroenterology 132* (2):551–61. doi: 10.1053/j.gastro.2006.11.040.

Wang, Y., J.D. Hoenig, K.J. Malin, S. Qamar, E.O. Petrof, J. Sun, *et al.* 2009. 16S rRNA gene-based analysis of fecal microbiota from preterm infants with and without necrotizing enterocolitis. *ISME J 3* (8):944–54. doi: 10.1038/ismej.2009.37.

Wang, Y., Y. Liu, A. Sidhu, Z. Ma, C. McClain, and W. Feng. 2012. Lactobacillus rhamnosus GG culture supernatant ameliorates acute alcohol-induced intestinal permeability and liver injury. *Am J Physiol Gastrointest Liver Physiol 303* (1):G32–41. doi: 10.1152/ajpgi.00024.2012.

Wang, Z., E. Klipfell, B.J. Bennett, R. Koeth, B.S. Levison, B. Dugar, *et al.* 2011. Gut flora metabolism of phosphatidylcholine promotes cardiovascular disease. *Nature 472* (7341):57–63. doi: 10.1038/nature09922.

Watanabe, T., M. Tada, H. Nagai, S. Sasaki, and M. Nakao. 1998. *Helicobacter pylori* infection induces gastric cancer in mongolian gerbils. *Gastroenterology 115* (3):642–8.

Wegener, M., J. Schaffstein, U. Dilger, C. Coenen, B. Wedmann, and G. Schmidt. 1991. Gastrointestinal transit of solid-liquid meal in chronic alcoholics. *Dig Dis Sci 36* (7):917-23.

Wei, M., R. Shinkura, Y. Doi, M. Maruya, S. Fagarasan, and T. Honjo. 2011. Mice carrying a knock-in mutation of Aicda resulting in a defect in somatic hypermutation have impaired gut homeostasis and compromised mucosal defense. *Nat Immunol 12* (3):264–70. doi: 10.1038/ni.1991.

Weingarden, A.R., C. Chen, A. Bobr, D. Yao, Y. Lu, V.M. Nelson, *et al.* 2014. Microbiota transplantation restores normal fecal bile acid composition in recurrent *Clostridium difficile* infection. *Am J Physiol Gastrointest Liver Physiol 306* (4):G310–9. doi: 10.1152/ajpgi.00282.2013.

Weir, T.L., D.K. Manter, A.M. Sheflin, B.A. Barnett, A.L. Heuberger, and E.P. Ryan. 2013. Stool microbiome and metabolome differences between colorectal cancer patients and healthy adults. *PLoS One 8* (8):e70803. doi: 10.1371/journal.pone.0070803.

Wells, J.E. and P.B. Hylemon. 2000. Identification and characterization of a bile acid 7alpha-dehydroxylation operon in Clostridium sp. strain TO-931, a highly active 7alpha-dehydroxylating strain isolated from human feces. *Appl Environ Microbiol 66* (3):1107–13.

Wentworth, J.M., G. Naselli, W.A. Brown, L. Doyle, B. Phipson, G.K. Smyth, *et al.* 2010. Pro-inflammatory CD11c+CD206+ adipose tissue macrophages are associated with insulin resistance in human obesity. *Diabetes 59* (7):1648–56. doi: 10.2337/db09-0287.

Wesemann, D.R., A.J. Portuguese, R.M. Meyers, M.P. Gallagher, K. Cluff-Jones, J.M. Magee, *et al.* 2013. Microbial colonization influences early B-lineage development in the gut lamina propria. *Nature 501* (7465):112–5. doi: 10.1038/nature12496.

Westerbeek, E.A., A. van den Berg, H.N. Lafeber, J. Knol, W.P. Fetter, and R.M. van Elburg. 2006. The intestinal bacterial colonisation in preterm infants: A review of the literature. *Clin Nutr* 25 (3):361–8. doi: 10.1016/j.clnu.2006.03.002.

Whelan, K. and E.M. Quigley. 2013. Probiotics in the management of irritable bowel syndrome and inflammatory bowel disease. *Curr Opin Gastroenterol* 29 (2):184–9. doi: 10.1097/MOG.0b013e32835d7bba.

Wiest, R., M. Lawson, and M. Geuking. 2014. Pathological bacterial translocation in liver cirrhosis. *J Hepatol* 60 (1):197–209. doi: 10.1016/j.jhep.2013.07.044.

Willing, B., J. Halfvarson, J. Dicksved, M. Rosenquist, G. Järnerot, L. Engstrand, *et al.* 2009. Twin studies reveal specific imbalances in the mucosa-associated microbiota of patients with ileal Crohn's disease. *Inflamm Bowel Dis* 15 (5):653–60. doi: 10.1002/ibd.20783.

Wilson, K.H. 1983. Efficiency of various bile salt preparations for stimulation of *Clostridium difficile* spore germination. *J Clin Microbiol* 18 (4):1017–9.

Wlodarska, M., B. Willing, K.M. Keeney, A. Menendez, K.S. Bergstrom, N. Gill, *et al.* 2011. Antibiotic treatment alters the colonic mucus layer and predisposes the host to exacerbated Citrobacter rodentium-induced colitis. *Infect Immun* 79 (4):1536–45. doi: 10.1128/IAI.01104-10.

Wong, A.P., A.L. Clark, E.A. Garnett, M. Acree, S.A. Cohen, G.D. Ferry, and M.B. Heyman. 2009. Use of complementary medicine in pediatric patients with inflammatory bowel disease: Results from a multicenter survey. *J Pediatr Gastroenterol Nutr* 48 (1):55–60. doi: 10.1097/MPG.0b013e318169330f.

Wu, G.D., J. Chen, C. Hoffmann, K. Bittinger, Y.Y. Chen, S.A. Keilbaugh, *et al.* 2011. Linking long-term dietary patterns with gut microbial enterotypes. *Science* 334 (6052):105–8. doi: 10.1126/science.1208344.

Wu, N., X. Yang, R. Zhang, J. Li, X. Xiao, Y. Hu, *et al.* 2013. Dysbiosis signature of fecal microbiota in colorectal cancer patients. *Microb Ecol* 66 (2):462–70. doi: 10.1007/s00248-013-0245-9.

Wu, S., K.J. Rhee, E. Albesiano, S. Rabizadeh, X. Wu, H.R. Yen, *et al.* 2009. A human colonic commensal promotes colon tumorigenesis via activation of T helper type 17 T cell responses. *Nat Med* 15 (9):1016–22. doi: 10.1038/nm.2015.

Yan, F. and D.B. Polk. 2002. Probiotic bacterium prevents cytokine-induced apoptosis in intestinal epithelial cells. *J Biol Chem* 277 (52):50959–65. doi: 10.1074/jbc.M207050200.

Yang, L., F. Francois, and Z. Pei. 2012. Molecular pathways: Pathogenesis and clinical implications of microbiome alteration in esophagitis and Barrett esophagus. *Clin Cancer Res* 18 (8):2138–44. doi: 10.1158/1078-0432.CCR-11-0934.

Yang, L., X. Lu, C.W. Nossa, F. Francois, R.M. Peek, and Z. Pei. 2009. Inflammation and intestinal metaplasia of the distal esophagus are associated with alterations in the microbiome. *Gastroenterology* 137 (2):588–97. doi: 10.1053/j.gastro.2009.04.046.

Yoon, Y.H. and H.Y. Yi. 2006. Surveillance report #75: Liver Cirrhosis Mortality in the United States, 1970–2003. Bethseda, MD: National Institute on Alcohol Abuse and Alcoholism.

Zenewicz, L.A., X. Yin, G. Wang, E. Elinav, L. Hao, L. Zhao, and R.A. Flavell. 2013.

IL-22 deficiency alters colonic microbiota to be transmissible and colitogenic. *J Immunol* 190 (10):5306–12. doi: 10.4049/jimmunol.1300016.

Zhang, F., W. Luo, Y. Shi, Z. Fan, and G. Ji. 2012. Should we standardize the 1,700-year-old fecal microbiota transplantation? *Am J Gastroenterol 107* (11):1755; author reply p.1755-6. doi: 10.1038/ajg.2012.251.

Zhu, L., S.S. Baker, C. Gill, W. Liu, R. Alkhouri, R.D. Baker, and S.R. Gill. 2013. Characterization of gut microbiomes in nonalcoholic steatohepatitis (NASH) patients: A connection between endogenous alcohol and NASH. *Hepatology 57* (2):601–9. doi: 10.1002/hep.26093.

Zimmer, J., B. Lange, J.S. Frick, H. Sauer, K. Zimmermann, A. Schwiertz, *et al.* 2012. A vegan or vegetarian diet substantially alters the human colonic faecal microbiota. *Eur J Clin Nutr 66* (1):53–60. doi: 10.1038/ejcn.2011.141.

CHAPTER 11

Probiotics and the Microbiome

GREGOR REID

11.1. A LONG HISTORY OF TIME

CROSBY, STILLS, AND NASH might have indoctrinated the phase, "It's been a long time coming" into modern psyche from their 1969 song, but it neatly reflects the field of probiotics.

Indeed, it is the microbiome that is the new concept, not probiotics, yet the two are interconnected. Reports of cheese-making date back at least six millenia (Salque *et al.* 2013), but the inducers of fermentation were not known then. The discovery of microbes was not apparent until Antonie van Leeuwenhoek's observation of organisms under the microscope in 1676, and the link with health conferred by fermented food was made by Elie Metchnikoff in 1907. This, in human terms, is a long history of time.

The realization that microbes are universal on the planet and the human body has opened up such dramatically different perspectives on life, its potential remains to be determined. But, why has it taken so long? In 1985, we proposed that the *Lactobacillus* species could be used as probiotics to prevent infection (Reid *et al.* 1985), but serious consideration of the concept took another 20 or so years. The question is why do radically different ideas take so long to become accepted? One answer is the intellectual stubborn nature of humans (Arbesman 2012), defensive in the face of change, and as physicist Max Planck stated, "New scientific truth does not triumph by convincing its opponents and making them see the light, but rather because its opponents eventually die, and a new generation grows up that is familiar with it."

Fortunately, Planck is not always right, and the emergence of probiotics to prevent infection has occurred without my demise.

Sometimes paradigm shifts require a number of factors to coalesce. As microbiology has evolved, so too has the identification of organisms and our ability to nurture them, the recognition that broad spectrum antibiotics can lead to a range of problems for the host through disruption of bacterial populations, and the appreciation for factors produced by beneficial microbes in maintaining homeostasis and/or health. In addition, side effects and poor efficacy of antimicrobials, certainly in the management of urogenital infections in women, led to increased interest in "alternative" therapies (Reid 1999). Studies with fermented food showed that metabolic products could prevent the growth of many pathogens (Barefoot and Nettles 1993), while the realization that dysbiotic events were responsible for many diseases led to the idea of replenishing the beneficial microbes in an attempt to regain homeostasis (Schiffrin and Blum 2002). Thus, no single moment caused the explosion of probiotic approaches that exists today (Hudson 2014).

11.2. RATIONALE FOR MODULATING HEALTH

The advent of pharmaceutical agents, rather than foods, to treat and prevent disease is a concept from the past one hundred or so years. The creation of the Food and Drug Administration in 1903 attests to the belief that chemical compounds needed to be better regulated in order to prevent harm or delude people into purchasing "snake oil" with no efficacy. While Metchnikoff may have promoted fermented foods for long life, such products cannot prevent or treat diseases like measles, whooping cough, polio, diphtheria, and tuberculosis which were major causes of death and morbidity prior to vaccination and the advent of hygiene procedures. But the human preponderance for taking one successful concept then using it widely untested and unproven for other reasons, has resulted in massive abuse and over use of antibiotics. From use in livestock for weight gain, to treatment of viral infections, prophylaxis for dental procedures, the use in multiple consumer products, and trace amounts in the water and food supply (Marshall and Levy 2011; Dinsbach 2012; Choutet 2003; AUPA 2011) the use of antibiotics has left a trail of drug resistant organisms, disrupted microbiotas, and long term ailments only now becoming understood (Blaser 2011; Srigley *et al.* 2013; Petschow *et al.* 2013; Reynolds and Finlay 2013; Cho *et al.* 2012).

In the mid-1980s, we rationalized that if lactobacilli were depleted and replaced by pathogens in the vagina and distal urethra as part of the process whereby pathogens caused urogenital infection, could we not restore health by administering back the lactobacilli (Reid *et al.* 1987; Bruce and Reid 1988)? This was viewed as a radical approach at the time, doomed to failure, and deemed by some as unnecessary given that infections were "easily" treated with antibiotics. Indeed, the concept of probiotics was not helped by people like Dr. Ron Atlas, the President of the American Society for Microbiology who stated that, "Probiotics may be today's snake oil, the liquid concoction of dubious or worthless medical value fraudulently peddled by hawkers from the backs of covered wagons during the settlement of the United States as a cure for innumerable ills," (Atlas 1999). Such inexcusable derisory and short-sighted comments likely contributed to a wider spectrum of researchers not taking probiotics seriously and delaying studies to better understand their mechanisms of action and efficacy. Nevertheless, the basis for the rationale only grew stronger, with a lean pipeline of new antimicrobial pharmaceuticals, increased microbial resistance, and no abatement of disease incidence.

Research on probiotic applications in humans in the late 1980s and early 1990s was primarily on *Lactobacillus rhamnosus* GR-1 (Cho *et al.* 2012; Bruce *et al.* 1992; Reid *et al.* 1992, 1994) and *L. rhamnosus* GG (Majamaa and Isolauri 1997; Pelto *et al.* 1998; Vanderhoof *et al.* 1999), and sporadic studies on *Bifidobacterium* and other organisms (Marteau *et al.* 1992; Möllenbrink and Bruckschen 1994). Encouraging support for the ability of exogenously administered microbes via ingestion or intravaginal administration to modulate health through countering disease laid the groundwork for an expanded interest in the field. Critics argued that clinical studies were small and that only large trials like those of new drugs would convince them that probiotics had merit. However, not all drug trials are large prior to products being made available, and even with large studies, there remain many drugs whose efficacy is debatable (Downing *et al.* 2014) or whose application to other diseases, genders, or age groups from those for which approval was initially granted, is questionable (Laughon 2014; Holdcroft 2007; Whitley and Lindsey 2009; Bongartz 2006). To that end, the strength of evidence for probiotics to prevent necrotizing enterocolitis as shown in meta-analyses (Deshpande *et al.* 2010; Ofek *et al.* 2014), even though sample sizes were not always large, is more convincing that many of the drugs used in the neonatal intensive care centers (Giacoia and Mattison

2005), and is such that to not using probiotics in this setting reflects badly on medical practice (Ofek 2014).

On the other hand, the growth of the "probiotic" industry in the past 5 years has resulted in thousands of products being sold around the world, the majority of which have never been tested in humans. This disturbing trend fuels the fire of critics, and does a disservice to the field. By definition, probiotics are "live microorganisms that, when administered in adequate amounts, confer a health benefit on the host," (Hill *et al.* 2014) and thus they need to be shown to modulate health and provide a benefit. Simply adding lactic acid bacterial strains to a capsule does not constitute proof, especially since strains can inhibit the activity of each other (Tejero-Sariñena *et al.* 2012). The net result of failure to adhere to standards of evidence (Reid 2005), is that regulatory agencies have recently clamped down on all products, rather than selectively approve those with excellent documentation. Taking this beyond common sense, the European Food Safety Authority banned the use of the word "probiotic", while the Food and Drug Agency of the United States halted clinical trials on probiotic foods, demanding they be registered as drugs if the intent is to treat, prevent, or cure disease (Anonymous 2012; FDA 2014). The net result of giving power to bureaucracies established to adjudicate chemicals, is that they are recalcitrant to change, lack the personnel who understand this new field, and hinder further research and development to the adversity of consumers and patients.

11.3. THE CONSTANTLY GROWING LIST AND HOW THE STRAINS FUNCTION

Nevertheless, probiotics are here to stay and many new formulations will emerge as a consequence of microbiome studies identifying dysbiosis associated with ill health. How they are regulated will depend on politicians and their willingness to create modern regulatory processes. Health Canada at least has shown some leadership in this area by creating a system that permits product claims if supported by sufficient human data (Health Canada 2014).

It is difficult to make a list of products that have sufficient human data to recommend, as not all results are published or easily accessible, and products will inevitably be inadvertently omitted. In addition, not all products are available in every country. Still, some published lists are available on well documented products (Skokovic-Sunjic 2013).

Basically, there are three categories: oral, gut, and urogenital health. For oral care, few products have been designed specifically and adequately tested for this purpose, but *Streptococcus salivarius* strains in BLIS formulations appear to counter caries and halitosis through bacteriocin production against pathogens like *Streptococcus mutans* (Burton et al. 2006, 2013).

The strains *L. rhamnosus* GR-1 and *L. reuteri* RC-14, sold by Chr Hansen through various distributors, were initially designed to reduce urinary tract infection (UTI) by inhibiting bacterial ascension, growth, and adhesion from the vagina to the bladder (Reid et al. 1987, 1990). This was based upon studies in the 1970s showing that the vaginal and perineal microbiotas are heavily colonized by the urinary pathogens prior to and during UTI (Bruce et al. 1973), while healthy controls had lactobacilli dominating. Given the problems of bacterial vaginosis (BV) and vulvovaginal candidiasis (VVC), the ability of these strains to prevent infections, cure BV, and enhance antimicrobial therapy (Martinez et al. 2009a, 2009b; Anukam 2006; Hummelen et al. 2011; Petricevic et al. 2008; Beerepoot 2012) has stimulated the introduction of other products. The growth and survival in milk of GR-1 provided an opportunity to develop and test its use in yogurt to modulate immune parameters. One study of patients with inflammatory bowel disease (IBD) showed an increase in regulatory T cells (Baroja et al. 2007). This would be expected to represent a positive change by reducing inflammation, but the study did not have sufficient patient numbers to test for clinical improvements. Likewise, application of probiotic yogurt to the prevention of allergies showed anti-inflammatory effects, this time systemically, and some patients had reduced symptoms, but the study was too small to reach statistical significance (Koyama et al. 2010). Of the other products purported to improve bladder and vaginal health, few have published human trial data, but *L. rhamnosus* Lcr35, with a similar genome to GR-1, is sold to restore lactobacilli to the vagina following antibiotic therapy for BV (Perticevic and Witt 2008). The product has also been used to reduce mortality and morbidity associated with necrotizing enterocolitis (Bonsante et al. 2013).

A number of babies suffer from colic, regurgitation, and constipation during the first three months of life. Administration of *L. reuteri* DSM 17938 has been shown to significantly improve these parameters. For example in one study, at 3 months of age, the mean duration of crying time improved using this product (38 versus 71 minutes; $P < 0.01$), the mean number of regurgitations per day (2.9 versus 4.6; $P < 0.01$), and the mean number of evacuations per day (4.2 versus 3.6; $P < 0.01$)(Indrio et al. 2014). This is intriguing and must relate to, as yet unidentified by-products of the strain.

Due to its commercial availability in the late 1980s ahead of most other probiotic strains, *L. rhamnosus* GG has been studied for many different disorders. However, supportive data is mostly to prevent diarrhea in children (Vanderhoof et al. 1999). Its use in preventing allergic reactions is somewhat mixed, with a study reporting that intake can reduce atopic dermatitis (Kalliomäki et al. 2003), another disagreeing (Kopp et al. 2008), while others showing reduced cumulative prevalence of eczema (Wickens et al. 2008), and administration in the second trimester of pregnancy reduced the severity of maternal allergic disease through increment of Th1 response, but not the incidence of childhood allergic sensitization or allergic diseases (Ou et al. 2012). A study also showed reduction in respiratory illness in day care centres (Kumpu et al. 2012). The effects at sites distant to the gut are intriguing and suggest immune effects or secretion of products across the gut epithelium influence pathogens and body functions elsewhere.

Irritable bowel syndrome (IBS) is a highly prevalent problem, with a number of symptoms and signs likely caused by multiple factors. The ability of probiotics to provide relief has been widely tested with several products shown to confer benefits. The finding that *Bifidobacterium infantis* 35624 reduced IBS symptom scores for abdominal pain/discomfort, bloating/distension, and bowel movement difficulty and normalized the inflammatory IL-10/IL-12 ratio (O'Mahony et al. 2005) led to its development by P&G worldwide. Subsequently, a double blind, placebo-controlled, parallel-designed study of 214 IBS patients showed that a daily capsule of *L. plantarum* 299v (DSM 9843) for 4 weeks reduced both pain severity and daily frequency, as well as bloating (Ducrotté et al 2012). As some IBS patients suffer from constipation, products that reduce food transit time could potentially be useful. As twice daily, two pots of yogurt containing *B. lactis* DN173-010 had been shown to improve transit times in adults (Agrawal et al. 2009); as has one pot per day (Guyonnet et al. 2009), the product was tested in 159 constipated children (defecation frequency less than times per week) for 3 weeks. The fermented dairy product did increase stool frequency, but not more so than the placebo (Tabbers et al. 2011).

Inflammatory bowel disease (IBD) is a more severe gastrointestinal condition, often requiring surgical intervention. Of the two types, Crohn's Disease (CD) has not been well managed by bacterial or yeast probiotics (Rolfe *et al.* 2006; Bourreille *et al.* 2013), for reasons that are still unclear, while Ulcerative Colitis (UC) therapy has had some success using three very different formulations. A meta-analyses showed overall risk ratios of 2.7 for inducing remission in active UC with *Bifidobacterium* supplemented fermented milk versus placebo or no additive treatment ($n = 2$); 1.88 with a commercial probiotic formulation VSL#3 versus placebo ($n = 2$); 1.08 for preventing relapses in inactive UC with *Escherichia coli* Nissle 1917 (Mutaflor) versus standard treatment ($n = 3$); 0.17 for preventing relapses in inactive UC/ ileo-anal pouch anastomosis patients with VSL#3 versus placebo; 1.21 for preventing endoscopic recurrences in inactive CD with *L. rhamnosus* GG versus placebo ($n = 2$); and 0.93 for preventing endoscopic recurrences in inactive CD with *Lactobacillus johnsonii* versus placebo ($n = 2$) (Jonkers *et al.* 2012). In children with active distal ulcerative colitis, rectal infusion of *L. reuteri* was found to be effective in improving mucosal inflammation and changing mucosal expression levels of some cytokines involved in the mechanisms of IBD (Olivia *et al.* 2012). These mixed results may in part be explained by differences in trial design and patient populations, but it is intriguing that single Gram-positive and Gram-negative probiotic strains are both effective when the might have been expected to induce different immunological reactions in the host. Plus, VSL#3, given at 3.6×10^{12} CFU VSL#3 $n = 77$) or placebo ($n = 70$), twice daily for 12 weeks, showed a 50% decrease in the Ulcerative Colitis Disease Activity Index at week 6, for VSL#3 (25; 32.5%) than the group given placebo (7; 10%) ($P = 0.001$). At week 12, there were 33 patients given VSL#3 (42.9%) who achieved remission, compared with 11 patients given placebo (15.7%) ($P < 0.001$) (Sood *et al.* 2009).

While the causes of IBD are not well understood, albeit there is some indication of shifts in the composition of the intestinal microbiota including lower diversity and increase in *E. coli* and *Fusobacterium* (Kostic *et al.* 2014), the disruption of the gut microbiota with antibiotic use has been well documented. The fact that bacterial and yeast probiotics, including *L. rhamnosus* GG, *Lactobacillus* casei DN-114 001 (Danactive) (Hickson *et al.* 2007), *L. acidophilus* CL1285 with *L. casei* LBC 8OR (BIO K Plus)(Gao *et al.* 2010), *Lactobacillus helveticus* R0052 and *Lactobacillus rhamnosus* R0011 (Lacidofil) (Foster *et al.* 2011), and *Saccharomyces cerevisiae* subsp. *boulardii* LYO (Florastor) (Kotowska *et al.* 2005) can reduce the incidence or help treat antibiotic-associated diarrhea indicates a collective effect of probiotics therapy. Of note, *S. boulardii* was not effective in a single-center, randomized, double-blind, placebo-controlled, parallel-group trial on 562 elderly (mean age 78–79) patients prescribed antibiotics or on antibiotic therapy for less than 48 hours (Pozzoni *et al.* 2012). This is interesting as the incidence was around 13%, which might be interpreted as being too low to impact, yet in a Turkish study at a military hospital, antibiotic-associated diarrhea was 9% (7/78) in the controls and fell to 1.4% (1/73) ($p < 0.05$) with *S. boulardii* prophylaxis (Can *et al.* 2006). A number of probiotic yogurts have *L. acidophilus* LA5 and *B. lactis* Bb12 (Yoptimal), with one study showing reduced duration of diarrhea when given with therapy for *Helicobacter pylori* (de Vrese *et al.* 2011). This disease is not prevalent in North America, but an interesting aspect of the study was that the same product pasteurized after fermentation and a milk acidified with lactic acid were not as effective, supporting the need for the viable probiotic organisms.

One serious and often fatal disease that has been responsible for many deaths in Canada is *Clostridium difficile* gastroenteritis. Indeed, in the Province of Quebec, a total of 619 people died from *C. difficile* over the course of 2010 and 2011. With almost 5,000 deaths in the past 10 years in this one area of Canada, invariably induced by antibiotic usage, the successful introduction of the commercial formulation Bio K+ into some hospitals to prevent the disease led to 2013 approval by Health Canada of this product use. This is an excellent example of four important points related to probiotics:

1. The company, encouraged by probiotic researchers and the publication of the FAO-WHO documents (FAO/WHO 2001a, 2001b), invested in R&D and performed studies to show that the product benefitted patients.
2. Hospitals were receptive to a food being used to prevent disease. This counters the antiquated stance of the European Food Safety Authority and the FDA who stipulate that only drugs can prevent disease, with the former banning the term probiotic for supposed lack of health evidence. This is also credit to Health Canada for being open to approving this food product for disease prevention.
3. It is important to translate research to those who need it the most, in this case patients at risk of death.
4. While antibiotics have saved many lives, their side effects are now being examined much more intensely, and questions must be raised about their widespread use, their efficacy, and methods that can replace some of their uses.

While probiotics have been useful to prevent *C. difficile* infection, they have been less effective in treating the condition. This has led to the more radical approach of fecal microbiota transplantation (FMT) in which donor feces from a healthy person is transferred via rectal or stomach infusion. With success rates around 90% in curing *C. difficile* infection (Smits *et al.* 2013; van Nood *et al.* 2013), this radical replacement therapy is now being tested for a range of conditions that appear to be affected by the gut microbiome. While not a probiotic per se (Hill *et al.* 2014), fecal transplant has evolved into other multiple strain therapies to manage chronic *C. difficile* infection. A 33 strain "RePOOPulate" composition (see previous chapter) has shown promise (Petrof *et al.* 2013a, 2013b), and as it is well defined and less variable between batches than feces, this product is deemed to be probiotic (Hill et a. 2014). Regulators in Canada have been open to allowing further use of RePOOPulate, whereas in the United States, the Food and Drug Administration has decided to regulate the treatment as an experimental drug. This, again, reflects on the antiquated system where the agency's framework, originating over 100 years ago, seems more important than grappling with new science and the growing realisation that drugs are not the savior of society. To define a normal microbial composition with food waste as a drug, and their decision to view yogurt as a drug if used to prevent disease, illustrates the ignorance of regulators and the need for legislation to change the system.

The world's first commercial probiotic introduced in the 1930s, a Japanese drink containing sweetened milk and *L. casei* Shirota, has been studied extensively. Widely available in Asia and Europe, but only partially introduced to North America, the product has been tested extensively in humans mostly in terms of its ability to modulate immunity. Two applications are particularly interesting: the ability to reduce the risk of recurrence of bladder cancer (Ohashi *et al.* 2002), and the possible reduction of nasal allergy symptoms in people exposed to Japanese cedar pollen (Tamura 2007). The former was based upon a retrospective study of 180 cases (mean age: 67 years, SD 10) of bladder cancer and 445 population-based controls. The adjusted odds ratios (OR) for smoking was 1.61 (95% confidence interval: 1.10–2.36), and for those having 10–15 years ago intake of fermented milk products 0.46 (0.27–0.79) for 1–2 times/week and 0.61 (0.38–0.99) for 3–4 or more times/week, respectively. It was suggested that an increase in natural killer cell activity might have increased the mucosa's ability to somehow prevent cancer cells recurring, but the mechanisms have not been verified.

TABLE 11.1. Probiotic Products and Strains with Documented Human Data Sold in Canada.

Brand Name*	Strain Contents	Effects Noted	References
BLIS Cultured Care	*Streptococcus salivarius* K12	Oral care, potentially including halitosis relief	Burton et al. 2006, 2013;
RepHresh Pro-B	*Lactobacillus rhamnosus* GR-1, *L. reuteri* RC-14	Improved urogenital health in women; gut and immune modulation	Bruce and Reid 1988; Bruce et al. 1992; Reid and Bruce 1992; Reid et al. 1997; Martinez et al. 2009a, 2009b; Anukam et al. 2006; Hummelen 2011; Petricevic 2008; Beerpoot et al. 2012; Baroja et al. 2007; Koyama et al. 2010
Provacare	*Lactobacillus rhamnosus* Lcr35	Prevention of necrotizing enterocolitis in infants	Petricevic and Witt 2008
BioGai drops	*Lactobacillus reuteri* DSM 17938	Treatment of colic and reduced crying in babies	Indrio et al. 2014
Culturelle	*Lactobacillus rhamnosus* GG	Prevention of diarrhea in children, reduction in respiratory illness, reduced cumulative prevalence of eczema; possible reduction in maternal allergy	Vanderhoof et al. 1999; Kalliomäki et al. 2003; Wickens et al. 2008; Ou et al. 2012
Align	*Bifidobacterium infantis* 35624	Relief from IBS symptoms	O'Mahony et al. 2005
TuZen	*Lactobacillus plantarum* 299v	Relief from IBS symptoms	Ducrotté et al. 2012
Activia yogurt	*Bifidobacterium lactis* DN173-010	Improved transit time in adults	Agrawal et al. et al. 2009
Mutaflor	*E. coli* Nissle 1917	Preventing relapses in inactive UC	Rolfe et al. 2006; Jonkers et al. 2012

(continued)

TABLE 11.1 (continued). Probiotic Products and Strains with Documented Human Data Sold in Canada.

Brand Name*	Strain Contents	Effects Noted	References
VSL#3	Lactobacillus acidophilus, Lactobacillus delbrueckii subsp. bulgaricus, Lactobacillus plantarum, Lactobacillus casei, Bifidobacterium infantis, Bifidobacterium longum, Bifidobacterium breve and Streptococcus salivarius subsp thermophiles	Decrease in the UC Disease Activity Index	Sood et al. 2009
Danactive	Lactobacillus casei DN-114 001	Prevention of antibiotic-associated diarrhea	Hickson et al. 2005
Bio K +	Lactobacillus acidophilus CL 1285 with Lactobacillus casei LBC 8OR	Prevention of antibiotic-associated diarrhea	Gao et al. 2010
Lacidofil	Lactobacillus helveticus R0052 and Lactobacillus rhamnosus R0011	Supplement for antibiotic-associated diarrhoea and acute gastroenteritis.	Foster et al. 2011
Florastor	Saccharomyces cerevisiae subsp boulardii LYO	Prevention of antibiotic-associated diarrhea	Kotowska et al. 2005; Can et al. 2006
Yoptimal	Lactobacillus acidophilus LA5 and Bifidobacterium lactis Bb12	Prevention of antibiotic-associated diarrhea	de Vrese et al. 2011
Yakult	Lactobacillus casei Shirota	Reduces the risk of bladder cancer	Tamura et al. 2007

Such a Th1 response could also be active in alleviating allergies. In a study of birch-pollen-specific IgE, the Shirota strain decreased the IgE and reduced the number of CD16(+)/CD56(+) cells in peripheral blood mononuclear cells (Snel et al. 2011).

11.4. PROBIOTICS AND MICROBIOME MYSTERIES

Of note, few probiotics appear to function by significantly altering the existing microbiome of the recipient (McNulty et al. 2011; Macklaim 2014). This may be a feature of current technical limitations in detection, or a reflection of the colonization resistance of the host, or the administration of strains in a condition not suitable to colonize a niche. But the inability to colonize the host has so far not prevented certain probiotics from conferring benefits. It had long been assumed that the probiotic strains had to adhere to host cells and colonize in order to be functional, but while temporary adhesion to surfaces and mucins may aid in the organisms growing and affecting the host, persistence rarely happens. There are likely many reasons for this colonization resistance, including entrenched commensals better able to compete for space and nutrients. It seems hard to imagine that immune tolerance excludes the incoming probiotic strains, as the species are already part of the autochthonous microbiota and no obvious genetic characteristics differentiate them from the probiotic. Studies are ongoing to attempt to select probiotic strains with adhesins, like pili (Douillard et al. 2013), in some cases through genetic engineering, but regulatory authorities appear to prefer strains that do not persist in the host.

If a probiotic is not able to persist, then it must confer its effects within the few days or weeks that it remains in the body. Metabolomic studies in humans and animals have shown that this indeed does occur (Endo et al. 2013; van Baarlen et al. 2009), with production of SCFAs being one mediator. In the near future, the combination of metagenomics, metabolomics, and transcriptomics will allow a deep understanding of the organisms present and what they are doing at a given time. The expansion of the probiotic field, so easily seen by the rapid increase in product sales and peer-reviewed publication, should be measured primarily by the impact made to human and animal health. The dramatic effects in saving the lives of babies and children illustrates the power of the concept, but our inability to take these products to developing countries cannot be condoned. Thanks to high throughput 16s rRNA sequencing and other tools, we are realizing how integral microbes are to life. Harnessing their power (Reid *et al.* 2014) and understanding their mechanisms of action (Reid *et al.* 2011), will make it possible to target specific probiotics for specific conditions, ultimately in somewhat of a personalized manner. Exciting times lie ahead.

11.5. REFERENCES

Agrawal, A., Houghton, L.A., Morris, J., Reilly, B., Guyonnet, D., Goupil Feuillerat, N., *et al.* 2009. Clinical trial: The effects of a fermented milk product containing *Bifidobacterium* lactis DN-173 010 on abdominal distension and gastrointestinal transit in irritable bowel syndrome with constipation. *Aliment Pharmacol Ther.* Jan; *29*(1): 104–14.

Anonymous. 2012. http://www.anh-europe.org/news/efsa-deliberately-blind-to-good-probiotic-science

Anukam, K.C., Osazuwa, E., Osemene, G.I., Ehigiagbe, F., Bruce, A.W., and Reid, G. 2006. Clinical study comparing probiotic *Lactobacillus* GR-1 and RC-14 with metronidazole vaginal gel to treat symptomatic bacterial vaginosis. *Microbes Infect.* *8*(12–13): 2772–2776.

APUA. 2011. Triclosan. http://www.tufts.edu/med/apua/consumers/personal_home_21_4240495089.pdf

Arbesman, S. 2012. http://blogs.hbr.org/2012/11/why-do-great-ideas-take-so-lon/

Atlas, R. 1999. Probiotics—snake oil for the new millennium? *Environ Microbiol.* Oct; *1*(5): 377–82.

Barefoot, S.F. and Nettles, C.G. 1993. Antibiosis revisited: Bacteriocins produced by dairy starter cultures. *J Dairy Sci.* Aug; *76*(8): 2366-79.

Baroja, M.L., Kirjavainen, P.V., Hekmat, S., and Reid, G. 2007. Anti-inflammatory effects of probiotic-yogurt in inflammatory bowel disease patients. *Clin Experimental Immunol.* *149*: 470–9.

Beerepoot, M.A., ter Riet, G., Nys, S., van der Wal, W.M., de Borgie, C.A., de Reijke, T.M., *et al.* 2012. Lactobacilli vs antibiotics to prevent urinary tract infections: A randomized, double-blind, noninferiority trial in postmenopausal women. *Arch Intern Med.* May 14; *172*(9): 704–12.

Blaser, M. 2011. Antibiotic overuse: Stop the killing of beneficial bacteria. *Nature.* Aug 24; *476*(7361): 393-4.

Bongartz, T., Sutton, A.J., Sweeting, M.J., Buchan, I., Matteson, E.L., Montori, V. 2006. Anti-TNF antibody therapy in rheumatoid arthritis and the risk of serious infections and malignancies: Systematic review and meta-analysis of rare harmful effects in randomized controlled trials. *JAMA.* May 17; *295*(19): 2275–85.

Bonsante, F., Iacobelli, S., and Gouyon, J.B. 2013. Routine probiotic use in very preterm infants: retrospective comparison of two cohorts. *Am J Perinatol.* Jan; *30*(1): 41–6.

Bourreille, A., Cadiot, G., Le Dreau, G., Laharie, D., Beaugerie, L., Dupas, J.L., *et al.* 2013. Saccharomyces boulardii does not prevent relapse of Crohn's disease. *Clin Gastroenterol Hepatol.* Aug; *11*(8): 982–7.

Bruce, A.W. and Reid, G. 1988. Intravaginal instillation of *lactobacilli* for prevention of recurrent urinary tract infections. *Can. J. Microbiol.* *34*: 339–343.

Bruce, A.W., Chadwick, P., Hassan, A., and VanCott, G.F. 1973. Recurrent urethritis in women. *Can Med Assoc J.* Apr 21; *108*(8): 973–6.

Bruce, A.W., Reid, G., McGroarty, J.A., Taylor, M., and Preston, C. 1992. Preliminary study on the prevention of recurrent urinary tract infections in ten adult women using intravaginal *lactobacilli*. *Int. Urogynecol. J.* 3: 22–25.

Burton, J.P., Chilcott, C.N., Moore, C.J., Speiser, G., and Tagg, J.R. 2006. A preliminary study of the effect of probiotic *Streptococcus salivarius* K12 on oral malodour parameters. *J Appl Microbiol.* Apr; *100*(4): 754–64.

Burton, J.P., Drummond, B.K., Chilcott, C.N., Tagg, J.R., Thomson, W.M., Hale, J.D., *et al.* 2013. Influence of the probiotic *Streptococcus salivarius* strain M18 on indices of dental health in children: A randomized double-blind, placebo-controlled trial. *J Med Microbiol.* Jun; *62*(Pt 6): 875–84.

Can, M., Beşirbellioglu, B.A., Avci, I.Y., Beker, C.M., and Pahsa, A. 2006. Prophylactic Saccharomyces boulardii in the prevention of antibiotic-associated diarrhea: A prospective study. Med Sci Monit. Apr; *12*(4):PI19–22.

Cho, I., Yamanishi, S., Cox, L., Methé, B.A., Zavadil, J., Li, K., *et al.* 2012. Antibiotics in early life alter the murine colonic microbiome and adiposity. *Nature.* Aug 30; *488*(7413): 621–6.

Choutet P. 2003. Antibiotic use: Excesses and consequences. *Rev Prat.* Sep 30; *53*(14): 1527-32.

de Vrese, M., Kristen, H., Rautenberg, P., Laue, C., and Schrezenmeir, J. 2011. Probiotic *lactobacilli* and bifidobacteria in a fermented milk product with added fruit preparation reduce antibiotic associated diarrhea and Helicobacter pylori activity. *J Dairy Res.* Nov; 78(4): 396–403.

Deshpande, G., Rao, S., Patole, S., and Bulsara, M. 2010. Updated meta-analysis of probiotics for preventing necrotizing enterocolitis in preterm neonates. *Pediatrics.* *125*(5): 921–30.

Dinsbach, N.A. 2012. Antibiotics in dentistry: Bacteremia, antibiotic prophylaxis, and antibiotic misuse. *Gen Dent.* May-Jun; *60*(3): 200-7.

Douillard, F.P., Ribbera, A., Kant, R., Pietilä, T.E., Järvinen, H.M., Messing, M., *et al.* 2013. Comparative genomic and functional analysis of 100 *Lactobacillus rhamnosus* strains and their comparison with strain GG. *PLoS Genet.* *9*(8):e1003683.

Downing, N.S., Aminawung, J.A., Shah, N.D., Krumholz, H.M., and Ross, J.S. 2014. Clinical trial evidence supporting FDA approval of novel therapeutic agents, 2005–2012. *JAMA.* Jan 22–29; *311*(4):368–77.

Ducrotté, P., Sawant, P., and Jayanthi, V. 2012. Clinical trial: *Lactobacillus plantarum* 299v (DSM 9843) improves symptoms of irritable bowel syndrome. *World J Gastroenterol.* Aug 14; *18*(30): 4012–8.

Endo, H., Niioka, M., Kobayashi, N., Tanaka, M., and Watanabe, T. 2013. Butyrate-producing probiotics reduce nonalcoholic fatty liver disease progression in rats: New insight into the probiotics for the gut-liver axis. *PLoS One.* May 16; *8*(5): e63388.

FAO/WHO 2001b. Guidelines for the evaluation of probiotics in food. ftp://ftp.fao.org/es/esn/food/wgreport2.pdf

FAO/WHO. 2001a. Evaluation of health and nutritional properties of powder milk and live lactic acid bacteria. Food and Agriculture Organization of the United Nations and World Health Organization Expert Consultation Report. http://www.who.int/foodsafety/publications/fs_management/en/probiotics.pdf

FDA. 2014. http://www.regulations.gov/#!documentDetail;D=FDA-2010-D-0503-0016

Foster, L.M., Tompkins, T.A., and Dahl, W.J. 2011. A comprehensive post-market review of studies on a probiotic product containing *Lactobacillus helveticus* R0052 and *Lactobacillus rhamnosus* R0011. *Benef Microbes.* Dec 1; *2*(4): 319–34.

Gao, X.W., Mubasher, M., Fang, C.Y., Reifer, C., and Miller, L.E. 2010. Dose-response efficacy of a proprietary probiotic formula of *Lactobacillus* acidophilus CL1285 and *Lactobacillus* casei LBC80R for antibiotic-associated diarrhea and *Clostridium difficile*-associated diarrhea prophylaxis in adult patients. *Am J Gastroenterol.* Jul; *105*(7): 1636–41.

Giacoia, G.P. and Mattison, 2005. D.R. Newborns and drug studies: The NICHD/FDA newborn drug development initiative. *Clin Ther.* Jun; *27*(6): 796–813.

Guyonnet, D., Woodcock, A., Stefani, B., Trevisan, C., and Hall, C. 2009. Fermented milk containing *Bifidobacterium* lactis DN-173 010 improved self-reported digestive comfort amongst a general population of adults. A randomized, open-label, controlled, pilot study. *J Dig Dis.* Feb; *10*(1): 61–70.

Health Canada. 2014. http://www.hc-sc.gc.ca/fn-an/label-etiquet/claims-reclam/probiotics_claims-allegations_probiotiques-eng.php

Hickson, M., D'Souza, A.L., Muthu, N., Rogers, T.R., Want, S., Rajkumar, C., et al. 2007. Use of probiotic *Lactobacillus* preparation to prevent diarrhoea associated with antibiotics: Randomised double blind placebo controlled trial. *BMJ.* Jul 14; *335*(7610): 80

Hill, C., Guarner, F., Reid, G., Gibson, G.R., Merenstein, D.J., Pot, B., et al. 2014. Expert consensus document: The International Scientific Association for Probiotics and Prebiotics consensus statement on the scope and appropriate use of the term probiotic. *Nat Rev Gastroenterol Hepatol.* Jun 10.

Holdcroft, A. 2007. Integrating the dimensions of sex and gender into basic life sciences research: Methodologic and ethical issues. *Gend Med. 4 Suppl B*: S64-74.

Hudson, E. 2014. Market for probiotics. http://uk.amiando.com/eventResources/0/G/DveLYOiZk70uuA/Ewa_Hudson.pdfmarket for probiotics

Hummelen, R., Changalucha, J., Butamanya, N.L., Koyama, T.E., Cook, A., Habbema, J.D.F., et al. 2011. Reid G. Effect of 25 weeks probiotic supplementation on immune function of HIV patients. *Gut Microbes.* 2(2): 80–5.

Indrio, F., Di Mauro, A., Riezzo, G., Civardi, E., Intini, C., Corvaglia, L., et al. 2014. Prophylactic use of a probiotic in the prevention of colic, regurgitation, and functional constipation: A randomized clinical trial. *JAMA Pediatr. Mar; 168*(3): 228–33.

Jonkers, D., Penders, J., Masclee, A., and Pierik, M. 2012. Probiotics in the management of inflammatory bowel disease: A systematic review of intervention studies in adult patients. *Drugs.* Apr 16; *72*(6): 803–23.

Kalliomäki, M., Salminen, S., Poussa, T., Arvilommi, H., and Isolauri, E. 2003. Probiotics and prevention of atopic disease: 4-year follow-up of a randomised placebo-controlled trial. *Lancet.* May 31; *361*(9372): 1869–71.

Kopp, M.V., Hennemuth, I., Heinzmann, A., and Urbanek, R. 2008. Randomized, double-blind, placebo-controlled trial of probiotics for primary prevention: No clinical effects of *Lactobacillus* GG supplementation. *Pediatrics.* Apr; 121(4): e850-6.

Kostic, A.D., Xavier, R.J., and Gevers, D. 2014. The microbiome in inflammatory bowel disease: Current status and the future ahead. *Gastroenterology.* May; *146*(6): 1489–99.

Kotowska, M., Albrecht, P., and Szajewska, H. 2005. Saccharomyces boulardii in the prevention of antibiotic-associated diarrhoea in children: A randomized double-blind placebo-controlled trial. *Aliment Pharmacol Ther.* Mar 1; *21*(5): 583–90.

Koyama, T., Kirjavainen, P.V., Fisher, C., Anukam, K., Summers, K., Hekmat, S., and Reid, G. 2010. Development and pilot evaluation of a novel probiotic mixture for the management of seasonal allergic rhinitis. *Can J Microbiol.* 56: 730–738.

Kumpu, M., Kekkonen, R.A., Kautiainen, H., Järvenpää, S., Kristo, A., Huovinen, P., et al. 2012. Milk containing probiotic *Lactobacillus rhamnosus* GG and respiratory illness in children: A randomized, double-blind, placebo-controlled trial. *Eur J Clin Nutr.* Sep; *66*(9): 1020–3.

Laughon, M.M., Avant, D., Tripathi, N., Hornik, C.P., Cohen-Wolkowiez, M., Clark, R.H., et al. 2014. Drug labeling and exposure in neonates. *JAMA Pediatr.* Feb; 168(2): 130–6.

Macklaim, J.M., Clemente, J., Knight, R., Gloor, G.B., and Reid, G. 2014. Effect of antimicrobial and probiotic therapy on the vaginal microbiome. Submitted to *Sci Trans Med.*

Majamaa, H. and Isolauri, E. 1997. Probiotics: A novel approach in the management of food allergy. *J Allergy Clin Immunol.* Feb; *99*(2): 179–85.

Manzetti, S. and Ghisi, R. 2014. The environmental release and fate of antibiotics. *Mar Pollut Bull.* Feb 15; *79*(1-2): 7-15.

Marshall, B.M. and Levy, S.B. 2011. Food animals and antimicrobials: Impacts on human health. *Clin Microbiol Rev.* Oct; *24*(4): 718-33.

Marteau, P., Pochart, P., Bouhnik, Y., Zidi, S., Goderel, I., and Rambaud, J.C. 1992. Survival of *Lactobacillus* acidophilus and *Bifidobacterium* sp. in the small intestine following ingestion in fermented milk. A rational basis for the use of probiotics in man. *Gastroenterol Clin Biol.* 16(1): 25–8.

Martinez, R.C., Franceschini, S.A. Patta, M.C., Quintana, S.M., Candido, R.C., Ferreira, J.C., et al. 2009b. Improved treatment of vulvovaginal candidiasis with fluconazole plus probiotic *Lactobacillus rhamnosus* GR-1 and *Lactobacillus reuteri* RC-14. *Letts Appl. Microbiol.* 2009; *48*(3): 269–74.

Martinez, R.C., Franceschini, S.A., Patta, M.C., Quintana, S.M., Candido, R.C., Ferreira, J.C., et al. 2009a. Improved cure of bacterial vaginosis with single dose of tinidazole (2g) and *Lactobacillus rhamnosus* GR-1 and *Lactobacillus reuteri* RC-14: A randomized, double-blind, placebo-controlled trial. *Can J Microbiol.* *55*(2): 133–138.

McNulty, N.P., Yatsunenko, T., Hsiao, A., Faith, J.J., Muegge, B.D., Goodman, A.L., et al. 2011. The impact of a consortium of fermented milk strains on the gut microbiome of gnotobiotic mice and monozygotic twins. *Sci Transl Med.* Oct 26; *3*(106): 106ra106.

Möllenbrink, M. and Bruckschen, E. 1994. Treatment of chronic constipation with physiologic Escherichia coli bacteria. Results of a clinical study of the effectiveness and tolerance of microbiological therapy with the *E. coli* Nissle 1917 strain (Mutaflor). *Med Klin (Munich)*. Nov 15; *89*(11): 587–93.

Ofek Shlomai, N., Deshpande, G., Rao, S., and Patole, S. 2014. Probiotics for preterm neonates: What will it take to change clinical practice? *Neonatology. 105*(1):64–70.

Ohashi, Y., Nakai, S., Tsukamoto, T., Masumori, N., Akaza, H., Miyanaga, N., *et al.* 2002. Habitual intake of lactic acid bacteria and risk reduction of bladder cancer. *Urol Int. 68*(4): 273–80.

Oliva, S., Di Nardo, G., Ferrari, F., Mallardo, S., Rossi, P., Patrizi, G., *et al.* 2012. Randomised clinical trial: The effectiveness of *Lactobacillus reuteri* ATCC 55730 rectal enema in children with active distal ulcerative colitis. *Aliment Pharmacol Ther.* Feb; *35*(3): 327–34.

O'Mahony, L., McCarthy, J., Kelly, P., Hurley, G., Luo, F., Chen, K., *et al.* 2005. *Lactobacillus* and *Bifidobacterium* in irritable bowel syndrome: Symptom responses and relationship to cytokine profiles. *Gastroenterology.* Mar; *128*(3): 541–51.

Ou, C.Y., Kuo, H.C., Wang, L., Hsu, T.Y., Chuang, H., Liu, C.A., *et al.* 2012. Prenatal and postnatal probiotics reduces maternal but not childhood allergic diseases: A randomized, double-blind, placebo-controlled trial. *Clin Exp Allergy.* Sep; *42*(9): 1386–96.

Pelto, L., Isolauri, E., Lilius, E.M., Nuutila, J., and Salminen, S. 1998. Probiotic bacteria down-regulate the milk-induced inflammatory response in milk-hypersensitive subjects but have an immunostimulatory effect in healthy subjects. *Clin Exp Allergy. 28*(12): 1474–9

Petricevic, L. and Witt, A. 2008. The role of *Lactobacillus* casei *rhamnosus* Lcr35 in restoring the normal vaginal flora after antibiotic treatment of bacterial vaginosis. *BJOG.* Oct; *115*(11): 1369–74.

Petricevic, L., Unger, F.M., Viernstein, H., and Kiss, H. 2008. Randomized, double-blind, placebo-controlled study of oral to improve the vaginal flora of postmenopausal women. *Eur J Obstet Gynecol Reprod Biol.* Nov; *141*(1): 54–7.

Petrof, E.O., Claud, E.C., Gloor, G.B., and Allen-Vercoe, E. 2013a. Microbial ecosystems therapeutics: a new paradigm in medicine? *Benef Microbes.* Mar 1; *4*(1): 53–65.

Petrof, E.O., Gloor, G.B., Vanner, S.J., Weese, S.J., Carter, D., Daigneault, M.C., *et al.* 2013b. Stool substitute transplant therapy for the eradication of *Clostridium difficile* infection: 'RePOOPulating' the gut. *Microbiome.* Jan 9; *1*(1): 3.

Petschow, B., Doré, J., Hibberd, P., Dinan, T., Reid, G., Blaser M., *et al.* 2013. Probiotics, prebiotics, and the host microbiome: The science of translation. Ann, N.Y. Acad Sci. 2013 Dec; 1306: 1–17.

Pozzoni, P., Riva, A., Bellatorre, A.G., Amigoni, M., Redaelli, E., Ronchetti, A., *et al.* 2012. Saccharomyces boulardii for the prevention of antibiotic-associated diarrhea in adult hospitalized patients: A single-center, randomized, double-blind, placebo-controlled trial. *Am J Gastroenterol.* Jun; *107*(6): 922–31.

Reid, G. 1999. The potential of probiotics. *Alternative Therapies in Women's Health.* November, 91–94.

Reid, G. 2005. Food and Agricultural Organization of the United Nation and the WHO.

The importance of guidelines in the development and application of probiotics. *Curr Pharm Des.* 11: 11–6.

Reid, G., Bruce, A.W., and Taylor, M. 1992. Influence of three day antimicrobial therapy and *Lactobacillus suppositories* on recurrence of urinary tract infection. *Clin. Therapeutics. 14* (1): 11–16.

Reid, G., Bruce, A.W., McGroarty, J.A., Cheng, K.J., and Costerton, J.W. 1990. Is there a role for *lactobacilli* in prevention of urogenital and intestinal infections? *Clin. Microbiol. Rev.* 3(4): 335–344.

Reid, G., Chan, R.C.Y., Bruce, A.W., and Costerton, J.W. 1985. Prevention of urinary tract infection in rats with an indigenous *Lactobacillus* casei strain. *Infect. Immun.* 49(2): 320–324.

Reid, G., Cook, R.L., and Bruce, A.W. 1987. Examination of strains of *lactobacilli* for properties which may influence bacterial interference in the urinary tract. *J. Urol.* 138: 330–335.

Reid, G., Millsap, K., and Bruce, A.W. 1994. Implantation of *Lactobacillus* casei var *rhamnosus* into the vagina. *The Lancet.* 344: 1229.

Reid, G., Nduti, N., Sybesma, W., Kort, R., Kollmann, T.R., Adam, R., *et al.* 2014. Harnessing microbiome and probiotic research in sub-Saharan Africa: Recommendations from an African workshop. *Microbiome.* 2:12. DOI: 10.1186/2049-2618-2-12.

Reid, G., Younes, J., van der Mei, H.C., Gloor, G.B., Knight, R., and Busscher, H.J. 2011. Microbiota Restoration: Natural and supplemented recovery of human microbial communities. *Nat. Rev. Microbiol.* 9(1): 27–38.

Reynolds, L.A. and Finlay, B.B. A case for antibiotic perturbation of the microbiota leading to allergy development. *Expert Rev Clin Immunol.* 2013 Nov; 9(11):1019–30.

Rolfe, V.E., Fortun, P.J., Hawkey, C.J., and Bath-Hextall, F. 2006. Probiotics for maintenance of remission in Crohn's disease. *Cochrane Database Syst Rev.* Oct 18; (4): CD004826.

Salque, M., Bogucki, P.I., Pyzel, J., Sobkowiak-Tabaka, I., Grygiel, R., Szmyt, M., *et al.*2013. Earliest evidence for cheese making in the sixth millennium BC in northern Europe. *Nature.* Jan 24; *493*(7433): 522-5. doi: 10.1038/nature11698. Epub 2012 Dec 12.

Schiffrin, E.J. and Blum, S. 2002. Interactions between the microbiota and the intestinal mucosa. *Eur J Clin Nutr.* Aug; 56 Suppl 3: S60-4.

Skokovic-Sunjic, D. 2013. http://www.isapp.net/Portals/0/docs/News/clinical%20guide%20probiotics%20canada%202013.pdf

Smits, L.P., Bouter, K.E., de Vos, W.M., Borody, T.J., and Nieuwdorp, M. 2013. Therapeutic potential of fecal microbiota transplantation. *Gastroenterology.* Nov; 145(5): 946–53.

Snel, J., Vissers, Y.M., Smit, B.A., Jongen, J.M., van der Meulen, E.T., Zwijsen, R., *et al.* 2011. Strain-specific immunomodulatory effects of *Lactobacillus plantarum* strains on birch-pollen-allergic subjects out of season. *Clin Exp Allergy.* Feb; 41(2): 232–42.

Sood, A., Midha, V., Makharia, G.K., Ahuja, V., Singal, D., Goswami, P., and Tandon, R.K. 2009. The probiotic preparation, VSL#3 induces remission in patients with mild-to-moderately active ulcerative colitis. *Clin Gastroenterol Hepatol.* Nov; 7(11):1202–9, 1209.e1.

Srigley, J.A., Brooks, A., Sung, M., Yamamura, D., Haider, S., and Mertz, D. 2013. Inappropriate use of antibiotics and *Clostridium difficile* infection. *Am J Infect Control.* Nov; 41(11): 1116–8.

Tabbers, M.M., Chmielewska, A., Roseboom, M.G., Crastes, N., Perrin, C., Reitsma, J.B., *et al.* 2011. Fermented milk containing *Bifidobacterium* lactis DN-173 010 in childhood constipation: A randomized, double-blind, controlled trial. *Pediatrics.* Jun; 127(6): e1392–9.

Tamura, M., Shikina, T., Morihana, T., Hayama, M., Kajimoto, O., Sakamoto, A., *et al.* 2007. Effects of probiotics on allergic rhinitis induced by Japanese cedar pollen: Randomized double-blind, placebo-controlled clinical trial. *Int Arch Allergy Immunol.* 143(1): 75–82.

Tejero-Sariñena, S., Barlow, J., Costabile, A., Gibson, G.R., and Rowland, I. 2012. In vitro evaluation of the antimicrobial activity of a range of probiotics against pathogens: Evidence for the effects of organic acids. *Anaerobe.* Oct; *18*(5): 530–8.

van Baarlen, P., Troost, F.J., van Hemert, S., van der Meer, C., de Vos, W.M., de Groot, P.J., *et al.* 2009. Hooiveld GJ, Brummer RJ, Kleerebezem M. Differential NF-κB pathways induction by *Lactobacillus plantarum* in the duodenum of healthy humans correlating with immune tolerance. *Proc Natl Acad Sci USA.* Feb 17; *106*(7): 2371–6.

van Nood, E., Vrieze, A., Nieuwdorp, M., Fuentes, S., Zoetendal, E.G., de Vos, W.M., *et al.* 2013. Duodenal infusion of donor feces for recurrent *Clostridium difficile.* *N Engl J Med.* Jan 31; *368*(5):407–15.

Vanderhoof, J.A., Whitney, D.B., Antonson, D.L., Hanner, T.L., Lupo, J.V., and Young, R.J. 1999. Lactobacillus GG in the prevention of antibiotic-associated diarrhea in children. *J Pediatr.* Nov; *135*(5):564–8.

Whitley, H. and Lindsey, W. 2009. Sex-based differences in drug activity. *Am Fam Physician.* Dec 1; *80*(11): 1254–8.

Wickens, K., Black, P.N., Stanley, T.V., Mitchell, E., Fitzharris, P., Tannock, G.W., *et al.* 2008. A differential effect of 2 probiotics in the prevention of eczema and atopy: A double-blind, randomized, placebo-controlled trial. *J Allergy Clin Immunol.* Oct; 122(4): 788–94.

CHAPTER 12

Considering the Microbiome as Part of Future Medicine and Nutrition Strategies

EMMA ALLEN-VERCOE

12.1. INTRODUCTION

THE purpose of *The Human Microbiome Handbook* is to provide an overview of current knowledge as it pertains to the human microbiome. It demonstrates that a few areas of health research have received such a surge in interest over the last decade. Moreover, while this handbook provides a current review of our understanding, the field is advancing at an astonishingly rapid rate. These are undoubtedly exciting times, since until recently modern medicine has considered human beings to be strictly human; our microbial passengers have been ignored—or worse—persecuted. It is my hope that this book has highlighted the very many aspects of our human biology and physiology that are influenced—or even controlled—by our microbial symbionts.

This chapter considers the current outlook for microbiome research, particularly as it pertains to the gut microbial ecosystem, and predicts areas where this research will be leveraged to benefit health in the near future.

12.2. MINING THE HUMAN MICROBIOTA FOR NEW DRUGS

What defines a healthy gut and why do some people seem to be more susceptible to GI infection than others? It is well known that people who have recently suffered microbial ecosystem depletion through, for example, antibiotic use or acute enteroviral infection are more susceptible to further gut infection during their convalescence (Croswell *et al.* 2009; Stecher *et al.* 2010). There are several reasons for this susceptibility, but the reduced ability for competitive exclusion of pathogens by a depleted microbiota has always been considered as a primary cause (Malago 2014). However, more recently there has been a growing appreciation for the role of the gut microbiota in maintaining homeostasis in the GI tract, through protective effects that include the secretion of chemical signals that modify pathogen behavior.

Microbes within an ecosystem interact dynamically and ecosystem cohesion may rely on microbial chemical "conversations" that inform ecosystem members of, for example, food substrate availability or type, and cross-feeding availability (El Aidy *et al.* 2013). Such chemical signals may also act as a signal for pathogens—both autochthonous opportunistic species as well as allochthonous species—to refrain from expression of virulence determinants, since this energetically expensive exercise is less likely to be fruitful for these pathogens in the face of an intact, protective microbial ecosystem. Antunes *et al.* (2014), demonstrated this principle recently by screening members of the normal gut microbiota for antvirulence activity against the well-studied food-borne pathogen, *Salmonella enterica*. By measuring expression of the *S. enterica* virulence global regulator, *hilA*, it was found that the spent culture supernatants of particular members of the *Lachnospiraceae* family in particular had repressive activity that was afforded by the secretion of an as-yet uncharacterized small molecule metabolite by these common gut microbial species.

This finding likely only scratches the surface of the potentially prophylactic chemical repertoire secreted by the healthy human microbiota, a pharmacopeia that is as-yet relatively untapped. The major barrier to this area of drug discovery lies in a general inability to culture many of our microbial symbionts; however, there are now several efforts underway to both bring recalcitrant species into *in vitro* study (reviewed in Allen-Vercoe 2013). In the future, we should expect to see an expansion in the development of drugs mined from gut microbial ecosystems.

12.3. PROTECTING THE GUT MICROBIOTA FROM COLLATERAL DAMAGE DURING ANTIBIOTIC EXPOSURE

The overuse of antimicrobial drugs has received a lot of recent attention, from the point of view that the targeted pathogens have evolved widespread resistance to these drugs, minimizing their effectiveness and creating fears of a return to the preantibiotic era when a simple puncture wound could lead to a life-threatening infection (Barriere 2015). Unfortunately, antibiotic resistance is not the only consequence of antimicrobial overuse, and there is now a growing realization that the collateral damage inflicted on the microbiota during antibiotic therapy is taking a toll on our health. Several studies have now conclusively shown that the gut microbial ecosystem changes profoundly during antibiotic administration, and that there may not be a recovery to the preantibiotic state, particularly if broad-spectrum antibiotics, or combinations of such, are used (Antunes *et al.* 2011; Arboleya *et al.* 2015; Cotter *et al.* 2012; Iapichino *et al.* 2008; Jernberg *et al.* 2007; Mangin *et al.* 2010; O'Sullivan *et al.* 2013). The missing microbiota hypothesis, as set out by Blaser and Falkow, also posits that because some aspects of the microbiota are inherited (through, for example, the processes of birth and breastfeeding), the ecosystem damage wreaked by antimicrobial use may compound over generations (2009).

The solution to both antibiotic resistance and collateral damage issues is to simply stop the use of antimicrobials; however, antibiotics are life-saving drugs when used appropriately, and an important weapon in the fight against infectious disease. Another strategy, therefore, is to find ways to protect the healthy microbiota during treatment. Many broad-spectrum antibiotics are given as oral preparations, and this fact as well as their pharmacology means that the gut microbiota, of all the host-associated microbes, is usually under the greatest threat during treatment. This is well illustrated by the common onset of diarrhea during a course of oral, broad-spectrum antimicrobials, which reflects a sudden change to the microbial ecology of the gut microbiota and a concomitant upset of the normal physiological homeostasis (Varughese *et al.* 2013). Part of the issue is that, if pharmacology allows, it is convenient to supply most antimicrobials by mouth for systemic absorption; however, most targeted infections are not found in the gut itself. Another problem is that for some infections where pathogenic biofilms are a component of the disease, such as otitis media, antibiotic doses have to be higher than the minimum inhibitory concentrations to be effective (Belfield *et al.* 2015), with potentially even greater collateral damage.

In the future, antibiotic administration will be much more carefully targeted. For example, treatment of ear or tooth infections may be carried out using topical applications of drugs that are less likely to accumulate to damaging concentrations in the GI tract (Dohar *et al.*

2006; Purucker *et al.* 2001). The necessity for prophylactic treatment as a routine part of surgical procedures will be more carefully evaluated (Young and Khadaroo 2014). Broad spectrum antimicrobials may be used only in emergency situations, with greater attention paid to rapid diagnostics allowing more targeted, narrow-spectrum antibiotics to be used (Spellbuerg *et al.* 2015). Alternatively, broad-spectrum antibiotics may be delivered orally in conjunction with compounds designed to maintain the antimicrobial in an inactive form until absorbed, to reduce damage to the gut microbiota from direct contact.

12.4. MICROBIAL ECOSYSTEM THERAPEUTICS

A greater understanding of the role of a damaged gut microbiota in disease has led to a surge in interest in the use of probiotics, defined as "live micro-organisms which, when administered in adequate amounts, confer a health benefit on the host," (Hill *et al.* 2014). There are many probiotics now on the market, although only a minority has had proposed beneficial effects clinically proven, and even then the effects are moderate at best (McFarland 2014). Eventually, probiotics may prove to be very useful, for example in extending remission in some types of inflammatory bowel disease, or for reducing the severity of traveler's diarrhea (Ghouri *et al.* 2014; Sarowska *et al.* 2013). Yet there are limitations to their effectiveness because, from an ecology point of view, the addition of a single or small group of similar species to the enormous diversity of the human gut is unlikely to have a dramatic effect on the ecosystem as a whole. Furthermore, because the gut microbiota is a cohesive ecosystem that can be thought of as a microbial "organ", the addition of incidental microbes in the form of probiotics does not add to the ecosystem; probiotics are unable to colonize the gastrointestinal tract and have an effect on the host only while they transit through the gut (Gonzalez-Rodriguez *et al.* 2013; Mills *et al.* 2011).

The principle of probiotic use is sound, and because the practice is generally regarded as safe, there is little reason for patients not to try it. But to view probiotics solely as a therapeutic regimen for one particular indication may exclude a greater potential. With the combined knowledge shared in *The Human Microbiome Handbook*, we have become aware of the ecological nature of the human microbiome. One particular direction involves using the combination of experimental and clinical evidence to identify the steps in development of an ecosystem rich in beneficial microbes. Alternatively, in the future, we could leverage the accumulating knowledge of the human gut microbiota to discover novel probiotic species or to create whole probiotic ecosystems. We are only beginning to understand how this is possible. Perhaps we should turn our attention to the microbiotas of individuals from varied geographical and cultural backgrounds, which traditionally are considered to be very healthy, often with higher than globally average numbers of centenarians. An expansion of the concept of probiotic use will require both time and further experimentation, yet more importantly, may result in a shift of the microbial-based medical mindset from one of treat and cure to adapt and restore.

To a certain extent, steps have already been made toward this goal. In the treatment of recurrent *Clostridium difficile* infection, where fecal transplant is rapidly emerging as an effective intervention (as discussed Chapter 3), concerns about the safety of using stool as medicine have driven us to try to determine the microbial components that are missing from the colons of patients, and then to effect a treatment by replacing these components in a defined way (Lawley *et al.* 2012; Perez-Cobas *et al.* 2014; Petrof *et al.* 2013; Shahinas *et al.* 2012; Shankar *et al.* 2014). Our prototype therapeutic, "RePOOPulate", or Microbial Ecosystem Therapeutic (MET)-1 is an example of this approach, where a 33-strain ecosystem, rich in Firmicutes, was applied to *C. difficile* patients (Petrof *et al.* 2013). *C. difficile* infection is known to correlate with a reduction in Firmicutes and a concomitant increase in Proteobacteria (Fuentes *et al.* 2014), and thus our defined ecosystem was introduced to try to redress this balance. Although only a pilot study, MET-1 rapidly cured two patients with severe, recurrent *C. difficile* infection; furthermore, 16S rRNA gene profiling of patient stool during the 6-month period after treatment revealed signatures that identified with MET-1 components, indicating that, unlike traditional probiotics, the delivered ecosystem was able to colonize for at least this long in the patients (Petrof *et al.* 2013). MET-1 was designed with microbial ecology in mind; the 33-strain mixture was derived from a single healthy donor (Petrof *et al.* 2013). We believe this to be important because these selected strains had formed part of a cohesive ecosystem in the donor. In other words, the gut environment of the donor had selected a groups of strains that could work together efficiently. Further work is underway to create more complex ecosystems from a series of different healthy donors with differing lifestyles (for example, various dietary practices), recognizing that different ecosystems may be optimal for diverse recipients.

Studies of the gut microbiotas of individuals from cultural backgrounds not typically exposed to widespread antibiotic exposure may help us to determine diversity loss in the Western world (Grzeskowiak *et al.* 2012; Schnorr *et al.* 2014) and could be instrumental in developing METs to restore the "missing microbiota". Understanding the host-microbiota cross talk that allows a given microbial ecosystem to work optimally within its host is a current research goal, and already bioinformatics approaches are being used to try to understand microbiota function in the context of disease (Collison *et al.* 2012). In the future, this stream of research will allow for the rational design of METs for use in the gut as well as other body sites. With accumulated knowledge, we may discover treatment or prevention regimens for a wide range of diseases.

12.5. PREDICTING THE INFLUENCE OF XENOBIOTICS ON THE HUMAN MICROBIOTA

Diet, so far, is the greatest known modulator of the gut microbiota (Dore and Blottiere 2015); microbes come into contact with and are influenced by the food we eat during the process of digestion, and the colon is essentially a specialized chamber where food substrates that are indigestible through the actions of human enzymes and processes can be broken down by the microbiota through anaerobic fermentation, a highly complex activity (Louis *et al.* 2007). As such, the food that we eat is more than food for our human selves, and we should consider our gut microbiota as an organ that takes part in the digestive process.

Recently, however, research on the effects of certain food additives on the colonic microbiota has brought to light some disturbing oversights. While xenobiotics such as food additives are rigorously tested for safety, in the past these toxicity assays have rarely, if ever, taken into account the effects of these additives on the gut microbiota. Some artificial food additives, such as sweeteners and emulsifying agents, have now been shown to affect the balance of microbes within the gut (Chassaing *et al.* 2015; Palmnas *et al.* 2014; Suez *et al.* 2014), and in the case of some sweeteners, may actually contribute to a microbiota reminiscent to that seen in metabolic disease (Palmnas *et al.* 2014; Suez *et al.* 2014).

In the same way that food additives have been overlooked as gut microbiota modulators, many of the drugs we consume have likewise rarely been tested for their effects on the gut microbiota (Li and Jia 2013). Pharmaceutical companies invest billions of dollars in drug discovery, and the added burden of testing for microbiome-associated effects (where every individual may be different) seems like an impossible achievement. However, drugs such as metformin, used to treat people with type-2 diabetes, serve as a good example of the role of the gut microbiota in modulation of pharmacological effects—this drug has been shown to directly affect the metabolic pathways of the microbiota, influencing the growth of some microbes over others, perhaps explaining why some individuals cannot tolerate the medication because of diarrheal side-effects (Lee and Ko 2014).

In the future, food additives and drugs will require more vigorous safety profiling, with predictions of effects on microbiota types from a wide range of individuals in addition to standard toxicology assessments. This will allow much more accurate assessments of detriment versus benefit and may alter the way that new and existing food additives and drugs are used or introduced.

12.6. LEVERAGING MICROBIOME KNOWLEDGE TO OPTIMIZE NUTRITION STRATEGIES

Simplistically, gaining nutrition from foods takes place via two pathways: (1) directly, through the actions of human enzymes and binding factors on the food and subsequent absorption of the breakdown/bound products through the gut; and (2) indirectly, through the actions of the microbiota on foods to yield host-absorbable substrates and metabolites. Until fairly recently, the second pathway has been generally ignored, however, there are important consequences of this pathway to nutrition.

At its most extreme, the gut microbiota is associated with malnutrition in both infancy and old age, with changes in the microbiota correlating with poor absorption of nutrients (Claesson *et al.* 2012; Ghosh *et al.* 2014; Kane *et al.* 2015; Lakshminarayanan *et al.* 2014; Subramanian *et al.* 2014). In childhood malnutrition, poor development of the gut microbiota, perhaps because of lack of exposure to a diverse diet, has been implicated in the disease (Subramanian *et al.* 2014). The gut microbial ecosystem becomes resistant to compositional change as successions in various taxa naturally decrease with age (Valles *et al.* 2014), and therefore a poorly developed microbial ecosystem may persist through childhood and contribute to malnutrition even in the face of dietary intervention.

At the other end of the scale, obesity and metabolic syndrome are now understood to be associated with the microbial content of the gut, and studies of identical twins discordant to obesity implicate certain microbial taxa in the disease (Goodrich *et al.* 2014). Two recent studies highlight the importance of the gut microbial ecosystem in obesity. The first of these was a trial of the effectiveness of fecal transplant, as donated from a healthy, lean individual, on metabolic disease in obese men (Vrieze *et al.* 2012). In this study, a reduction in insulin dependence was noted in the obese recipients who received the lean donor's stool, compared to those who received their own stool back as a control. The second study is a case report of a woman of average BMI who received a stool transplant from her obese daughter to treat a *C. difficile* infection, and though this patient was cured of her infection, she went on to gain significant weight in the months following the procedure, potentially as a consequence of receiving an obese-type microbiota (Alang and Kelly 2015).

In the future, the use of microbiome-modulating therapies to treat these conditions may become a reality, with a greater understanding of the development of the microbiota, as well as the influence of diet on these microbes. Such therapies may range from directed prebiotic therapy, using food starches targeted to specific microbial groups to stimulate their growth and effect more efficient digestion (Scott *et al.* 2015), to full MET strategies as above, to replace or modify ecosystems that are contributing to metabolic disease or malnutrition.

Future nutritional therapies need not be confined to disease management. Along with a dawning recognition that everyone has a unique gut microbial ecosystem, there is an opportunity for food manufacturers to capitalize on personalized nutrition. For example, it may become possible to determine optimal prebiotic foods from an assessment of gut microbiota profiles on an individual basis; armed with this knowledge, a person may be able to select food at the supermarket that is compatible with his or her gut microbiota, and to understand which food substrates might be the most optimal for their microbial symbionts.

12.7. SUMMARY

As was predicted thousands of years ago with the advent of Chinese traditional medicine, wellbeing originates in the gut (Li *et al.* 2009). This was echoed over 100 years ago by Élie Metchnikoff who postulated that microbes may be key to a longer and healthier life. Although much time has passed, we are now playing a form of catch-up to best understand and appreciate the involvement of our trillions of microbial passengers. Thankfully, this revolution is not limited to microbiology but is now widespread in medicine and incorporating numerous studies once considered unimaginable. As this book was published, researchers began to demonstrate the use of microbes to alleviate allergies to peanuts as well as in the remediation of psychiatric conditions. While the data is still scant and more work needs to be performed, these two studies alone demonstrate how microbes have transcended their initial denouncements as solely pathogens, and have become an integral part of our health and medicine. In the future, greater attention will be paid to our microbial symbionts and leverage their beneficial activities. In doing so, it is anticipated that our view of health will be expanded such that we no longer focus on our human selves, but rather on ourselves as human/microbial superorganisms that can maintain our wellbeing through support of all our biological systems, physiological, metabolic, immunological, neurological, endocrinological, and finally, microbial.

12.8. REFERENCES

Alang, A. and C. Kelly. 2015. Weight Gain After Fecal Microbiota Transplantation. *Open Forum Infectious Diseases 2*, 1–2.

Allen-Vercoe, E. 2013. Bringing the gut microbiota into focus through microbial culture: recent progress and future perspective. *Current opinion in microbiology 16*, 625–629.

Antunes, L.C., *et al.* 2011. Effect of antibiotic treatment on the intestinal metabolome. *Antimicrobial agents and chemotherapy 55*, 1494–1503.

Antunes, L.C., *et al.* 2014. Antivirulence activity of the human gut metabolome. *mBio 5*, e01183-01114.

Arboleya, S., *et al.* 2015. Intestinal Microbiota Development in Preterm Neonates and Effect of Perinatal Antibiotics. *The Journal of pediatrics 166*, 538–544.

Barriere, S.L. 2015. Clinical, economic and societal impact of antibiotic resistance. *Expert opinion on pharmacotherapy 16*, 151–153.

Belfield, K., R. Bayston, J.P. Birchall, and M. Daniel. 2015. Do orally administered antibiotics reach concentrations in the middle ear sufficient to eradicate planktonic and biofilm bacteria? A review. *International journal of pediatric otorhinolaryngology 79*, 296–300.

Blaser, M.J., and S. Falkow. 2009. What are the consequences of the disappearing human microbiota? Nature reviews. *Microbiology 7*, 887–894.

Chassaing, B., *et al.* 2015. Dietary emulsifiers impact the mouse gut microbiota promoting colitis and metabolic syndrome. *Nature 519*, 92–96.

Claesson, M.J., *et al.* 2012. Gut microbiota composition correlates with diet and health in the elderly. *Nature 488*, 178–184.

Collison, M., *et al.* 2012. Data mining the human gut microbiota for therapeutic targets. *Briefings in bioinformatics 13*, 751–768.

Cotter, P.D., C. Stanton, R.P. Ross, and C. Hill. 2012. The impact of antibiotics on the gut microbiota as revealed by high throughput DNA sequencing. *Discovery medicine 13*, 193–199.

Croswell, A., E. Amir, P. Teggatz, M. Barman, and N.H. Salzman. 2009. Prolonged impact of antibiotics on intestinal microbial ecology and susceptibility to enteric Salmonella infection. *Infection and immunity 77*, 2741–2753.

Dohar, J., *et al.* 2006. Topical ciprofloxacin/dexamethasone superior to oral amoxicillin/clavulanic acid in acute otitis media with otorrhea through tympanostomy tubes. *Pediatrics 118*, e561–569.

Dore, J. and H. Blottiere. 2015. The influence of diet on the gut microbiota and its consequences for health. *Current opinion in biotechnology 32C*, 195–199.

El Aidy, S., P. Van den Abbeele, T. Van de Wiele, P. Louis, and M. Kleerebezem. 2013. Intestinal colonization: how key microbial players become established in this dynamic process: microbial metabolic activities and the interplay between the host and microbes. *BioEssays: news and reviews in molecular, cellular and developmental biology 35*, 913–923.

Fuentes, S., *et al.* 2014. Reset of a critically disturbed microbial ecosystem: faecal transplant in recurrent Clostridium difficile infection. *ISME J 8*, 1621–1633.

Ghosh, T.S., *et al.* 2014. Gut microbiomes of Indian children of varying nutritional status. *PLoS One 9*, e95547.

Ghouri, Y.A., *et al.* 2014. Systematic review of randomized controlled trials of probiotics, prebiotics, and synbiotics in inflammatory bowel disease. *Clinical and experimental gastroenterology 7*, 473–487.

Grzeskowiak, L., *et al.* 2012. Distinct gut microbiota in southeastern African and northern European infants. *Journal of pediatric gastroenterology and nutrition 54*, 812–816.

Gonzalez-Rodriguez, I., L. Ruiz, M. Gueimonde, A. Margolles, and B. Sanchez. 2013. Factors involved in the colonization and survival of bifidobacteria in the gastrointestinal tract. *FEMS Microbiol Lett 340*, 1–10.

Goodrich, J.K., *et al.* 2014. Human genetics shape the gut microbiome. *Cell 159*, 789–799.

Hill, C., *et al.* 2014. Expert consensus document. The International Scientific Association for Probiotics and Prebiotics consensus statement on the scope and appropriate use of the term probiotic. *Nature reviews. Gastroenterology & hepatology 11*, 506–514.

Iapichino, G., *et al.* 2008. Impact of antibiotics on the gut microbiota of critically ill patients. *Journal of medical microbiology 57*, 1007–1014.

Jernberg, C., S. Lofmark, C. Edlund, and J.K. Jansson. 2007. Long-term ecological impacts of antibiotic administration on the human intestinal microbiota. *ISME J 1*, 56–66.

Kane, A.V., D.M. Dinh, and H.D. Ward. 2015. Childhood malnutrition and the intestinal microbiome. *Pediatric research 77*, 256–262.

Lakshminarayanan, B., C. Stanton, P.W. O'Toole, and R.P. Ross. 2014. Compositional dynamics of the human intestinal microbiota with aging: implications for health. *The journal of nutrition, health & aging 18*, 773–786.

Lawley, T.D., *et al.* 2012. Targeted restoration of the intestinal microbiota with a simple, defined bacteriotherapy resolves relapsing Clostridium difficile disease in mice. *PLoS Pathog 8*, e1002995.

Lee, H. and G. Ko. 2014. Effect of metformin on metabolic improvement and gut microbiota. *Appl Environ Microbiol 80*, 5935–5943.

Li, H., M. Zhou, A. Zhao, and W. Jia. 2009. Traditional Chinese medicine: balancing the gut ecosystem. *Phytotherapy research: PTR 23*, 1332–1335.

Li, H. and W. Jia. 2013. Cometabolism of microbes and host: implications for drug metabolism and drug-induced toxicity. *Clinical pharmacology and therapeutics 94*, 574–581.

Louis, P., K.P. Scott, S.H. Duncan, and H.J. Flint. 2007. Understanding the effects of diet on bacterial metabolism in the large intestine. *Journal of applied microbiology 102*, 1197–1208.

Malago, J.J. 2014. Contribution of microbiota to the intestinal physicochemical barrier. *Beneficial microbes*, 1–17.

Mangin, I., A. Suau, M. Gotteland, O. Brunser, and P. Pochart. 2010. Amoxicillin treatment modifies the composition of Bifidobacterium species in infant intestinal microbiota. *Anaerobe 16*, 433–438.

McFarland, L.V. 2014. Use of probiotics to correct dysbiosis of normal microbiota following disease or disruptive events: a systematic review. *BMJ open 4*, e005047.

Mills, S., C. Stanton, G. F. Fitzgerald, and R.P. Ross. 2011. Enhancing the stress responses of probiotics for a lifestyle from gut to product and back again. *Microbial cell factories 10* Suppl 1, S19.

O'Sullivan, O., *et al.* 2013. Alterations in intestinal microbiota of elderly Irish subjects post-antibiotic therapy. *The Journal of antimicrobial chemotherapy 68*, 214–221.

Palmnas, M.S., *et al.* 2014. Low-dose aspartame consumption differentially affects gut microbiota-host metabolic interactions in the diet-induced obese rat. *PLoS One 9*, e109841.

Perez-Cobas, A.E., *et al.* 2014. Structural and functional changes in the gut microbiota associated to Clostridium difficile infection. *Frontiers in microbiology 5*, 335.

Petrof, E.O., *et al.* 2013. Stool substitute transplant therapy for the eradication of Clostridium difficile infection: 'RePOOPulating' the gut. *Microbiome 1*, 3.

Purucker, P., H. Mertes, J.M. Goodson, and J.P. Bernimoulin. 2001. Local versus systemic adjunctive antibiotic therapy in 28 patients with generalized aggressive periodontitis. *Journal of periodontology 72*, 1241–1245.

Sarowska, J., I. Choroszy-Krol, B. Regulska-Ilow, M. Frej-Madrzak, and A. Jama-Kmiecik. 2013. The therapeutic effect of probiotic bacteria on gastrointestinal diseases. Advances in clinical and experimental medicine: *Official organ Wroclaw Medical University 22*, 759–766.

Schnorr, S.L., *et al.* 2014. Gut microbiome of the Hadza hunter-gatherers. *Nature communications 5*, 3654.

Scott, K.M., J.M. Antoine, T. Midtvedt, and S. van Hemert. 2015. Manipulating the gut microbiota to maintain health and treat disease. *Microbial ecology in health and disease 26*, 25877.

Shahinas, D., *et al.* 2012. Toward an understanding of changes in diversity associated with fecal microbiome transplantation based on 16S rRNA gene deep sequencing. *mBio 3*.

Shankar, V., *et al.* 2014. Species and genus level resolution analysis of gut microbiota in Clostridium difficile patients following fecal microbiota transplantation. *Microbiome 2*, 13.

Spellberg, B., J. Bartlett, R. Wunderink, and D.N. Gilbert. 2015. Novel approaches are needed to develop tomorrow's antibacterial therapies. *American journal of respiratory and critical care medicine 191*, 135–140.

Stecher, B., *et al.* 2010. Like will to like: abundances of closely related species can predict susceptibility to intestinal colonization by pathogenic and commensal bacteria. *PLoS Pathog 6*, e1000711.

Subramanian, S., *et al.* 2014. Persistent gut microbiota immaturity in malnourished Bangladeshi children. *Nature 510*, 417–421.

Suez, J., *et al.* 2014. Artificial sweeteners induce glucose intolerance by altering the gut microbiota. *Nature 514*, 181–186.

Valles, Y., *et al.* 2014. Microbial succession in the gut: directional trends of taxonomic and functional change in a birth cohort of Spanish infants. *PLoS genetics 10*, e1004406.

Varughese, C.A., N.H. Vakil, and K.M. Phillips. 2013. Antibiotic-associated diarrhea: a refresher on causes and possible prevention with probiotics--continuing education article. *Journal of pharmacy practice 26*, 476–482.

Young, P.Y., and R.G. Khadaroo. 2014. Surgical site infections. *The Surgical clinics of North America 94*, 1245–1264.

Vrieze, A., *et al.* 2012. Transfer of intestinal microbiota from lean donors increases insulin sensitivity in individuals with metabolic syndrome. *Gastroenterology 143*, 913–916 e917.

Index

16S rRNA, 9, 13, 30, 33, 39, 44–46, 60, 63, 65, 71, 218, 220, 233, 243, 254, 278, 315, 320, 323, 338, 351, 358
3, 5, 3′-triiodothyronine, 187
4-ethylphenylsulfate (4EPS), 116, 238
4-hydroxyphenylacetate, 83
4-hydroxyphenyllactate, 83
4-hydroxyphenylpropionate, 83
4- hydroxyphenylpyruvate, 83
4-ethylphenol, 83
5-bromocytosine, 247
5-chlorocytosine, 247
5-halocystosine, 247
5-methylcytosine, 247
7-α-dehydroxylase, 77, 321
Acetate, 19, 21, 43, 74–76, 78–80, 82, 83, 86, 103, 147–149, 157, 172–173, 175–177, 181, 190, 205–206, 210–212, 237–239, 283
Acetylation, 80–81, 107, 122, 175, 210, 236, 239, 245, 265
Acidaminococcus fermentans, 5
Acinetobacter lwoffii F78, 245
Acquapendente, Fabricus, 271
ACTH, 116
Actinobacteria, 10, 12–13, 22, 36, 51, 54, 73, 189, 192, 220, 227, 229, 239, 287, 289, 292
Actinomycetales, 25

Adenosine diphosphate (ADP), 236
Adenosine triphosphate (ATP), 75, 78, 88, 200, 254
Adipocytes, 80, 176, 178, 181, 194–195, 197, 204, 211–212, 242, 297
Adipose tissues, 105, 176–177, 179, 180–181, 187, 195–196, 198, 200–201, 205, 207, 211, 312
Afferent neurons, 111, 119, 120, 126, 128, 223
Akkermansia, 7, 45–47, 51, 198, 204, 284, 306
Akkermansia spp, 284
Alistipes, 8, 36, 44–46
Alkaline phosphatase, 199, 201–202, 204, 206–208
Alkaliphilus, 46
Allergy, 29, 33, 68–69, 157, 160, 162–163, 167, 169, 257–258, 265, 335–336, 342–345
Ammonia, 43, 82, 89, 98–99, 101, 103, 106–107, 215, 228, 265, 290, 297
Amphiregulin, 240
Anaerobic bacteria, 1, 3–4, 184
Anaerococcus, 8, 192
Anaerotruncus, 8, 45, 58, 284
Anaerotruncus coliohominis, 284
Angiogenesis, 137, 321

359

Anthocyanin, 92, 94, 98
Antibiotic-associated diarrhoea (AAD), 48–49
Antibiotics, 15, 35, 42, 48–49, 57, 59, 61–62, 90, 135, 147, 155, 157, 159, 162, 184, 191, 199–200, 215–217, 219, 225, 230, 233–234, 255, 268, 272–275, 277–278, 281, 286, 288, 290–291, 293, 297, 303–304, 307, 313, 319, 328–329, 334, 339–342, 345, 349–350, 355–356
Antigen presenting cells (APCs), 141
Antimicrobial, 5, 6, 13, 18, 20–21, 27, 41, 74, 94–95, 113, 132, 135, 137–139, 161, 164, 166, 168–170, 188, 215, 272, 277, 280, 293, 296, 321, 328–329, 331, 342, 344–345, 348–350, 355, 357
Antimicrobial—associated molecular patterns (MAMPs), 135
Antimicrobial peptides, 21, 132, 135, 161, 164, 166
Antimicrobial peptides (AMPs), 132
Anxiety, 28–29, 54, 111–113, 115–116, 118, 122, 124–128, 238
Apoptosis, 25, 80, 196, 203, 239, 248, 251, 256, 259–260, 291, 324
Arabinogalactan, 76
Arabinoxylan, 76, 99, 105–106
Archaea, 10, 12, 17, 241, 254, 260, 267
Archaed, 55
Arginine, 236, 262
Aryl hydrocarbon receptor (AhR), 151, 238–240
Atherosclerosis, 90, 102, 189–193, 201, 205–207, 210, 244–245, 259, 261–262, 266
Atlas, Ron, 329, 339
Atopic dermatitis, 68, 154, 166, 332
Atopobium, 40
Autism, 7–8, 28, 31, 109, 115, 124–125, 127, 237, 255–256, 258, 261–262, 264–265
Autism Spectrum Disorder (ASD), 115–116, 307, 312
Autoimmunity, 32, 132, 161, 202, 269
Avenanthramides, 92

B cells, 119, 132, 134, 140, 143–145, 168, 180, 240, 244
Bacterial vaginosis, 331, 339, 342–343
Bacterial vaginosis (BV), 331
Bactericidal/permeability increasing protein (BPI), 199
Bacteriology, 1, 3, 205, 258
Bacteroides fragilis, 4, 15, 44, 53, 141, 225, 238, 253, 290
Bacteroides thetaiotaomicron, 36, 72
Bacteroidetes, 10, 12, 13, 14, 22, 24, 26, 31, 36, 39, 44, 46, 50, 51, 52, 53, 54, 57, 58, 73, 94, 95, 174, 189, 192, 217, 220, 225, 227, 228, 229, 255, 276, 284, 287, 289, 292, 293, 294
Balb/c mice, 113, 115, 316
Barrett's oesophagus, 27, 294, 324
β-glucosidase, 77
β-glucuronidase, 77
Bifidobacterium, 6–7, 15, 18, 23–26, 32, 36, 40, 68–69, 81, 92, 94, 128, 184, 225, 234, 268, 271, 280, 284, 329, 333, 342–343, 357
Bifidobacterium animalis, 23, 31, 314
Bifidobacterium breve, 100, 337
Bifidobacterium infantis, 126, 141, 288, 332, 336–337
Bifidobacterium lactis, 336, 339, 341, 345
Bifidobacterium longum, 41, 64, 67, 125, 128, 288, 291, 337
Bile Acid Metabolism, 171, 188, 285
Bilophila wadsworthia, 8, 188, 221, 224–225
Body Mass Index (BMI), 47, 76, 354
Borrelia burgdorferi, 244
Botulism, 1
Brain derived neurotrophic factor (BDNF), 114–115, 122, 127
Branched-chain fatty acids (BCFA), 83
Breast cancer, 240
Breast milk, 10, 41, 67, 287
Butyrate, 20–21, 24–25, 31–32, 36, 43, 47, 50–53, 55, 64, 67, 75–83, 88, 98–101, 103–107, 122, 128, 147, 149–150, 156, 163–164, 168–169

Butyrate (continued), 171–175, 177–181, 204, 207–208, 212, 237–239, 249, 254, 256–259, 261, 270, 279, 283–285, 289, 305, 307, 314, 340
Butyrivibrio, 44, 47, 147
Butyrivibrio fibrisolvens, 147
Butyrovibrio crossotus, 284
Butyrylation, 236

Caco-2, 101, 152
Cadaverine, 90
Campylobacter, 21, 126, 244
Candida albicans, 96, 100
Capnocytophaga gingivalis, 242
Capsaicinoids, 92
carbohydrate response element binding protein (ChREBP), 297
Carboxylation, 90, 237
Carcinogenesis, 30, 88, 100, 102, 105, 248, 258, 260, 264, 266, 290–292, 309, 313, 316, 319
Carcinogenic N-nitroso compounds (NOC), 91, 106
Carcinoma, 181, 200, 240, 242–243, 246–257, 259–260, 262–263, 289, 292–294, 301–302, 306, 309, 312
CARD15, 22
CARD9, 158
Cardiovascular, 22, 27, 49, 87, 90–91, 97, 99, 103, 106–107, 172, 190, 193, 210–211, 226, 239, 244–245, 249, 259–261, 323
Cardiovascular Disease (CVD), xi, 22, 49, 87, 90, 97, 172, 244, 249
Carnitine, 90–91, 102, 189–193, 206, 209, 245, 261
Catenibacterium, 44, 46
Cathelicidins, 137, 139, 161, 165–166, 170
CCL8, 157
CD14, 50, 198, 207
CD39, 253–254, 261, 265
CD4 T-cells, 82, 132, 134, 137, 140–141, 149, 167, 236, 245, 252–253, 256, 259, 263, 306
CD41, 245
CD8 T-cells, 134, 140, 142, 252, 260, 263

Cecum, 15, 67, 74, 184, 215
Centers for Disease Control and Prevention (CDC), 217
Cephalosporins, 49, 272
c-Fos, 121
Chemokine receptor, 141
Chemotaxis, 20, 137, 180
Chenodeoxycholic acid, 182–183
Chinese Hamster Ovary (CHO), 81
Chlamydia psittaci, 244
Chloramphenicol, 5
Cholesterol, 44, 78, 95, 136, 151, 167, 173, 181–182, 189–191, 193, 197, 207, 240
Cholic acid, 182, 189
Chromatin, 99, 122, 129, 175, 235–237, 239, 247, 257
Chromosomes, 235, 247
Chylomicrons, 181
Cirrhosis, 89, 208, 227–231, 234, 295–296, 300, 303, 310, 321, 324
Citrobacter, 139
Citrobacter rodentium, 81, 127, 139, 269, 324
Citrullination, 236, 262
c-Jun N-terminal kinase (JNK), 195
Clarithromycin, 48
Claudin-1, 198
Claudin-3, 198
Clindamycin, 49, 272
Clostridial toxins (tcdA, TcdB), 274–275, 321
Clostridium, 23, 25, 66, 83, 184, 192, 229–230, 269, 280, 296, 300, 304
Clostridium bolteae, 7–8
Clostridium clostridioforme, 57, 284
Clostridium cluster, 10, 14–15, 20, 24, 43–44, 46, 58, 268, 274, 276, 285
Clostridium coccoides, 293, 295
Clostridium difficile, 5, 41, 48, 66, 69, 215–216, 272, 300, 302–303, 305–315, 317–324, 334, 343–345, 351, 356–358
Clostridium histolyticum, 158, 255
Clostridium leptum, 53, 94, 221, 321
Clostridium lituseburense, 158

Clostridium perfringens, 287
Coabundance gene groups, 222
Coagulase negative staphylococci, 40
Colitis, 112, 138–139, 141–142, 148, 167–168, 180, 203, 207, 216, 221, 230, 273, 290–291, 299, 301–302, 304, 307, 310–311, 317, 320, 322, 324, 343, 345–355
Colitogenic bacteria, 53
Collinsella, 192
Colon, 5, 7, 8, 14–15, 19–20, 24, 42–43, 59, 64, 73, 76–80, 82–83, 87–88, 90–91, 97–98, 100–106, 138, 148, 157–158, 162–163, 169, 172, 175, 181, 184, 200, 212, 214–215, 229, 248, 256, 259, 262, 265, 288, 290–291, 305, 308, 310, 316–317
Colon cancer, 31, 78, 80, 103, 175, 212, 257–258, 262, 288, 290, 306, 317
Colonocytes, 19, 21, 77–78, 88–89, 103, 173
Colorectal cancer, 24, 29–30, 33, 91, 101–102, 128, 257, 262, 265, 288–289, 301, 303–304, 308, 314–315, 322–324
Colorectal carcinoma (CRC), 289
Coping Checklist (CCL), 118
Coprobacillus, 45, 58
Coriobacteridae, 289
Corynebacterium, 8, 27, 41
C-reactive protein (CRP), 56, 58, 282
Crohn's disease, 21, 28, 31–32, 52, 67, 70, 105, 138, 161, 163, 166–167, 169, 212, 219, 221–222, 231–234, 238, 278–280, 301, 304, 306–307, 311, 313, 316–317, 320, 324, 332, 339, 344
Crohn's disease (CD), 44, 53, 138, 219, 221, 278–280, 332
Crotonylation, 236
C-type lectin, 137–138
Cyclic adenosine monophosphate, 254
CYP7A1, 183–186
Cystathionine beta-synthase, 89
Cystathionine gamma lyase, 89
Cysteine, 87, 89, 137, 251, 279
Cytokine, 137

Defensins, 136–139, 160–163, 167–168, 280, 296
Dendritic cells (DCs), 82, 111, 119, 134, 141, 163, 166, 194, 251, 263, 270, 304, 319, 321
Deoxycholic acid, 183, 187, 238
Dermitis, 332
Desulfitobacterium, 192
Desulfomonas, 25, 87
Desulfomonas spp, 87
Desulfovibrio, 7, 25, 46, 51, 57, 87, 192, 201, 205, 231, 238, 279, 319
Desulfovibrio alaskensis, 192
Desulfovibrio desulfuricans, 192, 201, 205
Desulfovibrio spp, 87
Diabetes, 23–24, 30–32, 44, 49–51, 62, 65–66, 68–69, 74, 105–106, 125, 155–156, 160–161, 163–164, 166, 168–169, 172, 175, 179, 181–182, 187, 193–194, 196, 201–206, 208–211, 217, 237, 242, 249, 263, 282–284, 295, 299, 302, 307, 312, 318, 323, 353
Dimethylamine (DMA), 90
Dopamine, 111, 201
Duodenum, 12–13, 18, 42, 214, 345
Dysbiosis, 20–23, 25–28, 35–36, 38, 40–41, 49, 53, 59, 62, 116, 120, 163, 168, 177, 225–226, 228–229, 231–232, 234, 255, 265, 281, 286, 295, 301, 304, 308, 311–313, 320, 324, 330, 357

Eczema, 27, 332, 336, 345
ELDERMET, 58
Endocannabinoid system, 50, 66, 68
Endotoxemia, 24, 50, 62, 64–65, 197, 201–202, 204, 206–209, 232, 283, 285, 296, 299–300, 302, 310, 316
Enterobacter cloacae, 144
Enterobacteriaceae, 15, 21, 26, 221–222, 225, 229, 273, 278, 280–281, 288, 291
Enterococcus, 94
Enterococcus faecium, 39, 310

Enterocolitis, 5, 27, 32, 40, 70, 285–286, 298, 300, 302–305, 308–310, 312, 314–316, 322, 329, 331, 336, 340
Entero-pathogens, 48
Enterotype, 8, 15–16, 20, 28, 42, 44, 46, 51, 60, 72, 97, 234, 256, 324
Epigenetic, 80, 110, 122, 124, 127–129, 175, 200, 202, 209, 235–246, 248–252, 254–266, 270
Epigenome, 235, 237, 252, 261, 263
Epinephrine, 117
Epiregulin, 240
Erysipelotrichales, 222
Erysipelotrichi, 298
Escherichia coli, 6, 21, 25–26, 31, 39, 41, 51, 77, 92, 99, 117, 150, 158, 161, 184, 221–222, 230, 238, 271, 278, 280, 284, 287, 289, 310, 315, 320, 333, 343
Esophagitis, 27, 30, 294, 324
Eubacterium, 6, 10, 15, 25, 36, 43, 45, 58, 92–93, 184, 192, 238–239, 248, 268, 311
Eubacterium aerofaciens, 248
Eubacterium hallii, 285
Eubacterium ramulus, 93
Eubacterium rectale, 20, 24, 44–46, 50, 55, 238, 239
Eubacterium spp, 91

Faecalibacterium prausnitzii, 20, 21, 33, 43, 53, 57, 70, 72, 77, 158, 162, 167, 169, 220–221, 233–234, 238–239, 279, 284, 320, 322
Farnesoid X Receptor (FXR), 151, 184, 185–187, 209
Fast-acting-induced adipocyte factor (FIAF), 176, 181, 242, 297
Fecal microbiota transplantation (FMT), 267–268, 270–278, 281–282, 285, 288, 291, 298, 335
Fermentation, 20, 39, 42–43, 50, 54–55, 58–59, 66–67, 71, 74–77, 82–83, 87, 89, 91–92, 94, 95, 99, 101, 103–107, 111, 122, 125, 127, 156, 162, 169, 171, 177, 189, 201, 205, 215, 227, 230, 262, 327, 334, 352

Fibroblast growth factor (FGF), 185–186, 201, 206
Firmicutes, 10, 12, 13, 21–22, 24–26, 31, 36, 39, 43–46, 50, 53–54, 57–58, 73, 94–95, 156, 158, 174, 184, 189, 192, 221, 225, 227, 229, 238–239, 254–255, 276, 278–279, 284, 289, 292–294, 351
Flavin mono-oxygenase (FMO), 90, 189, 279
Fluorescent in situ hybridisation (FISH), 54
Fluoroquinolones, 49, 272
Foam cell, 90, 191
Food and Drug Administration (FDA), 328, 330, 335
Formyl peptide receptors (FPRs), 148
Formylation, 236–237
Free Fatty Acid Receptor (FFA), 79–82, 102–103, 106, 148, 176, 187, 206, 208–210, 261–262, 306
Functional gastrointestinal disorders (FGIDs), 52–54
Fusobacteria, 12–13, 39, 222, 228, 294
Fusobacteriaceae, 222
Fusobacterium, 4, 279, 290, 312, 333
Fusobacterium nucleatum, 27, 290, 302, 321
Fusobacterium varium, 6, 221, 223, 316

Gamma aminobutyric acid (GABA), 113–114, 121, 123, 125, 164, 238
Gas gangrene, 1
Gemella, 292
Gemella asacchrolytica, 8
GLP-1, 79–80, 102, 176, 178, 180, 187–188, 198, 208, 212
Glucagon-like peptide-1 (GLP-1), 178
Gluconeogenesis, 78, 173, 180
Glucose transporter type 4 (GLUT4), 196
Glutamine, 58, 89
Glutathione, 279
Glycans, 64, 72, 95, 283, 290
Glycine, 90, 184, 190, 245, 320
Glycoside hydrolases, 92, 172

Glycosylation, 236
Gnotobiotic mice, 32–33, 43, 72, 262, 342
GPR109a, 80, 82, 99, 105–106, 148, 169
GPR41, 23, 79, 98, 102, 106–107, 148, 161, 176–180, 182, 201, 206, 208–210, 212
GPR43, 23, 79, 98, 102, 148–149, 161, 166, 176–182, 201, 206, 208, 210, 212, 314
GR-1, 331
Graft-versus-hostdisease (GVHD), 138
Guar gum, 76, 104
Gut-associated lymphoid tissue (GALT), 19, 132, 134–135, 137, 139, 140–141, 143, 145, 269
Gut-brain axis, 53, 109–111, 116–121, 123–125

Hadza, 36–38, 70, 357
Hall, Wendel, 2
HDL, 198
Health Canada, 330, 334, 341
Helicobacter, 334
Helicobacter hepaticus, 269, 291, 316
Helicobacter pylori, 13, 31, 244–245, 257–265, 289, 299, 301, 304–306, 309, 313–314, 316–317, 319–321, 323, 334, 340
Hemostasis, 328
Hepatic encephalopathy, 89, 98, 106, 228, 230, 296–297, 300, 318
High fat diet, 64, 91, 194, 197
Histamine, 137, 157
Histidine, 236
Histone, 80–81, 99, 105, 107, 122, 128, 175–176, 202, 210–211, 235–239, 245, 249, 256, 258, 262–263
Histone deacetylases (HDAC), 80–81, 122, 175–176, 180, 202, 239
Histone deacetylases acetylation (HDAC), 175
Homocysteine, 245
Hospital Anxiety and Depression Scale (HADS), 118–119
HT-29 cell line, 87–88, 259
Human Microbiome Project, 9–10, 16, 30

Hydrogen sulfide (H_2S), 21, 25, 87–88, 97–100, 102–104, 137, 221, 225, 233, 284, 290
Hydroxylation, 33, 95, 103, 183, 229, 236, 274, 309, 323
Hydroxymethylation, 237
Hygiene hypothesis, 27, 33, 155, 251
Hyperlipidaemia, 242
Hypersensitivity, 223, 230
Hypertension, 27, 89, 98, 242, 249
Hypobromous acid, 247
Hypochlorous acid, 247
Hypothalamic-pituitary-adrenal (HPA), 116
Hypothalamic-pituitary-adrenal axis (HPA), 110, 116–117, 123
Hypoxia, 240, 248

IEC-6 cells, 78
IFN-γ-inducible protein 10 (IP-10), 153
IFN-γ-inducible protein 10 (IP-10), 153
IgA, 70, 119, 132, 134–135, 140, 143–144, 145, 161–164, 167–168, 270
IgE, 144, 157, 162–163
IgG, 132, 144
Interleukin 12, 81, 236, 253, 310, 332
Interleukin 13, 153
Interleukin 17, 142, 155, 308–309
Interleukin 1β, 81, 152–153, 196, 306
Interleukin 4, 153, 236, 253
Interleukin 5, 153
Interleukin 6, 56, 58, 81, 153, 179, 194, 236, 240, 253, 259, 280, 294
Interleukin 8 receptor, 158
Interleukin 18, 56
Interleukin 22, 142, 151, 270, 325
Interleukin 23 receptor, 158
Ileum, 6, 14, 19, 77, 89, 183, 185, 214, 221, 229, 280
Immunoglobulin, 134, 144, 161, 168–169, 269, 296
Indican, 86
Indirubin, 239
Indole, 83, 86–87, 99, 104, 150–151, 160, 165–166, 238–239
Inducible nitric oxide synthase (iNOS), 132

Inflammation, 18, 22, 24–27, 29, 36, 38, 47, 50, 53, 55–56, 58, 62, 64–66, 68, 71, 81–82, 88, 101, 103, 105–106, 112, 121, 125, 132, 140–142, 148–152, 154–155, 157–158, 160–161, 163–169, 176, 178–179, 185, 188–189, 194–199, 201, 203–206, 208–212, 217, 221, 224, 226, 228, 230, 232, 238–239, 241, 243, 247–249, 251, 253–254, 258–259, 261, 264–265, 269–270, 278–281, 283, 285, 296, 289–290, 293, 295–296, 298, 300, 307, 311, 313, 318–320, 324, 331, 333
Inflammatory Bowel Disease (IBD), 21–22, 52–53, 70, 134, 142, 150, 152, 155, 157–159, 217, 219–221, 231, 278–282, 301, 306–307, 310–311, 313, 320–321, 324, 331–333
Insulin sensitivity, 176–78, 181–182, 184–185, 187–188, 196, 198, 201, 204, 211, 285, 307, 322, 358
Interferon gamma (IFN-γ), 80, 105, 138, 153, 155, 196, 236, 245, 253, 280
Interferon regulatory factor (IRF), 195
Intestinal gluconeogenesis (IGN), 180
Intestinal intraepithelial lymphocytes (IELs), 142
Intraepithelial lymphocytes (IEL), 129, 132, 134, 142, 161–162, 164, 313
Irritable Bowel Syndrome (IBS), 27, 29, 31, 52–55, 65, 67–70, 74, 76, 82, 101, 103, 107, 120, 208, 223–225, 230, 232, 263, 271, 301, 308, 332, 336, 341, 345
Isoleucine, 83
IκB kinase (IKK), 195

Jejunum, 6, 14, 18–19, 214
Junctional adhesion molecule 1 (JAM-1), 198

Kanamycin, 5
Keratinocytes, 154
Klebsiella, 158, 192, 287
Klebsiella oxytoca, 26

L cells, 102, 178
Lachnospiraceae, 52
Lachnospiraceae, 52, 228, 254, 273–274, 281, 348
Lactic acid, 330
Lactobacillus, 6, 8, 24–25, 28, 32, 40, 46, 54, 83, 92, 94, 112, 125, 184, 225, 227, 238, 268, 294–295, 306, 316, 327, 339, 341–342, 344–345
Lactobacillus acidophilus, 106, 248, 261–262, 337
Lactobacillus brevis, 27, 32
Lactobacillus casei, 100, 280, 291, 309, 313, 333, 337, 341, 343, 344
Lactobacillus delbrueckii, 337
Lactobacillus gasseri, 284, 290, 302
Lactobacillus helveticus, 128, 333, 341
Lactobacillus johnsonii, 156, 166, 169, 291, 333
Lactobacillus plantarum, 336–337, 344
Lactobacillus reuteri, 23, 31–32, 156, 336, 342–343
Lactobacillus rhamnosus, 113, 120–121, 128, 323, 329, 336–337, 340–342
Lactococcus, 227, 295
Lactose, 23, 76, 107
Lactulose, 76, 97, 297
Lamina propria, 21, 81, 119, 132, 134, 140–141, 143, 198, 200, 270, 304, 314, 321, 323
L-carnitine, 90
LDL, 95, 181, 190–191, 197–198, 227
Leptotrichia, 41
Leuconostoc, 227, 279, 295
Leuconostocaceae, 279
Lipogenesis, 78, 173, 181, 205
Lipolysis, 176, 181, 197, 204–205, 208, 275, 317
Lipopolysaccharide (LPS), 18, 24, 50, 81, 136–138, 153, 157, 178, 194–199, 201, 207, 235, 240, 251, 283, 290, 294
Lipoprotein lipase (LPL), 181, 226, 242, 297

Lipoproteins, 18, 95, 104, 106, 181, 190, 226–227, 231, 242, 275, 297–298, 310, 317
Listeria monocytogenes, 138
Lithocholic acid, 169, 187, 290
Low-density lipoprotein–cholesterol (LDL), 95
Lymphocytes, 149
Lysine, 80, 236

M cells, 119
Macrophage Inflammatory Protein (MIP-1α), 148
Macrophage inflammatory protein 1α (MIP-1α), 148
Major Hostological Complex (MHC), 139
Malonylation, 236
Maternal Immune Activation (MIA), 115
Megacolon, 216
Megasphaera, 5, 228
Megasphaera elsdenii, 5
Mesenchymal, 240
Mesenteric lymph nodes, 120, 141, 144–145, 198, 296
Metabolism, 16, 19, 23, 25, 27, 29, 31, 33, 41–42, 46, 49–52, 58, 60, 62–63, 66–68, 74–75, 78–79, 83–86, 89–90, 92–93, 95, 97, 99–107, 151, 153, 162–163, 171, 176–185, 187–194, 196, 200–201, 203, 206–212, 227, 234, 237, 239–240, 244–245, 249, 255, 258, 261–262, 264, 266, 268, 274, 279, 282–285, 291, 297, 305, 312, 316, 318, 323, 357
Metabolomics, 16, 90, 152, 262, 265, 278, 282, 338
Metagenomic, 10, 16, 22, 30, 33, 39, 43, 50–52, 60, 67, 69, 71, 104, 142, 161, 166, 206, 232, 233, 267, 278–279, 282, 284, 306, 318, 338
Metaproteomics, 16, 306
Metatranscriptomics, 16
Metchnikoff, Elie, 270–271, 315, 327–328, 354

Methionine, 87, 242, 245, 264
Methylation, 80, 122, 176, 236–238, 242–249, 255–257, 259, 261–262, 264–266
Methylcellulose, 76
Methyltransferase, 88, 104, 236, 244–245, 247
Methyltransferase, 88, 104, 236, 244–245, 247
Microbial-associated molecular patterns (MAMP), 135, 138
microRNA (miRNA), 237, 242, 244, 252, 257, 262–264
Monocarboxylate transporter (MCT), 77, 101
Monocyte chemoattractant protein 1 (MCP-1), 153
Monocytes, 137, 194, 203, 319
Morganella morganii, 145
MUC1, 247
MUC19, 157
Mucin, 18, 20–21, 46, 51, 82, 87–88, 135, 142, 162, 198, 204, 231, 247, 285, 306, 310, 338
Multiple sclerosis, 28, 251, 254–255, 258, 260–262
Muricholic acid (MCA), 182, 209
Mycobacterium, 14
Myeloid differentiation primary response gene 88 (MyD88), 156, 254

Necrotizing, 329, 331
Necrotizing enterocolitis (NEC), 286
Neisseria, 12–14
Neomycin, 3, 5, 113
Neonatal, 329
Neuroinflammation, 253, 265
Neutrophils, 135, 137, 142, 148, 169, 179, 243, 247, 270
Nitric oxide, 27, 29, 30, 99–100, 132, 290, 306
Nitrosamine, 90
N-nitroso compounds (NOC), 91
NOD, 22, 29, 31, 128, 138, 141, 144–145, 158, 161, 165, 182–183, 209, 244, 269, 280, 290–291, 296, 301, 303–304, 316–317, 320

NOD2, 138
Nonalcoholic steatohepatitis, 49, 51, 72, 194, 202, 205, 209, 225, 232, 233–234, 295, 299, 306, 322, 325
Nonalcoholic fatty liver disease (NAFLD), 49, 51–52, 60, 68, 91, 97, 99, 187, 208, 225–227, 229, 232, 294, 315, 318, 340
Nonalcoholic Steatohepatitis A, 49, 51, 194, 197–198, 225, 227, 229, 295–298
noncoding RNAs (ncRNA), 237, 242
Norepinephrine (NE), 114
Nuclear Factor Kappa B (NF-κB), 55–56, 69, 81, 96, 105, 115, 138, 148, 150, 153, 166, 179–180, 188, 194–197, 200–211, 240, 244, 280, 294, 307, 314, 345
Obesity, 6, 22–23, 28–29, 31–34, 42, 49–51, 56, 62–64, 66, 68, 70–71, 79, 80, 95, 103, 105, 155, 172–175, 178–179, 181–182, 185, 187, 193–194, 198–199, 201–212, 217, 219, 226, 230, 232, 234, 242, 249, 257, 264, 282–285, 295, 298, 302, 306, 308, 312–313, 315, 318–323, 353–354
Odoribacter, 279
Oligofructose, 76, 100, 107
Oligosaccharides, 83

p38 mitogen-activated protein kinase (MAPK), 153, 195, 250
Palmitate, 78
Pancreatic β-cell, 24, 178
Paneth cells, 135, 137–138, 160–163, 167, 169, 280, 296, 321
Parabacteroides, 8, 44–46, 58
Paraprevotella, 44, 46
Paraventricular nucleus (PVN), 111
Paromomycin, 5
Pasteur, Louis, 26, 131, 167
Pasteurellacaea, 222
Pasteurization, 155
Pathogens, 328–329, 331
Pattern Recognition Receptors (PRR), 119, 253, 290

PBMC cells, 81, PCR-denaturing gradient gel electrophoresis (PCR-DGGE), 54, 67
p-cresol, 58, 83, 86–87, 97–98, 100, 103–104
Penicillin, 1, 135, 184
Peptide YY (PYY), 23, 79, 176, 178–180, 198
Peptidoglycan, 137–138, 164, 254, 283, 290, 303
Peroxisome proliferator activated receptor γ (PPAR- γ), 29, 36, 48, 78, 181, 187, 197–198, 298–299, 311, 327
Peroxisome proliferator activated receptor γ (PPARγ), 181
Peroxisome proliferator-activated receptor gamma coactivator (PGC)-1α, 181
Peyer's patches, 19, 119, 132, 134, 141, 145, 163, 269
Pharmacopeia, 348
Phascolarctobacterium, 279
Phenotype, 36, 50, 95, 100, 113, 115, 121–122, 149, 163, 165, 172, 174, 189, 191, 200, 202–203, 219, 221, 230, 234–235, 237, 240, 245, 252, 259, 268, 285, 305–306
Phenylacetate, 83
Phenylacetyglutamine, 58
Phenylalanine, 83, 85
Phenyllactate, 83
Phenylpropionate, 83
Phenylpyruvate, 83
Phosphatidylcholine, 90–91, 106–107, 189, 210–211, 227, 244, 323
Phosphoinositide 3-kinase, 197, 254
Phospholipase A2, 138, 166
Phospholipid, 136, 190, 231, 233, 310
Phosphorylation, 188, 195, 199, 201, 236, 248, 252, 313
Piperidine, 90
piwi-interacting RNA (piRNA), 237
Placebo, 332–333
Planck, Max, 327–328
Platelet derived growth factor, 240
Polycystic ovarian disease, 242
Polyphenols, 43, 74, 92–98, 100–106

Polysaccharide A, 121, 141, 253
Porphyromonas, 8, 12–13, 17, 26, 242, 257, 260–262
Post-transcriptional modification, 235
Poth, Edgar, 1, 3
Prebiotics, 43, 61, 101, 111, 125, 225, 302, 304, 308, 319, 341, 343, 356
Pregnane X receptor (PXR), 151, 169
Prevotella, 10, 12–13, 15–17, 20–21, 24, 26, 40, 41, 44, 46–47, 50–51, 57, 192, 220, 228, 242, 255, 260, 268, 292, 294, 296
Probiotics, 327, 329–330, 332, 335
Proinflammatory, 17, 21, 24, 49, 55, 81, 112, 121, 137–138, 142, 153, 158, 162, 170, 175–176, 180, 188, 194–195, 199, 203, 240, 252, 283, 293, 297, 317
Prophylaxis, 6, 302, 328, 334, 340–341
Proprionate, 19, 43, 75–80, 82–83, 86, 98, 103, 122, 147, 149, 157, 161, 172–181, 201, 205, 207, 211, 237, 238, 261, 283
Proprionylation, 236
Prostaglandin, 153
Proteobacteria, 10, 12–14, 22, 24, 26, 36, 39, 41–42, 51–52, 54, 73, 192, 220, 228, 238–239, 255, 276, 278, 281, 287, 289, 292, 294, 298, 351
Proteus, 184, 192, 287, 297
Providencia, 192
Pseudomembranous colitis, 271, 272, 305
Pseudomonas, 184, 287
psoriasin, 154
Putrescine, 90
Pyrrolidine, 90

Qsec sensor kinase, 117

Ralstonia, 287
Reg IIIα, 138
Regulatory T-cell (Treg), 60, 70, 82, 106, 140–142, 149, 151, 155, 157–158, 168, 210, 216, 236, 251–254, 261–263, 269–270, 300, 320–321
RelA/p50, 240

RePOOPulate, 277, 335, 351
Resveratrol, 92, 95, 99
Riboflavin, 279
Ribosylation, 236
Rice, 63, 76, 125, 154, 164, 202, 228, 307, 331, 336, 343
Riegel, Gordon, 2
Rifaximin, 276, 278, 296–297, 310
RNase 7, 154
Rome criteria, 54
Roseburia, 20–21, 25, 43–46, 55, 58, 77, 192, 221, 228, 238–239, 268, 279, 284, 297, 305
Roseburia intestinalis, 285
Ruminococcus, 8, 10, 15, 20, 36, 46, 221

Saccharomyces, 18, 333–334
Saccharomyces boulardii, 271, 339–340, 342–343
Saccharomyces cerevisiae, 333, 337
S-adenosylhomocysteine, 245
S-adenosylmethionine, 245
Saliva, 16, 26–27, 33, 262, 331, 336–337, 340
Salmonella, 71, 77, 96, 153, 163,
Salmonella enterica, 56, 348, 356
Salmonella typhimurium, 25, 48, 161
Sarcopenia, 55
SCFA, 17, 19–21, 23, 25, 32, 43, 47, 58, 70, 74–82, 89, 94, 96, 105, 122, 146–150, 157, 172–182, 209–210, 215, 237–239, 242, 249, 255, 269–270, 279, 283, 289–290, 320, 338
Segmented filamentous bacteria (SFB), 145
Selenomonas, 12
Sepsis, 40, 67, 168, 202, 321
Serine, 195, 236
Serotonin, 111, 127–128, 238–239, 257
Short heterodimer partner (SHP), 185
short interfering RNA, 237
Signal transducer and activator of transcription 3 (STAT3), 236, 244, 247–248
Silencing RNA, 179, 237

skatole, 83, 107, 165
SLC5A8, 78, 99, 103, 106
SMCT-1, 78
Smith, Louis D.S., 3
Specific pathogen free (SPF) mice, 113, 117
Spink, Wesley, 2
Spirochaetes, 10, 12, 294
Squamous cell carcinoma, 240, 242–243, 259–260, 262–263
Staphylococcus aureus, 23, 27, 154, 162, 254
Staphylococcus aureus δ-toxin, 154
Staphylococcus epidermidis, 26–27, 39, 154
Streptococcus, 12–14, 18–19, 25–28, 31–32, 39, 41, 243, 284, 289, 292, 294, 301, 331, 336–337, 340
Streptococcus bovis, 15, 25, 27–28, 289
Streptococcus gallolyticus, 25, 28, 289, 301
Streptococcus mitis, 13, 27, 106, 243
Streptococcus salivarius, 27, 331, 336–337, 340
Streptomycin, 48, 152, 184
Succinylation, 236
Sulfurtransferase, 88
Sulphur-reducing Bacteria, 87
Sumoylation, 236
Sutterella, 7, 8, 13

T cell receptor (TCR), 139, 252
T cells, 331
T helper 1 (Th1), 81, 100, 102, 104, 106, 140, 142, 149, 155–156, 166, 236, 252, 253, 256, 262, 269, 310, 332, 335
T helper 2 (Th2), 140–141, 153, 157, 170, 236, 252–253, 262
T helper 17 (Th17), 140–142, 149, 155–156, 166, 236, 252–253, 258–259, 261, 269, 309, 313, 316
Taurocholic acid, 203, 230, 274
Tauromuricholic acid, 185
Tempol, 185, 201–202
Tenericutes, 12, 39, 243
TGR5, 184, 187–188, 206, 208, 210, 212

Thiol S-methyltransferase (TMT), 88
Thiosulfate sulfurtransferase (TST), 88
Threonine, 236
Thyroxine, 187
TLR3, 154
TMAO, 90–91, 189–193, 244–245
Toll-Like Receptor, 18, 22, 26, 148, 153–154, 157, 194–198, 203–204, 240–241, 244, 253–256, 265, 286, 290, 298, 307, 310, 319
Transcription, 53, 69, 80–82, 141, 151, 166, 175, 180–181, 185, 195–196, 235, 237–240, 244, 247, 250–255, 257, 261, 280, 297, 317
Translation, 237, 250, 256, 262, 343
Translocation, 21, 50, 81, 118, 138, 143, 145, 197–199, 205, 208, 226, 228, 230, 240, 262, 283, 295–296, 306, 316, 321, 324
Treponema, 27, 44, 47
Trimethylamine (TMA), 90–91, 189, 202, 212, 227, 244, 298
Trimethylamine-N-oxide (TMAO), 90
Tryptamine, 151, 238–239
Tryptophan, 86, 107, 150–151, 163, 165, 170, 238–240, 260, 265–266
Tumour necrosis factor-alpha (TNF-α), 56, 58, 81, 138, 148, 150, 153, 179, 194, 196–197, 240, 250, 280, 293–294, 297–298, 307, 339
Tyrosine, 83–84, 236, 246, 252

Ubiquitination, 236, 246
Ulcerative colitis, 21, 28, 33, 52, 99–102, 104–105, 139, 164, 220, 231–234, 278, 299, 301–302, 310, 314, 316–317, 319–320, 322, 333, 343
Ulcers, 21, 27
Urea, 89, 215, 246, 330
Uremia, 87
Urinary Track Infection (UTI), 331
Urogenital, 239, 243, 249, 328–330, 336, 344

Vaccination, 155, 299, 328
Vagus nerve, 111, 120–122, 125
Valine, 83

Valproic acid (VPA), 115
van Leeuwenhoek, Antoine, 327
Vancomycin, 6–7, 48, 184, 211, 264, 274–276
Vascular endothelial growth factor, 195, 240
Vegetarian diet, 44, 46, 280, 325
Veillonella, 12–14, 18–19, 23, 41, 54, 222, 228, 255, 294
Veillonellaceae, 222, 255
Verrucomicrobia, 10, 36, 227
Very low-density lipoproteins (VLDL), 181

Vitamin D Receptor (VDR), 151
Vulvovaginal candidiasis, 331, 342

Western Diet, 7, 22, 268, 290
Wheat dextrins, 76, 101, 104

X receptor, 151
Xenobiotic, 51, 217, 240, 243, 352
Xylanibacter, 44, 47, 268

Yokenella, 192

Zona occuldens (ZO), 198